The Radio Years

The Radio Years

A HISTORY OF BROADCASTING
IN NEW ZEALAND

VOLUME ONE

Patrick Day

AUCKLAND UNIVERSITY PRESS
in association with the
BROADCASTING HISTORY TRUST

First published 1994

Auckland University Press
University of Auckland
Private Bag 92019
Auckland

Publication is assisted by the Historical Branch, Department of Internal Affairs.

Cover illustration: Lois White *On the Air*, c. 1935, Manawatu Art Gallery.
Photograph provided by Auckland City Art Gallery.

ISBN 1 86940 094 1

Typeset in Christchurch by Deadline Typesetting Ltd
Printed in Wellington by GP Print Ltd

To my wife Barbara

The isle is full of noises
Sounds and sweet airs, that give delight and hurt not.
Sometimes a thousand twangling instruments
Will hum about mine ears, and sometimes voices.

William Shakespeare, *The Tempest*

Contents

Abbreviations		*viii*
Preface		*ix*
	Introduction	*1*
1	Radio and State Control	*6*
2	The Birth of Broadcasting	*34*
3	The Radio Broadcasting Company	*63*
4	Disputes and Growth	*94*
5	The Extension of Broadcasting and the Fall of the RBC	*121*
6	New Zealand Broadcasting Board	*148*
7	Victory for the B Stations and Labour	*182*
8	Government Controlled Broadcasting	*213*
9	Broadcasting in the War Years	*249*
10	New Zealand Broadcasting Service	*283*
	Broadcasting and New Zealand	*314*
Appendix 1	Radio Licence Ownership Statistics	*321*
Appendix 2	Map Showing Placing of Stations by the Coverage Commission	*322*
Appendix 3	Stations Licensed 1922–1961	*323*
Notes		*328*
Bibliography		*341*
Index		*345*

ABBREVIATIONS

AJHR	*Appendices to the Journals of the House of Representatives*
A&P	Agricultural & Pastoral Association
ATL	Alexander Turnbull Library
APRA	Australasian Performing Right Association
AWA	Amalgamated Wireless (Australasia) Limited
BBC	British Broadcasting Company to 1 January 1927, then British Broadcasting Corporation
DSIR	Department of Scientific & Industrial Research
FOL	Federation of Labour
GPO	General Post Office
HC	Harris Collection
HMV	His Master's Voice
M	Mackay papers
N	Nash papers
NA	National Archives
NBS	National Broadcasting Service
NCBS	National Commercial Broadcasting Service
NPA	Newspaper Proprietors' Association
NZART	New Zealand Amateur Radio Transmitters
NZBB	New Zealand Broadcasting Board
NZBC	New Zealand Broadcasting Corporation
NZBS	New Zealand Broadcasting Service
NZCDC	New Zealand Cooperative Dairy Company
NZCER	New Zealand Council for Educational Research
NZEF	New Zealand Expeditionary Force
NZPA	New Zealand Press Association
NZPD	New Zealand Parliamentary Debates
NZRU	New Zealand Rugby Union
ORA	Otago Radio Association
P&T	Post and Telegraph
RBC	Radio Broadcasting Company of New Zealand Limited
RNZ	Radio New Zealand
RNZSA	Radio New Zealand Sound Archives
RR	*Radio Record*
S	Scrimgeour papers
TAB	Totalisator Agency Board
UPA	United Press Association
WEA	Workers' Education Association

Preface

The history of New Zealand broadcasting falls into two periods, divided by two developments of the early 1960s — the introduction of television and the removal of the long-standing administration of broadcasting as a government department. This volume deals with the first of these periods, when broadcasting was the preserve of radio. These radio years saw the development of a new form of communication and its use to provide new genres of entertainment, new varieties of information and education and new forms of community and national understanding. It is a story that stretches back to the 1890s and the work of early researchers. The first broadcasting stations opened in the 1920s, but the 1930s is the decisive decade. During these years radio became a common household possession and broadcasting grew to become a national institution, under direct government control. Broadcasting assumed a prominence in New Zealand society which has never been lost. This history offers an account of this growth and acceptance, and of the actions taken by many individuals to further and to influence the nature of New Zealand broadcasting.

This history was commissioned by the Broadcasting History Trust. Without the trustees' foresight and assistance, it would not have been completed. I am grateful for the continuing support of the five trustees — Hugh Rennie, Elizabeth Alley, Brian Corban, Richard Rowntree and Beverley Wakem. I am also grateful for their scrupulous non-intervention. I have been assisted by the Victoria University of Wellington and especially its Stout Research Centre, where I was housed for a year while researching and writing. I thank the director at the time, James Collinge, for his assistance. For my time at the centre and for my conversations on broadcasting and other topics with my colleagues there, I am grateful. I also wish to acknowledge the assistance of the University of Waikato, which has given me periods of research leave to work on this history. I am indebted to Elizabeth Caffin and Gillian Kootstra for their careful attention to my manuscript.

Broadcasting records are to be found in various collections, and I acknowledge the assistance I have received from the staff of the Alexander Turnbull Library, the General Assembly Library, the National Archives, the

Radio New Zealand Library, Stephen Riley and staff of Radio New Zealand Sound Archives, and George Newlands of the Telecom Museum and Archives.

Various individuals have been influential on this work and I am happy to acknowledge them. Particularly I thank C. E. Beeby, John Campbell, the late Tom Clarkson, Peter Coleman, Peter Downes, Brian Easton, John Farnsworth, Geoff Lealand, Arthur Olsson, Alwyn Owen, Noel Palmer, Jack Perkins, John Roberts, Claire Toynbee and Tony Wilson. This history is the better for their assistance.

Hamilton
April 1994

Introduction

Radio began as wireless telegraphy, used for Morse code communication. It is an innovation from the first years of the twentieth century that brought a new method of point-to-point communication. By the start of the 1920s the transmission of voice and music was possible and broadcasting, a new form of communication, had begun. Broadcasting is the deliberate dispersal of a signal to all and sundry; the word is a metaphor derived from agriculture, where it refers to the open-handed method of sowing seed. Broadcasting stations were soon transmitting in many countries including New Zealand, where the new medium was rapidly adopted. During the 1930s radio receivers became household possessions for most of the New Zealand population and broadcasting became a major source of entertainment and information. At first stations transmitted usually for two or three hours in an evening, and not every evening. By the end of the 1930s morning and afternoon sessions had been added, and radio listening was both the country's main form of leisure-time activity and a new means of communication from the government to the population. By the end of the 1950s stations were broadcasting continuously throughout the day — 2YA from 5.30am and the commercial stations from 6am — and carrying through to 11pm or midnight.

Radio was taken under government control soon after its invention, and treated as part of the telegraph and postal services. Broadcasting was another matter: the government's initial view was that it should not be involved except as a regulator and licenser. Small private stations were started in various towns. The official view soon became that broadcasting should be organised nationally, and these stations found themselves in competition with the Radio Broadcasting Company, a private company contracted by the government to transmit via 1YA, 2YA, 3YA and 4YA from the main centres. Government involvement increased, culminating in the Labour Government's 1936 decision that broadcasting be run by the state. Nearly all stations were taken over. The national stations (the four YA stations) became a government department. Local stations were purchased and transformed into commercial stations, coordinated through another government department. In 1943 the two departments were amalgamated and later renamed the New Zealand

Broadcasting Service. New Zealand broadcasting was run by this organisation until the 1960s; that is, for the rest of the period covered by this volume.

The interest and importance of broadcasting lies in its widespread acceptance by the population and in the way it became part of and changed people's way of life. Initially both radio and broadcasting were regarded as impractical fads, not to be taken seriously. This attitude passed and within fifteen years of the start of broadcasting in New Zealand, more than half of households possessed a set and radio ownership was continuing to grow rapidly. Radio sets were eventually put in the workplace, in hotels, in motorcars and other places. But most were in the family home. Broadcasting developed as a home-oriented form of entertainment and information. The convenience of the home-based radio sets and the popularity of the programmes was such that radio listening became an everyday activity for individuals of all ages. Within each centre the local radio station became a focus for entertainment and information and for community activity — and also a reliable means of communication during times of emergency.

Music was the basis of radio programming. It was made available to the bulk of the population for hours daily, in a way which few people had previously been able to experience. Debate over broadcasting practices centred at first on music. At issue was the purpose of broadcasting. It was argued that broadcasters had a duty both to music and to the cultural enhancement of the audience to concentrate on established classical music rather than on contemporary popular songs. To educate the audience in the appreciation of classical repertoire meant the deliberate neglect of contemporary music. This was never accepted by the general New Zealand audience.

Also at issue was the relationship between radio stations and their local communities. Elitist standards increasingly defined the YA stations. They developed as centrally controlled, coordinated national stations offering a quality entertainment explicitly defined as superior to that available locally. A role with the local community became more a feature of the private stations and, after their demise, the commercial stations. Even here, as record-playing became the dominant format for all stations, the music played was increasingly popular rather than local. Rather than responding to the local community, broadcasting practices redefined the nature of the community. Broadcasting assisted the development of new communities of interest that cut across their geographically defined predecessors.

Radio broadcasting developed a range of new entertainments. Ironically these were looked at askance by many broadcasters, who instead wished to concentrate on earlier forms of entertainment: the pop song should not replace the symphony, nor the radio serial the play. From the time a choice was available, unwavering through the radio years, the majority New Zealand audience chose not to use radio to educate itself in an appreciation of the classical fare. The eventual decision made by broadcasters was to develop different types of programming for different stations, to provide a range of entertainment and let the audience choose which station to tune to. Allied with this station specialisation was the political decision to link the majority programming with commercial broadcasting and to finance the other stations from the annual licence fee.

The programmes broadcast were new presentations of older forms of entertainment and instruction, along with new variants. The styles of radio entertainment quickly rose in popularity but, like broadcasting itself, also became topics of public discussion and dissension. Radio serials — a new genre — were avidly listened to, but were debated both for their content and, because most of them were imported, for their introduction of what were regarded as non-New Zealand values. In addition to these were programmes ranging from quiz shows and talent quests to broadcasts to schools and religious services. The programmes created new interests and changed both the nature of entertainment and the home lives of the audience. Broadcasting also brought new personalities to public prominence. Many announcers and entertainers gained a widespread popularity which sometimes perturbed their employers. While not necessarily as high in public esteem, senior broadcasting administrators also became well-known figures whose policies and actions were subject to constant scrutiny.

The broadcasting audience was an unexpected collective that grew in size beyond any voluntary group until it approximated the entire population. Soon commonplace, radio broadcasting was a revolutionary means of communication that reached straight into the family home, previously a sanctuary from outside influence. The audience was cohesive on various levels and represented a new grouping of both the local community and the nation — and eventually the world. The size of the audience and the nature of the communication were intensely debated. Although radio is inherently a two-way medium, broadcasting was developed as a one-way form of communication: broadcasters spoke and audiences listened. While this was similar to earlier entertainment and public communication, broadcasting was seen as changing the nature of the experience. Views varied considerably, from the understanding that broadcasting should be applauded as representing a new democratic era to the criticism that it should be deplored as ushering in a form of totalitarian manipulation. All sides of this debate saw the audience as a grouping that was so large as to make broadcasting a topic of great interest and importance.

On a political level, the growth and increasingly public nature of broadcasting saw governments become more and more involved in what they had originally regarded as beyond their concerns. New Zealand politicians were slow to see a political significance for radio. But during the 1930s broadcasting became part both of the political contest and of the terrain where that contest took place. The first Labour Government developed radio as a means of circumventing what it regarded as unfavourable newspaper publicity. Radio became a new way for representatives to communicate with the electorate, for rulers to talk to the governed. Parliamentary broadcasts were begun, a system of national news broadcasts, prepared through the Prime Minister's Department, was set up, and generally radio was used by the government to communicate directly with the people. That broadcasting should offer an independent source of news and commentary and should be administered in a manner other than by direct government control was increasingly considered, but not until the 1960s was there substantial change.

A study of broadcasting presents its own difficulties. Radio programmes are cast upon the air waves, received or not received and then generally lost beyond recall. Unlike newspapers, broadcasts are not delivered in any permanent form. During the 1930s it became possible to record, and a collection of sound archives was started, but only a tiny fraction of programmes has been preserved. A written history of broadcasting is thus presented in a medium other than the one under discussion, and deals with a topic of which the actual productions are largely unavailable for examination.

The written records of the various organisations, public and private, that have been involved in broadcasting are also no longer necessarily available. Often, material has unexpectedly survived. Radio New Zealand Sound Archives was in 1989 given a crate of papers that included most of the records of the Radio Broadcasting Company from 1927 to its close in 1931. Little information is available about the early years of the company. Some state organisations have been less than assiduous in record preservation. The Post & Telegraph Department was for many years closely involved in not just the regulation but also the day-to-day practice of broadcasting, but there is little record of this aspect of the (now abolished) department's activities. Other organisations have been more painstaking in their record collection and preservation. In particular the Broadcasting Archives — the records of the Broadcasting Corporation and its predecessors back to the New Zealand Broadcasting Board, are a rich resource — though they too are by no means complete. Many broadcasting employees are also authors, published and unpublished, who have contributed much to my understanding. Of the unpublished material, mention must be made of the work of the NZBC historian, J. H. Hall. There are few records from the private stations that existed prior to the election of the first Labour Government. Not all kept records and fewer still attempted to preserve them. There are exceptions: the prolific Ian Mackay of 2ZR Nelson has left much information on that station and on his involvement in later commercial broadcasting. However many details of many of our radio stations, particularly but not exclusively those of the early private stations, are no longer available.

What does remain clear is the sense of a new phenomenon permeating society. As listening to radio programmes became a favourite pastime, broadcasting became a topic of public interest. As radio was used as a medium for news and political communication, broadcasting came under increasing scrutiny. Broadcasting is an item that crops up frequently in parliamentary records, newspapers and the range of publicly available records. As it became part of our lives so it became part of public discourse. Its history is of the expectations for the new method of communication and of the delights and disappointments that followed. It is also a history not just of the growth of broadcasting but of the changing nature of the society that adopted the medium.

1
Radio and State Control

Broadcasting stations went on air in the 1920s. The introduction to broadcasting history begins, not with these first stations, but with the development and usage of radio itself. The technology that is radio and attitudes towards it grew over many years, before and after the start of these stations. At first radio's broadcasting characteristics were not appreciated and the medium was simply regarded as a new way to take a message from point to point. Within New Zealand, as in many countries, by the time broadcasting began, radio was already operating within legislation and under an official scrutiny that much influenced the early practice of broadcasting. In New Zealand both official and amateur attention focused on radio in the opening decade of the twentieth century. Before this New Zealand had been influential in the development of the new medium through the work of one of her more illustrious sons. As a prologue to New Zealand broadcasting, we begin with the first steps in radio and the work in this connection done by Ernest Rutherford.

FIRST STEPS IN RADIO

The pivotal experiment that began the development of radio was that of Heinrich Hertz in 1888. Hertz's experiment confirmed the electromagnetic theory first published in 1865 by the Scottish physicist, James Clerk Maxwell, which contended that light was a type of electromagnetic wave, and that other such waves existed. The theory remained untested until Hertz, in Germany, succeeded in producing and detecting the waves Maxwell had described. Hertz magnified the electrical oscillations produced by the discharge of a Leyden jar by passing them through an induction coil. The charge was then passed through rods on each end of the induction coil to hollow metal balls on the end of each rod, with a space between the balls just large enough to make an air gap. Hertz's achievement was to project the charge out from the space between the balls, known as the spark gap, to the detection apparatus — a loop with a similar gap. Hertz had thus demonstrated the existence of radio waves, or 'Hertzian' waves, as they were to be known for some years. His experimental confirmation of Maxwell's predictions spurred many researchers to follow his lead. What was implicit in his work, though by no means fully understood, was that signals could be carried on these waves and radiated in all directions without any intermediary means of conveyance — without wires.

The first major change to Hertz's apparatus was in the method of detection. Reasoning that no matter how good a conductor a piece of metal might be, its resistance would be much enhanced if it were shaved into little pieces, Edouard Branly in 1890 developed his coherer. It was a 6-inch-long glass tube full of iron filings. When struck by radio waves, the metal filings cohered, at which point their resistance to further waves was greatly reduced. When the filings were loosened by shaking or tapping, they could be used to register the reception of further waves. Oliver Lodge, the British physicist, developed from this the receiver most used in early apparatus. In a lecture in 1894 to the Royal Institution he demonstrated his coherer, which included a 'trembler' to mechanically shake the filings and restore their resistance.

ERNEST RUTHERFORD

Many people from different countries have been influential in the development of radio technology. Ernest Rutherford, destined to be his country's most eminent scientist, began his contribution towards radio in 1894 when, as a research student at Canterbury University College in New Zealand, he studied the high frequency magnification of iron — a problem then at the forefront of electrical technology. He was interested in the suggestion made by Tesla (developer of the Tesla coil, a high-frequency, high-voltage transformer), that it was possible to transport electrical power without wires.

As part of his research, Rutherford repeated Hertz's already famous experiment. While his method of generating radio waves was essentially a duplication of Hertz's work, his method of detection was novel. With the aim of detecting short current pulses, he developed the magnetic detector. This consisted of short, thin strips of steel — (Rutherford experimented with sewing needles, piano wire and other material). Around the steel strips he wound insulated fine copper wire. He found that the rapidly alternating current would magnetise the steel strips or, if they had already been magnetised, would demagnetise them. He further showed that the effect was only on the outer layer of the strips. This research, which Rutherford developed further at Cambridge, provided the 'essential physical basis of an oscillation detector, and therefore of a receiver for electric waves'.[1]

The early experiments with radio waves should not be confused with message transmission, even though a distinct signal was involved. Such apparatus was capable, in the language of modern information theory, of transmitting one 'bit' of information. It could be used to turn a switch and thus, for example, ring a bell — which is what Rutherford did in his later Cambridge experiments. In his New Zealand work, the reception of radio waves was indicated by a signal on a galvanometer registering the magnetisation of the steel strips of the detector.

At Canterbury Rutherford is recorded as transmitting and receiving over a distance of about 60 feet through various intermediary walls.[2] It is most unlikely that Rutherford did achieve these results, or that he even considered the implications of his experimental work for wireless telegraphy while he was a student in New Zealand. The only evidence for it is in a letter in the *Press* on 22 October 1937 from R. M. Laing, a fellow-student. Laing, writing after Rutherford's death, stated: 'I saw him send a message by Hertzian waves from one end to the other of Professor Bickerton's laboratory . . . through various intermediary walls.' There is no other corroborative evidence. Bickerton did not mention the matter in his report on Rutherford at the time of Rutherford's application for the Exhibition scholarship which was to take him to England. Bickerton's report is highly laudatory and such success is surely a point he would have included. Also, as Eve points out, Rutherford, in a letter from Cambridge, dated 8 December 1895, writes, 'I can get quite a large effect at twenty yards', as though this result was obtained for the first time.

In 1895 Rutherford moved to the Cavendish Laboratory at Cambridge University. J. J. Thomson, the Cavendish professor, encouraged his new student to continue his Canterbury research. It was at this time that Rutherford became aware of the implications of his work for communication technology. A man who sought and delighted in the practical application of scientific knowledge, he was in the company, at Cambridge, of people who were quick to see an application for the new technology. In particular Sir Robert Ball, to whom Rutherford was introduced, 'developed a tremendous interest in my experiments on the detection of electric waves for long distances'.[3] Sir Robert was the Royal Astronomer for Ireland, and this position meant that he was interested in the matter of communication with lighthouses and lightships. At Ball's instigation and with Thomson's encouragement, Rutherford now directly addressed the matter of the detection of radio waves. He realised his method was superior to Lodge's coherer, and with further modifications to his detector he was soon able to transmit and receive over a distance of half a mile. Thomson claimed in later years that Rutherford held the world record for distance of transmission and reception during his first few months at the Cavendish.[4] Rutherford's achievements soon gained him considerable attention. Within a few weeks of his arrival at Cambridge he was invited to demonstrate his experiments to the Cambridge Physical Society, and in September 1896 to the British Association for the Advancement of Science. His first paper to the Royal Society, in June 1896, also concerned his magnetic detector.

Rutherford's equipment could detect radio waves in a manner superior to anything so far employed. His research was later seized on and developed by others. As Rutherford wrote in 1910, 'This effect of electrical oscillations of altering the magnetism of iron is the basis of magnetic "detectors" developed by Marconi and others which have proved one of the most sensitive and reliable of receivers in radiotelegraphy.'[5]

Rutherford was interested in the application of his work to marine communication, and he was quick to see the financial possibilities. In a letter to his fiancée he wrote:

> I have every reason to hope that I may be able to signal miles, without connections, before I have finished. The reason I am so keen on the subject is because of its practical importance. If I could get an appreciable effect at ten miles I would probably be able to make a considerable amount of money out of it, for it would be of great service to connect lighthouses and lightships to the shore, so that signals could be sent any time.[6]

Thomson also appreciated and attempted to promote the development of Rutherford's apparatus, though his assistance may have been counter-productive. He sought the advice of Lord Kelvin, the Glasgow professor of physics. Although impressed, Kelvin questioned the long-term practicality of such a communication system and advised against any large-scale capital expenditure. Even if the advice had been to go ahead, it is not likely that Rutherford, Thomson or the Cavendish would have had access to the investment capital necessary to embark on the commercial development of a new means of communication.[7]

After his first few months at the Cavendish and before his demonstrations in 1896, Rutherford moved away from radio research. He was intrigued by the detection of radio waves and entranced by the fact that solid matter was as open air to their passage. However, for him the greater challenge was not the development of a practicable communication system, but the intellectual puzzle of the nature of matter raised by the new discoveries. His radio research was an early part of that investigation which was to lead, in 1911, to his announcement of atomic theory and introduction of 'the greatest change in our idea of matter since the time of Democritus'.[8]

GUGLIELMO MARCONI

Contemporaneously with Rutherford, Guglielmo Marconi, a young Italian, was also experimenting with radio waves. In spite of Thomson's claim that Rutherford was transmitting further than anyone else in 1896, it is likely that Marconi's work at least equalled Rutherford's. Marconi was not so interested in the theoretical physics that intrigued Rutherford. He had a different obsession — the development of a new communication system. He was to produce an effective communication device from the work of the early experimenters and take it from the laboratory into the commercial world. Marconi, three years younger than Rutherford, was born in Bologna in 1874. In his early twenties he developed a system of radio communication, and offered his invention to the Italian government. When they turned him down, he moved to England early in 1896 to seek support for his research. Marconi was no outsider to the British business world. His mother, who tirelessly supported her son in his radio interests, was of the Anglo-Irish Jameson family of brewers and distillers, and her family connections gave him the financial backing he needed.

In 1896, radio researchers sought to develop apparatus which automatically restored the magnetisation to the steel strips of the detector so that a continuous stream of signals could be received. In Rutherford's research the magnetisation was lost on receipt of the first signal and thus did not allow reception of further signals. Rutherford did find a method of overcoming this problem.[9] Similar solutions were reached independently by others, in particular Marconi, who designed the magnetic detector that gave the most practical service.[10] Marconi's other contributions to the development of radio included the earthed aerial, which much increased the strength of the received signal, and the introduction of the telegraph Morse code key, which turned radio into a communication system capable of transmitting messages.[11]

Marconi was determined to bring radio to public attention. He established a link between the East Goodwin lightship and the South Foreland lighthouse — a distance of 12 miles. In March 1899, when a ship ran aground on the Goodwin Sands, radio signals hastened the rescue of passengers and cargo. From this point the need for radio in lighthouses and lightships was recognised. On 28 March 1899, Marconi successfully sent a message across the English Channel.[12] In December 1901 he signalled across the Atlantic.[13] Marconi was succeeding in making radio a viable and well-known communication device. But there was still some time to go before it became financially profitable. By 1908, the Marconi Wireless Telegraph Company had spent £500,000 and still had paid no dividend.[14]

What Marconi in particular had developed was the communication system known as wireless telegraphy. The wonder of radio at its inception, and for at least the first generation who experienced it, was that it allowed the transmission of messages without using wires. This was so startling that it was long known by what it was not — 'the wireless'. Marconi's original system was not the radio of later years. The electrical generators then in existence, such as those used in the spark transmission systems, produced interrupted waves; voice and music transmission were not yet possible. The system was, however, well suited to the transmission of Morse code via a telegraph key — hence the term 'wireless telegraphy'. The initial systems, including Marconi's, were devised as ways of transmitting Morse code messages from point to point.

The equipment used by James Passmore in 1902 in his first successful attempt at wireless telegraphy communication. OTAGO EARLY SETTLERS' MUSEUM

AMATEUR RADIO

The professional development of radio interested many individuals besides Marconi, and competing systems emerged. Around the world there were also countless amateurs interested in the device. New Zealand had its share of enthusiasts.

The next New Zealand experimenter after Rutherford was another Canterbury College student, John Sissons St George Cooper. The sole reference to Cooper's experiments is in the minute book of the Canterbury Philosophical Institute. The main item on the agenda for the meeting of 7 June 1899 was an address by Cooper on 'wireless telegraphy illustrated by experiments'.

After Cooper, interest in radio in New Zealand was shared by various young men. It was an exciting area of research that was beginning to enter the public imagination. Various amateurs attempted, with gradually increasing success, to transmit and receive signals. One of these was James L. Passmore, a sixteen-year-old in 1901 when he 'seriously took up the study of things electrical'. During the Easter holidays of 1902 he successfully communicated over a distance of 100 yards. The equipment Passmore constructed for this initial foray now holds an honoured position in the Otago Early Settlers' Museum in Dunedin. By February 1903, Passmore's communication range was 250 yards; 4 months later it was 6 miles, and in 1904 he picked up signals from HMS *Powerful*, off the Otago Heads about 14 miles from Passmore: 'The next night the warship's operator sent a series of dots and dashes to Passmore who successfully received them in the presence of the yeoman of signals.'[15] No doubt the yeoman of signals was delighted to meet a confrère. The Royal Navy had begun equipping its ships with radio in 1900,[16] but in 1904 in New Zealand waters, opportunities for radio communication were few and far between.

Harris accords Passmore the honour of having 'built and operated the first wireless telegraphy set in the South Island, and as far as can be established, in New Zealand'. This may be so. But even if we ignore Cooper, there are other likely candidates. Among them are two young men, J. E. (Ernest) Strachan and Robert Scott, also from Otago. Writing in 1970, Strachan recalled that he and Scott 'were associated in all our enterprises — electrical experiments, long cycle trips, and countless discussions on science, philosophy, religion and the state of the world. "Wireless" came into that very naturally.'[17] Strachan and Scott 'began by making an induction coil throwing a two inch and a half spark. Then we made two Leyden jars out of jam jars and connected the apparatus up with two copper rods and brass balls as a spark gap . . . I do not suppose we would have got much further if we had not got hold of a journal describing Branly's coherer.'[18] Hall records that the pair began their experiments as early as 1899; in the summer of 1900–01 they repeated Hertz's work to the extent of sending a spark from one room to another, and by mid 1901 they could ring a bell at a distance of 100 yards from their transmitter. By 1902, Scott and Strachan were in communication with W. P. Huggins, an experimenter in Timaru who was achieving similar results to their own.

Teenage boys are not the most meticulous recorders of historical events. It is possible that there were other radio pioneers in New Zealand, for many people were quick to appreciate its potential. But the New Zealand government was to bring in a policy that would discourage experimentation by its citizens.

EARLY RADIO POLICY

Probably the first time radio was brought to the attention of a New Zealand politician was on 16 February 1902, when Premier Richard Seddon, about to depart for England, was given the London address of the 'inventor of the automatic lighthouse indicator by means of wireless telegraphy' — though whether Seddon took the matter any further is not recorded.[19]

The first official mention of radio in New Zealand is a *New Zealand Gazette* notice to mariners on 17 October 1902. It is a notification by the Marconi Wireless Telegraph Company of stations established by that company. None were in the southern hemisphere, let alone in New Zealand waters.

In spite of this slight acquaintance with radio, the New Zealand government in 1903 moved to take control of the medium. Briggs cites the British Wireless Telegraph Act of 1904 as 'the first of its kind in the world',[20] but it was in fact preceded by the New Zealand Wireless Telegraphy Act, passed on 26 September 1903. The first Australian equivalent, also called the Wireless Telegraphy Act, was passed in 1905. A major task for any government is the provision of means of communication, and this had been a particular preoccupation in New Zealand. The European settlers had always felt isolated, and desire for contact with the homeland had led all New Zealand governments to pay close attention to such concerns as postal and shipping services and the telegraph. Radio was recognised immediately by New Zealand's Liberal Government as an innovation in communication, and as such, worthy of government interest.

The Wireless Telegraphy Act of 1903 was a short one of four clauses that gave the government the right to receive and transmit messages by wireless telegraphy and to establish stations for that purpose. It also decreed that any person who 'erects, constructs or establishes any station or plant for the purpose of receiving or transmitting' without first obtaining official consent was liable to a maximum fine of £500 as well as forfeiture of their equipment. Albert Pitt, the Attorney-General, stated, 'The whole principle of the Bill is that the Government intend to acquire a monopoly of this system in the colony, just in the same way as has been done in regard to telegraph lines and telephones.'[21] The Act granted the government the right to control the use of radio and was a statement that the government intended, although not necessarily immediately, to exercise that right.

Radio, until well into the second decade of this century, was wireless telegraphy, Morse code communication without wires. Point-to-point communication was the only use envisaged for it. The nature of radio was by no means fully understood by experimenters, let alone by politicians and officials. It was realised that radio did transmit in all directions simultaneously, but it was not until the advent of radio telephony, with the transmission of voice and music, that radio's potential to broadcast communications deliberately to all and sundry was recognised as a distinct form of communication. In 1903 radio's omnidirectional transmission was seen as a problem in point-to-point communication — it made it difficult to preserve the confidentiality of private messages. It was also seen as an obstacle to delivering a message successfully to the intended recipient. This understanding of radio helped lead to the perceived need for total government control of the new technology. Sir Joseph Ward, the postmaster-general and electric telegraph commissioner, argued that

> Under the Marconi System one of the difficulties was that the messages could not be transmitted in a direct line from station to station . . . if a Marconi message was going to any point in the colony it would not go in a direct line but in a series of circles. It would therefore surely be seen that if private people were allowed to establish stations, as was suggested, even although not for hire or profit, it would be possible for many of these Marconigrams to be intercepted and never to reach their proper destination at all.[22]

One has a vision of a radio message as a single entity racing in ever-increasing circles until stolen by an unintended recipient.

There was already a fear that this new method of point-to-point communication could have consequences for existing systems. The Act protected the considerable government investment of the previous forty years in telegraph and telephone systems. Ward made it clear that the government was concerned not only that an outsider might make money from the new 'Marconi system of telegraphy', but that the new system might render the telegraph network obsolete.[23]

The need for a government monopoly received little debate in Parliament. There was a general acceptance of monopolistic control in New Zealand — a country which, by the early years of this century, had acquired a reputation for state involvement in activities properly left to private enterprise. This 'perfect mania for appealing to the state'[24] meant that the New Zealand government's monopoly control of radio was unexceptional. The policy of state involvement, coupled with the wish to protect previous state investment from future competition, made the Wireless Telegraphy Act a logical move by the Liberal Government.

The major objection to the proposed government monopoly was voiced in the Legislative Council by A. Lee Smith of Otago, who saw the proposal as 'a premium on what I might call scientific inertness'. The work of the Otago teenagers was a matter of public notice, and Smith contended that such experimentation should be encouraged. He also argued that rural people should be allowed the use of the new device. 'Take a large station in the country, for instance, where there are two houses, say, twenty miles from each other. What a great advantage it would be to the owner if he could by means of wireless telegraphy communicate from one place to the other. By this Bill that man would not be allowed to do it to the extent of even a quarter of a mile.'[25] It was a prescient suggestion, for the development of radio was soon to give good service among 'neighbouring' farms. But the government's actions discouraged the development of any indigenous use of radio.

While Smith's colleagues pointed out that licences could be applied for, they were not to know that licensing procedures would not be implemented. However, the objection which stopped a proposed amendment to the bill allowing individuals to experiment with radio came from Seddon. He raised his powerful voice against the amendment with the argument that anything less than full government control would allow foreign powers into New Zealand. This possibility, he said, had caused the British government to express its concern. 'If the British Government thought it necessary to send a communication of this nature to this colony respecting such a matter then our Government should do its duty to the colony and to the Empire as well.'[26]

Having passed the Wireless Telegraphy Act and thus protected its right to control any future development of radio in New Zealand, the government investigated the current state of radio — but it showed no haste to introduce the medium to New Zealand, nor any wish to assist or take part in its development. Nor did it make any immediate moves to curb the activities of the amateurs.

Discussions whereby New Zealand eventually set up international radio links first took place in 1904 between the New Zealand and Australian governments and representatives of the Telegraph Construction Company of the United States, which was then operating the De Forest and Shoemaker systems. J. W. Ormsby Hamilton, who negotiated on behalf of the American company, recalled that the negotiations collapsed because of the two governments' 'exacting demands'.[27] The exacting demands were, essentially, that the service did work, and that communication be established and maintained between the two countries. This was, at that time, beyond radio's capabilities. In 1904 the Post and Telegraph (P&T) Department also had unsuccessful negotiations with the Marconi Wireless Telegraph Company; it reported that 'advancement in the direction of making wireless telegraphy a commercial success appears to be slow'.[28]

Parliamentary indication of the government's plans for use of radio came on 18 October 1905 in a reply by Ward to a query whether the government would establish 'wireless telegraphic communication between the Chatham Islands and the main land'. Ward acknowledged that such a plan had been 'under consideration', but added that the government had in mind a link of longer distance: 'This up-to-date means of communication should be established between New Zealand and Australia.'[29] This was in advance of radio's capabilities. Two years later, on 3 February 1908, the first radio message was sent from New Zealand to Australia when the prime minister took advantage of the presence of British warships to send a message to his Australian counterpart. HMS *Pioneer*, in Wellington harbour, transmitted to HMS *Psyche*, stationed at Port Jackson. But this cross-Tasman transmission was not direct. HMS *Powerful*, stationed in the Tasman at the edge of its radio range, received the signal then steamed towards Port Jackson. Six hours later it was in radio contact with HMS *Psyche* and the prime minister's message was relayed.

In 1906 the Marconi Company completed stations at Queenscliff, Victoria and Devonport, Tasmania. The Australian and New Zealand postmasters-general decided at this stage to defer any further consideration of the proposed Australia–New Zealand link until these stations gave practical evidence of effective communication.

The other application of radio, as noted in the 1905 P&T Department report, was 'for communication with vessels at sea, and outlying islands, [where] the wireless-telegraph system could be brought into use with considerable advantage'. This was to be the first use of radio in New Zealand.

During a 1905–6 trip to Europe and the United States, Ward was courted by the British and American companies, and much increased his understanding of radio. In Rome he saw the Marconi military station in communication over 200 miles with its counterpart in Sardinia. He noted that both the Italian and British authorities had fitted their war vessels with the Marconi system. In London he inspected the Marconi system and that of the Lodge–Muirhead Wireless and General Telegraphic Syndicate. Ward sailed from London to New York on the *Majestic*, which was fitted with a Marconi system. He was able to observe the ship receiving daily news bulletins from Britain and America as well as communicating with steamers travelling east and west. In New York, Ward visited the United States navy yard and inspected the Shoemaker system, which the navy was fitting to its vessels. Ward returned to New Zealand appreciative of radio's potential for international communication and convinced of its practicality for marine communication, both merchantile and military.[30]

By this time the New Zealand government, along with most other governments, was aware of the strategic value of radio. Its indispensability in naval warfare had been demonstrated by the Japanese in the Russo-Japanese war. The New Zealand government was also aware, as was the New Zealand press, that radio offered an alternative to the service offered by sea cables. Radio could thus lessen New Zealand's reliance on the cable service — which was vulnerable in war and subject to the ravages of the sea — and on the cable companies, widely viewed as arrogant and extortionate.

On Ward's return, Parliament voted £2000 in the public works estimates for 1906–7 for wireless telegraphy — the first allocation of public funds for this purpose. There were no rapid results. Negotiations with the Australian government to establish radio links between the two countries did not lead to any immediate action. In general, New Zealand's policy was to wait and see: 'It will probably be found the best policy to withhold any considerable outlay . . . until something more definite is known . . . from scientific work now proceeding.'[31]

The Marconi Company gave the first public demonstration of radio in New Zealand at the end of 1906, as part of the Christchurch International Exhibition. Two Marconi ship spark transmitters were installed, one at the Exhibition in Hagley Park and the other at the Islington Meat Freezing Works, where a 220-volt direct current was available to operate the alternators. The Marconi engineer in charge was H. M. Dowsett, and the local operator assigned from the P&T Department was E. A. Shrimpton. Various New Zealanders gained their first interest in radio from this Exhibition, including Shrimpton, who in later years as a senior Post Office official and a keen amateur radio operator was to be of much influence on early New Zealand amateur operators and broadcasters.

The concentration on radio as a medium with particular application to maritime and military uses was given further impetus by the Radio-Telegraphic Convention, signed at Berlin on 3 November 1906. New Zealand ratified the convention the following year. The convention's major achievement was to overturn the Marconi Wireless Telegraph Company's insistence on communication with its own stations only and with no others. This attempt at monopoly, supported by the British and the Italians, was negated by the concerted efforts of other delegates, in particular the Americans and Germans. After resolving this issue, the protagonists united to bolster a military control of the new maritime communication system. The convention promoted the use of wireless telegraphy for maritime purposes, but with a primary orientation, as the British phrased it, 'to the supreme consideration of naval interests and national defence'.[32]

The emphasis on maritime and military communication meant that radio's commercial applications were neglected. It was argued that this was necessary because of radio's rudimentary development.[33] There is some strength in this contention, although it ignores the considerable commercial use of radio at that time in North America and Europe. But it does not acknowledge that the major block to the commercial use of radio was the opposition from the established and powerful cable companies.

A comprehensive scheme for radio coverage of the Pacific was first introduced in 1908 by the London-based Pacific Radio Telegraph Company. The Lodge–Muirhead system was the preferred type of radio, and the proposal was to use Fiji as the headquarters of a chain of stations linking Fiji, Samoa, the New Hebrides, the Caroline Islands, Tahiti, the Cook Islands, Fanning Island, New Guinea and the Sandwich Islands with Australia and New Zealand. The Pacific Cable Board supported the scheme, which safeguarded the Board's dominant position by declaring that 'Wireless telegraphy would not be used between Fiji, New Zealand and Australia, except in cases of emergency, as radio-telegraphy cannot compete profitably with an already completed cable.' The New Zealand government expressed cautious interest. But it also pointed out that the scheme did not cover all New Zealand's requirements; in particular the Chatham and Auckland Islands should be included.[34]

The proposal was eventually rejected at the Wireless Telegraphy Conference held in Melbourne in December 1909, though not for these particular objections. The decision at the Melbourne conference was against private ownership of the scheme and for state control. The earlier effort by the Marconi Wireless Telegraph Company to establish and control a communication network by allowing only their clients to lease sets and operators, had taken much international effort to overturn. The Australian and New Zealand governments were not to be led into an antipodean version of that proposal.

An alternative scheme was introduced at the Melbourne conference by representatives from Australia, New Zealand, Fiji and the Admiralty Islands. The conference passed resolutions calling for three high-power stations to be established at Suva in Fiji, at Sydney in Australia, and at Doubtless Bay in the north of New Zealand. The Sydney station would communicate with Doubtless Bay, and Doubtless Bay with Suva. Three medium-power stations were to be established at Tulagi in the Solomon Islands, Ocean Island in the Gilbert Archipelago, and Vila in the New Hebrides. The conference scheme was principally to link Australia and New Zealand, through Suva, with the international network. For the Australian and New Zealand governments, dominant at the conference, the linking with other Pacific islands was of lesser consideration, and the cost of the scheme was more in keeping with the modest payments the two governments were prepared to make.[35]

The New Zealand government decided to add to the Melbourne conference scheme. It proposed, in its own domain,

> Five wireless-telegraph stations to meet the requirements of defence, the navy, and the merchantile marine service in the Dominion. Provision is made for high-power stations at or near Doubtless Bay and Bluff, with a range of 1,250 nautical miles, and three medium-power stations, one at or near Gisborne, one at or near Cape Farewell, and one at Sumner.[36]

The government viewed the stations as electronic lighthouses for shipping, as well as for international communication. With the high-power stations at opposite ends of the country, they could be in touch with all approaching ships.

The question of whether the new system was to compete with the cable companies for commercial traffic was soon raised. The Australian government was prepared to leave the field to the cable companies, at least in the meantime. Alfred Deakin, the Australian prime minister, supported the opinion of his postmaster-general 'that there is no present justification for providing wireless stations to handle ordinary commercial telegraphic traffic'. Ward was more oblique: he noted that 'the stations on both sides of the Tasman could at any moment be used for such a purpose'.[37] However, when challenged on the point, Ward realised capitulation was necessary before the scheme could begin. The Pacific Cable Board chairman told Ward that 'The Imperial Parliament . . . [deemed] it necessary to obtain the assurance of the Commonwealth and Dominion governments that the proposed wireless-telegraph schemes of the Pacific will not compete commercially with the existing cable systems.' Ward replied clearly: 'It is not intended that the wireless telegraphic system in New Zealand is to compete with the existing cable system for commercial work between New Zealand and Australia. The wireless apparatus will be used for communicating with ships and for defence purposes only.'[38]

Thus, by the second decade of the twentieth century and before it possessed any radio equipment, the New Zealand government had decided to reserve radio for marine and government communication. The obvious commercial utility of radio was neglected in favour of the existing cable and telegraph systems. Its potential for use by New Zealand citizens was also neglected, for fear that this might imperil the government's preferred use for radio.

For some years, though, the government showed little interest in enforcing the provisions of the Wireless Telegraphy Act requiring would-be enthusiasts to obtain official consent before establishing a radio station. These provisions were regularly observed in the breach. Some individuals obtained a verbal agreement from their local telegraph inspector for their experiments; others took no notice of officialdom. Strachan continued with his experiments. In 1905 he moved to Lawrence, where he taught at the high school until 1912 and managed to enthuse various pupils and townspeople about radio.[39]

The less than assiduous application of the provisions of the Act was most apparent on 10 September 1908, when an intriguing public demonstration of radio took place in Dunedin. Rawson Stark, Stanton Hicks and Cyril Brandon, aged sixteen and seventeen, 'three clever boys' as reported by the *Otago Daily Times*, transmitted back and forth across the Otago Harbour, exchanging greetings between the mayors of Dunedin and West Harbour and sending a message, forwarded by telegraph, to the prime minister from the 'S. H. & B. Wireless Company'.[40] Stark and Hicks were school pupils who began their experiments in 1906. Their search for equipment led them to various city firms and to friendship with Brandon, an apprentice electrician. Stark, whose father was the city electrical engineer, had previously visited Strachan at Lawrence and no doubt gained considerable information from that source as well as from his home. The three young men had been transmitting between their homes and thus had established three-way communication between Ravensbourne, Andersons Bay and Caversham. Their demonstration was attended by various public figures. While the transmission was in breach of the 1903 Act, it was also one that could be publicly celebrated and recorded as well as being noticed on an official level. The message of greetings drew congratulations from the prime minister. This was a rare occasion on which the amateur use of radio received an appreciative response from officialdom.

THE GOVERNMENT, THE AMATEURS AND THE NAVY

From this point on the official response to amateur use of radio became antagonistic. In and around New Zealand in the first decade of this century the Royal Navy was the only official user of radio. Amateur and naval usage were to conflict, and this led to the government increasingly tiring of and acting against the activities of non-official users of radio. The incident which led to the policing of the regulations took place in November 1909 when HMS *Pioneer* was berthed in Timaru Harbour. When it radioed to HMS *Challenger* at Wellington, its signals were impeded by other transmissions. The offending apparatus was found the following day at the home of Alfred Hathaway. It was no large concern; Hathaway later wrote that 'as my power is derived from a single four-volt accumulator and a motor-car ignition coil, it was only while at the wharf, about one mile distant from my home, that the *Pioneer* was able to hear me'. The problem of interference was one which could be rectified. It was possible to tune out the offending signal, though Hathaway required somewhat more sophisticated equipment to do so. The investigating naval officer recommended 'that the authorities be informed, and a tuned system used'. The navy reaction was not antagonistic; they accepted that the two types of radio users, naval and amateur, could coexist. However, the government's response was to take the Timaru experimenter off the air entirely 'in order to avoid the possibility of interference with communications between men-of-war.' Two months later a similar denial of permission was communicated to two other Timaru residents who sought 'to erect an aerial at our respective houses in Timaru, and to attach our instruments which we are partly buying and partly making, so that we can continue our experiments, which are solely for our instruction, and not for the purposes of pecuniary gain'.[41]

Others who suffered a similar fate included D. G. (Toots) Mitchell and O. E. (Ted) Meinung, who became long-standing amateur radio enthusiasts. They unwittingly established communication with a warship at Nelson and brought on themselves 'an enforced period of inactivity'.[42] Mitchell, a telegraph cadet, had first become interested in radio while a pupil of Strachan's at Lawrence High School. Meinung, who had worked for ten years with Siemens, one of the German firms which in 1903 had merged to form Telefunken, had been in association with Mitchell since introducing himself after spotting the younger man's aerial. The two were to have the last word about being taken off the air. They both had long associations with the Dunedin private radio broadcasting station, 4XD. Mitchell was elected the 4XD inaugural secretary and operator. He remained with the station, and was station director at the time of his death in 1970.

A further casualty of the government's policy was A. E. H. (Hugh) Simpson, who first decided to make a radio set after seeing the Marconi

D. G. (Toots) Mitchell pictured late in his life at 4XD, Dunedin.
RADIO NZ

demonstration at the 1906 Christchurch exhibition. He has related his efforts and those of his colleagues John Bingham, Cecil Russell and Harold Wragge in these early years, during which they increased their range until they were in contact with ship stations. 'All this was illegal, of course,' Simpson blithely states.[43] At the time of a visit by HMS *New Zealand* to Lyttelton, Simpson and Bingham were put off the air and officially warned by the P&T Department of all the penalties under the Act. Naval ratings had established a portable station atop a Christchurch building to communicate with the ship, and Simpson and Bingham had contacted the portable station. Unfortunately for them the ship intercepted the transmission and reported to the authorities. Neither man was down permanently, and both were later involved in broadcasting.

Not all contact between amateurs and the navy caused strife. A group of boarders at Waitaki Boys' High School built a station with their rector's permission, and also listened to HMS *New Zealand*. The interest of the school was aroused by their success in picking up communications while the warship was at Melbourne and during her passage of the Tasman Sea. One of the messages intercepted was from General Godley inviting Captain Halsey to dinner at the military commandant's residence in Wellington. It happened that General Godley's inspection of the school cadets took place soon after, and he noticed the aerial on the foreshore. Initially sceptical of its efficiency, he was astounded when the rector quoted to him verbatim his own message to Captain Halsey. Now keenly interested, he insisted on seeing the record book of the club and offered to promote a wireless field section of the signallers' division of the cadet corps.[44]

There is no full record of the amateurs who were involved with radio in these early years. Other experimenters operated in various parts of the country, but the law made them unwilling to publicise their actions. Clarkson notes that 'Probably every keen would-be amateur at some time carried out clandestine operations, using the technique of that era, motor car spark coils and crystal detectors.'[45] As Clarkson further notes, the New Zealand law and the fear of detection forced amateurs to neglect the transmission potential of radio and to content themselves, at least in the meantime, with reception. Many felt the official attitude was working against the development of radio. Strachan argued that political and commercial interests were concerned to reduce amateur activities to the level of a 'harmless hobby'.[46]

THE FIRST GOVERNMENT RADIO STATIONS

In 1911 contracts were finally let for the installation of the New Zealand government's first radio stations. The plans were similar to those announced the previous year. There were to be two high-power stations at opposite ends of the country. The northern station, first planned for Doubtless Bay, was now to be a few miles away at Awanui. This station was to be 'able to maintain uninterrupted communication with Sydney day and night'. The other high-power station was still to be 'near Bluff'. The major change was in regard to the low-power stations. Five, instead of three, were to be purchased: two were to be erected 'at or near Gisborne and Christchurch', another was to be upon the summit of Mount Etako in the South Tinakori Range, near Wellington, and 'probably one or two government steamers will be similarly equipped'.[47]

The equipment was manufactured by the Australasian Wireless Company, which also built the high-power stations. The name of the company disguised the fact that, surprisingly, the government had bypassed the British manufacturers and chosen the Telefunken system of Germany. The government's final choice for the supplier of its radio stations lay between Telefunken and the Marconi Wireless Telegraph Company. The two had standardised their spark transmission systems, and had many features in common. The Telefunken system emitted a high-pitched, more musical note than the lower, raspy tone of the Marconi system, and was more easily heard in bad weather. It was favoured by most operators.[48] But the main reason for the decision in favour of Telefunken was the New Zealand government's insistence on a guarantee that the equipment would work. Only Telefunken was prepared to guarantee communication between the distant terminals.[49]

The government's first Telefunken station in New Zealand was the Wellington Radio Station. A low-power station, it was situated in the tower of the General Post Office. The low-power sets were 2.5-kW plants of the type used at sea and designated as first-class ship-stations. The Wellington station opened on 26 July 1911 and communicated at night over a normal range of 600 miles with those radio-equipped ships, generally the international steamers, which were within range. In 1911 the government began licensing ships for the installation of wireless plants; eight licences were issued in the first year. The Wellington station was closed down the following year when the station at Mount Etako, behind Wellington, was opened.

The department was pleased with the running of the station, noting that 'this branch of the service bids fair to become an indispensable adjunct to the department's telegraph system'. Wireless telegraphy was viewed as an extension of, rather than alternative to, the telegraph system. It was to be used where the wires would not go — to ships at sea. Other uses could be found, but for many years the government would not look. Meanwhile radio was a maritime means of communication that 'minimizes maritime dangers and anxieties, and, by keeping ships in touch with the shore or other ships renders unique service to society and commerce'. As a maritime communication, radio was soon valued for its strategic and commercial applications, along with its aid to marine safety.[50]

The 600-mile range of the Wellington Radio Station meant the government decided it required fewer stations. In September 1912 Robert Rhodes, postmaster-general in the two-month-old Massey ministry, announced that the government did not intend to proceed with the Gisborne station, and that 'the stations at Awanui, Auckland and Wellington will meet all the requirements of the North Island'.[51] Though the minister did not announce it at the time, the Sumner station was also cancelled. The government's policy was to equip as many ships but as few land stations as possible. Shore stations were conduits into the telegraph system and, because they interfered with one another, should be few.

With the opening of the Wellington Radio Station the Meteorological Office moved to take advantage of the new facility and became the second government agency to make use of radio. Arrangements were made for weather radio-telegrams to be forwarded to the office from warships and merchant vessels within range. The P&T Department provided this service without charge. Cable messages of Australian weather conditions had been available for many years, but with ships able to be reached two and even three days from port, this was the first peek into the Tasman and Pacific.

The Mount Etako station, Radio Wellington, opened on 14 October 1912

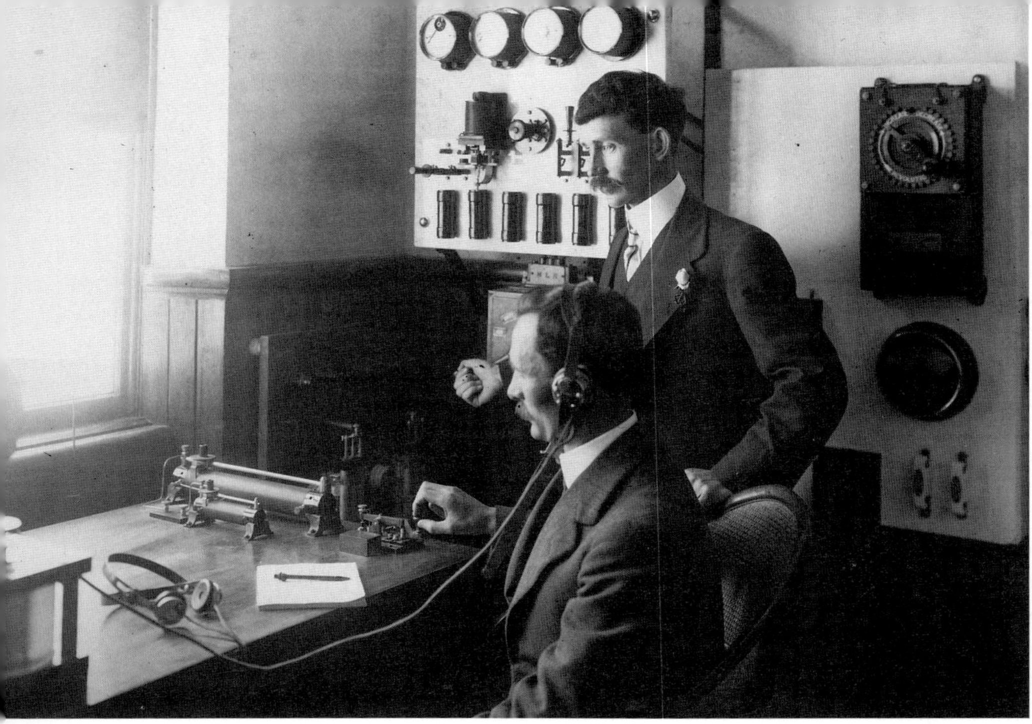

Interior of Mount Etako station, Radio Wellington, opened in 1912. ALEXANDER TURNBULL LIBRARY

on a site 985 feet above sea level, immediately behind the city of Wellington. Power came from an extension of the city's electric mains, and telegraph and telephone wires were laid between the station and the city. The station had the same 2.5-kW power output as the Wellington Post Office experimental station had, but its distinctive aerial and earthing systems and its exposed site enhanced its performance. The station building was bunker-like, built to provide a quiet interior and protection from high winds. It was the aerial system which was the dominant architectural feature for all the stations. At Mount Etako the aerial, which was supported by two masts of Oregon pine 150 feet high and 300 feet apart, was of a flat-top type, composed of four phosphor-bronze wires four feet apart. From the middle of the aerial four wires extended downwards to the building. The earth connections consisted of 300 copper wires radiating about 300 feet from the station, situated midway between the masts. Originally envisaged as having a range of around 300 miles during the day and double that distance at night, the completed station exchanged signals at night with vessels 600–1000 miles distant, depending on the atmospheric conditions, and almost every night with some of the Australian stations.[52] The Governor-General, Lord Islington, who opened the installation in a 70-mph breeze, noted that the station was intended to be part of an empire-wide hook-up and would be important for British trade and imperial interests as well as being vital in time of war.[53]

Ten days after the opening of Radio Wellington, another 2.5-kW station was established in Auckland on the roof of the new Chief Post Office. Its aerial system was supported by 40-foot tubular steel masts erected on the two hemispherical domes of the building. The masts were stayed to outriggers projecting from the corners of the building. As the tops of the domes were already some 35 feet above the roof level, the aerial system was a prominent feature in Queen Street. It covered a range slightly less than that of Radio Wellington. In June 1912 the third 2.5-kW plant was installed on the government cable steamer Tutanekai. A fourth plant was sent to the Chatham Islands where, with motive power generated by an oil engine, a station was opened on 18 September 1913. The last of the five plants was kept for spare parts.

The government's view of the appropriate use for radio was strengthened at the International Radio-Telegraphic Convention in London in June 1912, to which it sent two representatives. The Convention continued with the understanding that wireless telegraphy had a few select applications — communication between ships, lighthouses, outlying islands and shore. The topic of interference, which had been of particular concern to the New Zealand government, was also discussed at the Convention. The first radio transmitters and detectors were unable to confine their wavelength to any firm measure. Even when tuning systems were developed, they were neither immediately efficient nor rapidly adopted. By 1912 it was realised that the earlier fear that only one of all the transmitters within range could operate at any one time was no longer valid. But the problem of interference remained a serious one. The Convention recommended it be countered by regulation. It also recommended that ships be graded into three classes with hours of attendance prescribed, principally to ensure that distress signals were observed.

The New Zealand government implemented the Convention strategy against interference by requiring ships in its waters to use minimum power for their radios and to communicate only with the nearest station. Generally the new regulations requiring ships to have radios were undemanding and applied only to the larger passenger steamships registered in New Zealand and engaged in foreign or intercolonial trade. The regulations did not apply to steamships trading to the Chatham, Auckland, Campbell and Antipodes Islands. Those vessels required to have radio were placed in the third class: they had no fixed working hours, were not bound to perform any regular listening service and were not required to carry an emergency installation.[54] At the time there were twenty-two ship-stations registered in New Zealand. By the end of 1914 there were twenty-seven.

The government's resolve that it had fully met the New Zealand requirements for radio coverage was tested, but unshaken, by the SS *Suffolk* incident in August of 1913. The ship had been unable to find the entrance to Port Chalmers because of fog. Unable to communicate with the port, it had radioed to Wellington, the nearest station, which sent a telegram to Port Chalmers and despatched the local tug to the harbour mouth in

Awanui station opened in 1913. ALEXANDER TURNBULL LIBRARY

search of the ship. Edward Clark, the local MP, used the incident to ask the government to authorise a wireless telegraphy station at Port Chalmers or to subsidise radio equipment for the new tug, then under construction. Rhodes, however, was unmoved: 'All the requirements of the vessel were fully met by the existing station at Wellington. None but the most trifling additional advantage could be gained by having a wireless station at Port Chalmers.'[55]

The high-power stations at Awanui and Awarua, identical 30-kW stations, were both opened on 18 December 1913. Each was provided with a 400-foot steel tower, from which umbrella-shaped aerials spread from summit to base over an area of about 90 acres. They were impressive structures, complemented by similar stations in Australia and Western Samoa. Awanui communicated mainly with Radio Apia, the Samoan station. There were two Australian stations, one at Applecross in Western Australia and another at Pennant Hills in New South Wales. Awarua communicated chiefly with Pennant Hills.[56]

The two New Zealand stations were both sited on ground that had a flat, unobstructed outlook and soil of good conductivity — sites similar to Nauen, the Telefunken development station on a plain some 25 miles from Berlin. Telefunken sought such sites as likely to allow the stations to meet the performance tests specified.[57] Awarua was completed well ahead of schedule. With the Great War soon to break out, the German employees of Telefunken 'had no wish to linger on predictably foreign soil; the work was consequently speeded up to allow for their early departure'.[58]

Interior of Awanui station. ALEXANDER TURNBULL LIBRARY

THE BANNING OF AMATEUR RADIO

The government had shown little sympathy for New Zealand amateur experimenters when they had come to the attention of the Royal Navy. With its own radio system in operation, the government regarded amateurs as unwarranted eavesdroppers on the messages of others. This view was shared by both the Liberal government and its successor, the Massey ministry. The clamping down on illegal amateur plants, and the withdrawal of permission to experiment from those that had received a local authorisation, were followed by a 1912 instruction that all amateur aerials were to be taken down. The rationale for this decision — in effect a ban on all amateur use of radio — was that unauthorised wireless installations were interfering with official messages.

Not everyone observed the ban. Brenda Bell, a member of the family from Shag Valley in Otago that was so influential in the development of amateur radio, recalls 1912 as the year of her family's first successful listening, on a crystal set built by her father. Over the next two years, she recalls, 'We spent many a cold night muffled up in woollies playing with that crystal set'. Brenda and her brother Frank were just beginning a long association with radio. Frank was later to achieve the first two-way radio communication with a station in England. They and many others were undaunted by an official ban. Their understanding of what they were involved in overcame any feelings of intimidation: 'Many of us had a sense of awe, of dealing with things greater than we could understand. And we knew that throughout history many good men, and women too, had been ruined, tortured and burnt at the stake for dabbling in magic not one fraction as wonderful as this.'[59]

Three Wellingtonians, equally prepared to flout the law, were serendipitously and exceptionally to win official acceptance. The details are recorded by Arthur McClay, the youngest of the three and later the principal figure in Wellington's 2YK broadcasting station.[60] In 1911 McClay, then a fourteen-year-old schoolboy, decided to learn all he could about 'that wonderful science' of wireless telegraphy. He was shocked to learn his desire was illegal. McClay remembers walking down Britomart Street one evening in June 1911, and seeing in the dim light 'a young man hauling up an aerial which was attached to two tall trees. The leading wire from the aerial trailed into the basement room of a house. I was overcome with amazement and joy. Here was a man after my own heart.' McClay befriended the other enthusiast, Brian Robinson, a twenty-year-old electrician who made shipboard calls as part of his job, and who was on friendly terms with a number of ships' radio operators. Robinson, who had built a replica of a shipboard Telefunken set, introduced McClay to the wireless operators' fraternity and McClay soon had his own set at his parents' home at 13 Gordon Place, Newtown. McClay's neighbour and friend Guy Tinney, a Post Office landline Morse code telegraphist, also joined in the illegal wireless use. Their aerials were kept out of sight during the day; reception and transmission were only at night. Their activity went undiscovered until one stormy night in 1912 when McClay and Tinney picked up a ship's distress call for urgent help. Hearing no reply from Radio Wellington they assumed, correctly, that the gale force wind had put Radio Wellington's aerial out of action. Risking a possible conviction for illegal possession, they informed the P&T Department of the distress call. A successful rescue was mounted. Far from being prosecuted, the trio were thanked for their services and issued with licences to use a wireless telegraphy set. The licences were to be renewed annually and their equipment was subject to the inspection of the department.

Amateur use of radio became a matter of public discussion culminating in the debate on the passage of the Post and Telegraph Amendment Act 1913. Legislative control of radio remained at the start of 1913 with the provisions first seen in the 1903 Act, and repeated in the Post and Telegraph Act 1908. But curbs on amateurs were to be made stronger. The topic was raised in Parliament in August by Hugh Campbell, the Napier MP, after amateurs in his constituency had been taken off the air.

> In his district one boy had been fined and two others had been compelled to pull down their machinery. These boys had worked at the business until they could talk to each other over a distance of a mile, and he thought that instead of being blocked for fear of some wireless message drifting into their hands they should be encouraged to go on. They had managed this absolutely by themselves,

> and it was to the advantage of the country that a certain number of youths should train themselves in this manner.[61]

By 1913 the objections to the government's policy were phrased in terms of the service being given to the country by the amateurs. It was an important point. New Zealand had a growing need for wireless operators and they were not being trained. But it is notable that, a decade after the first parliamentary debate on radio in 1903, the understanding that radio could be used by individual citizens for their own purposes was no longer voiced. Radio was an official means of communication and the technology was not to be available to the average citizen.

The matter was raised in Parliament with respect to school instruction in wireless telegraphy by George Anderson, the member for Mataura. Strachan was science teacher at Gore High School, which was in his electorate. Armed with a letter from Strachan, Anderson argued that the government's ban on outside aerials hindered schools from teaching wireless telegraphy. A 'gentleman who was science master of a school in his district' had informed him that 'such apparatus as the Postmaster-General suggests would bear about as much resemblance to modern wireless as a pea-blower does to a Gatling gun, or a soap-bubble to a Zeppelin. For teaching purposes it would be inferior to the diagrams in a good text-book. As a signalling-device it would be less reliable than a penny whistle.' In fact Anderson, or Strachan, had exaggerated the dismal performance of the Gore High School radio apparatus, sans aerial. When the aerial, a district wonder, was dismantled, Strachan had shown the local P&T Department inspector the dismantled parts and then invited him to listen to Morse signals rolling in strongly from the corrugated iron roof of the school.[62]

Anderson's view was that 'what is wanted is not prohibition of experimental stations, but the licensing of them. At present the Government is making an enemy of all young wireless enthusiasts in the country.' He noted that not only the current but 'each Postmaster-General had by every means in his power taken upon himself to prevent the study of this important science, because it was said it interfered with the business of the office.'[63] The United States government allowed its citizens to own and use radio transmitters and receivers. Its policy was well known and Anderson raised the basic question of entitlement with respect to radio when he asked the New Zealand government to allow a similar access. Rhodes took the opportunity to declare the government's case:

> The opinion of the merchantile public using the public wireless system is strongly against allowing the indiscriminate use of wireless apparatus that could be used to tap their messages, and it is clearly opposed to public policy to allow of interference with the transmission of a distress-signal or call for help by wireless

> . . . The Dominion Government will not consent under any circumstances to private messages coming, by means of any regulation it makes, to the knowledge of experimenters in wireless telegraphy.

Far from accepting the United States as an exemplar, Rhodes used it to justify his policy:

> On the Californian coast it is extremely difficult for ship-stations to communicate with coast-stations from sun-set until about eleven p.m., owing to the large amount of 'talk' indulged in by amateurs. It has indeed been said that, outside the period of 'silence' which the United States Navy has had to enforce during the first fifteen minutes of each hour, should it unfortunately happen that a ship-station found it necessary to make the distress-signal the chances of the call for help being attended to are remote . . . The difficulty experienced in the United States in connection with amateur wireless installations interfering with the transmission of legitimate business is well known.[64]

The government made its policy clear in the Post and Telegraph Amendment Act 1913, in which it strengthened the provisions of the 1903 Act, making it now an offence to establish, without permission, any station *capable* of reception and transmission. The government's view was most openly voiced by Francis Bell in the Legislative Council. 'Private wireless stations are a nuisance wherever they are; and, frankly, it is not intended to permit them.'[65]

Permission to install a radio station was available via an Order in Council. The first of these was given in April 1914 to the Corporation of Canterbury College in connection with research the college was conducting with the British Association for the Advancement of Science. The permission was hedged with conditions which reflected the government's strong reluctance to allow any individual or organisation to trespass on what it had determined was its monopoly over radio. The station was to be used only for the reception and not for the transmission of signals. Its use was to be confined to members of the scientific teaching staff of the college and their assistants, and no telegraphic message received was to be communicated to any person other than an officer of the department. The department reserved the right to inspect the station and plant at any time, and to suspend its working during any period. There is no record of the Canterbury College station ever being installed. The project was stopped with the start of World War I.

WORLD WAR I

With the outbreak of war a military guard was placed at each of the government stations and the small number of amateur stations were closed. The possession of an unofficial radio came within the province of the military authorities. Private possession was illegal and the matter was staunchly policed, as the following parliamentary exchange indicates:

> Hon. Mr. Allen, Minister of Defence: By aid of the means adopted by the military authorities, assisted by the Postal Department, we are able to find out if any wireless messages other than those we are sending out are sent out of New Zealand. One private wireless plant was found and the man in possession of it was arrested. That man is still under arrest, and will not be released.
> An Hon. Member: Was he sending messages?
> The Hon. Mr. Allen: No.[66]

Punishment was not always as extravagant as this panic reaction, which occurred early in the war. However, illegal use of radio was always treated seriously. Two years later, when a sixteen-year-old boy was caught, he and his father were prosecuted and convicted, the boy for erecting and the father for aiding to erect an illegal wireless. They were fined £2 and £10 respectively.[67]

Although amateur use of wireless was banned during the war, it continued. By this time the simple technology of the crystal set was known, and there were many clandestine owners and users in New Zealand. Probably there were also owners with more sophisticated receivers. Gisborne, where half a dozen or more amateurs worked under cover of darkness, is one of the few areas where some record survives of this illegal activity. Before the end of the war three of the Gisborne group, Ivan O'Meara, P. R. (Percy) Stevens and R. J. (Bob) Patty, had brought in the Morse signals of Apia using only a crystal receiver but with an aerial 300 feet long, which was hidden away in a tall pine tree by day, extended at night and then rolled up again about 2am.

The ban on amateur usage and the strict telegraph censorship over inland telegrams, cable messages and radio-telegrams at the start of the war led to a large reduction in radio usage as it had previously been practised. But as some avenues closed, others opened: the war considerably enlarged the number of New Zealanders competent in radio operation. They served in signals units of the army and as wireless operators at sea, in fighting ships and with the merchant navy. Even a few aviators learned Morse code and the intricacies of spark transmission operation so as to relay details of troop positions while flying behind enemy lines.

The Dominion Radio College was set up in Auckland and Wellington to train men in wireless telegraphy. Considerable assistance in the form of equipment and experienced operators as teachers was given to the college by Amalgamated Wireless Company of Australasia Limited (AWA), a company which began in 1913. (The amalgamation was of the Marconi and Telefunken interests.) The company thus began a long association with Australian and New Zealand radio and broadcasting.[68]

At the same time, the New Zealand government and people had reason to be grateful for the skill and readiness shown by self-trained amateurs. As Brenda Bell recalls:

> When I was in England during the war a team of New Zealand radio men came over on loan to the British Government for highly secret service in connection with radio. And several were men who years ago belonged to our regular correspondence. Men who began as boys dabbling with a new kind of hobby which took all their pocket money and all their time and kept them awake at night, but taught them geography as no schoolmaster could and trained them to serve their country as badly needed specialists.[69]

Radio operators and technicians were among the first needed when the war began for, at its outbreak, the radio link between Awanui and Apia was broken when the German staff at Apia disabled the station there. The First Expeditionary Force from New Zealand which occupied Samoa included four radio operators and two radio engineers, all volunteers from the P&T Department, who brought spare wireless apparatus from the department. By late October 1914 Awanui and Apia were again in communication.

Radio was soon valuable in the New Zealand conduct of the war. At the time of departure of the main body of the New Zealand Expeditionary Force, Radio Wellington was a listening post for the Royal Navy. The evening after the supposed secret sailing from Wellington, Clive Drummond, one of the station telegraphists and later a widely known broadcasting announcer, picked up a message passing between two ships of a German naval squadron in the Pacific. When decoded in Melbourne, the contents indicated that the Germans were aware of the convoy's departure. The convoy was recalled and held in Wellington to await a stronger escort. Once the force had sailed Private W. C. Falconer, a telegraphist from Eltham who was acting as wireless operator on the *Arawa*, picked up an alarm call from Cocos Island that led to the destruction of the German naval raider *Emden* by HMAS *Sydney*.[70]

Drummond left New Zealand for Mesopotamia on war service as Sapper Drummond. He was a member of the First New Zealand Wireless Troop, a group of sixty linesmen and telegraphists from the P&T Department.[71] In Mesopotamia the troop, together with the Australian Wireless Squadron, took over the lines of communication from the Royal Engineers. The two units were subsequently united, and named the Anzac Wireless Squadron. The squadron maintained the lines of communication throughout the Mesopotamian campaign.

Drummond was exceptional in that he had considerable pre-war radio experience as a staff member of Radio Wellington. But his experience was matched during the war by that of many other New Zealanders. The 1914–18 years opened up many more opportunities for radio enthusiasts than in the previous decade. These opportunities were not only due to the wartime requirements for skilled operators and technicians, but were also part of the change in radio over the years since the passage of the 1903 Wireless Telegraphy Act, from a futuristic dream to a practical means of communication. All this development was increasingly within the orbit of official sanction. That there could be other uses for radio was not conceded by the New Zealand government. The ban on all private use of radio that came with the start of the war was no sudden change of policy with the new wartime necessities. It was the end-point of a development of official control over the medium that had been stated in principle when the government first became aware of radio, and had been put into practice over the years since the passage of the 1903 Wireless Telegraphy Act. In the pre-war years there had been tension between this attitude and the desires of amateur enthusiasts. The war greatly increased the number of radio enthusiasts. The coming years were to see the tension also increase, as developments in radio led to a growing conflict between the government's limited vision for radio and its citizens' enthusiasm for a widening use for the new medium.

2
The Birth of Broadcasting

THE BEGINNING OF BROADCASTING

The change from wireless telegraphy to wireless telephony, from morse code to voice transmission, did not gain pace until after World War I. However its origins date from 1900 in the work of the Canadian, Reginald Aubrey Fessenden. Fessenden's revolutionary aim was continuous wave transmission on which voice or music could be superimposed as variations or modulations. This was in contrast to the series of bursts of the interrupted wave transmission that was the spark transmission system. While Fessenden soon achieved success, it was not until 1906, when General Electric devised and constructed the alternating current generator he needed, that he was able to broadcast.

The other critical innovation for broadcasting was the vacuum tube, or valve, developed from the work in England of John Ambrose Fleming and in America of Lee De Forest. In 1905 Fleming returned to work he had done fifteen years earlier as scientific advisor to Thomas Edison. Beginning with a regular incandescent light bulb and retaining a vacuum in the bulb, he sealed a metal plate inside the bulb in such a way that it would act as a valve, thus rectifying the signals or allowing them to flow in one direction only. This produced an audible sound. From this De Forest developed his 'audion' by the addition of a third element or grid in the tube, which much increased its power as a detector and amplifier.

When De Forest formed the De Forest Radio Telephone Company in 1907, he had a vision of what was to come. 'My present task (happy one) is to distribute sweet melody broadcast over the city and sea,' he wrote in his diary on 5 March 1907.[1] This is among the first uses of the word 'broadcast' in its modern sense. In 1909 he stated 'I look forward to the day when opera may be brought into every home. Someday the news and even advertising will be sent out over the wireless telephone.'[2]

The translation of broadcasting into a business plan based around the selling of radio sets to consumers began in November 1916 with a famous memo from David Sarnoff of American Marconi to his superiors. The idea was considered preposterous at the time but in 1920, after the formation of the Radio Corporation of America (RCA) and its takeover of *Wireless Age*, the American Marconi periodical, Sarnoff resubmitted his plan. He also proposed that *Wireless Age* should publish the forthcoming radio programmes. This time RCA accepted Sarnoff's plan, to its great financial advantage. RCA was not the first organisation to begin regular broadcasts. That honour is contested by two stations — the Netherlands station PCGG in The Hague, and the Westinghouse station KDKA of Pittsburg run by Frank Conrad. Both began regular transmissions in 1920. But RCA was well placed to benefit from the broadcasting boom that swept the US. It began selling radio sets in 1922, and sales in the first three years exceeded $80 million.[3]

England lost the lead in radio broadcasting in 1920 when the British government stopped the experimental broadcasts from the Marconi Laboratories at Chelmsford. British broadcasting started again in 1922 with the Marconi Company's resumption of experimental programmes and with the formation of the British Broadcasting Company, a cartel of radio receiver manufacturers, with John Reith as general manager.[4] By that time radio broadcasting was spreading rapidly, and New Zealand had its own first stations.

NEW ZEALAND POSTWAR RADIO EXPERIENCE AND POLICY

After the war the New Zealand radio amateurs were ready with new enthusiasm and new technology. Strachan was on the wharf at Lyttelton when one of his pre-war associates returned from service as a ship's wireless operator. 'Wait till you hear these vacuum tubes', the homecomer shouted from his ship's rail, 'You can play tunes on them.' One signaller recalled, 'We all learned much from our wireless experience in France and returned full of energy for fresh experiments. The valve had been invented — we had used them in France — and most of us managed to stow a few in our inside pockets to take back with us.'[5]

In the early postwar years most valves entered New Zealand in a similar surreptitious fashion. It was some time before they were freely available. This situation was not peculiar to New Zealand. In the US in 1920, for instance, they were still unavailable to the public.[6] The military control of radio that was a feature of the war years was only slowly relinquished. In New Zealand the censorship restrictions on the use of the telegraph, cable and wireless telegraph were removed in December 1918.[7] Further removal of restrictions was not offered, and in particular amateur usage of transmitters and receivers remained outlawed. The government returned to the pre-war use of radio. Nothing further was proposed.

In September the following year, after one parliamentarian, J. C. Thomson, suggested New Zealand and the Falkland Islands were the only countries in the world where citizens were not allowed radio licences, the government reconsidered the matter.[8] Its decision, appropriately announced by the minister of defence, was not to remove the embargo 'at present'.[9] However it was becoming obvious that New Zealand could not continue to outlaw all amateur, non-official use of radio. The banning of amateur usage was increasingly regarded as a denial of public access. International attitudes and a rising discontent at home meant change was inevitable. The first change was signalled by Gordon Coates, the postmaster-general, on 11 August 1920. The Act, which provided for licensing only ship stations, would be joined by legislation 'to license and suitably regulate the operation of wireless stations of all kinds'. However Coates also announced a need to safeguard against 'the promiscuous propagation of wireless signals capable of interfering with legitimate traffic.'[10] The new legislation came in the Post and Telegraph Amendment Act 1920. While other than official and marine use of radio was now permitted, the government separated the reception and transmission capabilities of radio. The existing legitimate users of radio for transmission were allowed to continue, but new licensees could use radio only for reception. Amateurs were allowed to apply for licences but were permitted only to receive signals. Thus the separation of radio's reception and transmission capabilities, introduced elsewhere as part of a developing pattern of broadcasting, was introduced politically in New Zealand.

Applications from amateurs were submitted immediately, but as no regulations had been gazetted, no licences could be issued. In April 1921, with the necessary regulations still non-existent, the P&T Department began to issue provisional permits. By the end of the year over 300 had been issued, authorising the erection and operation of receiving stations. Coates still would not move on the matter of transmission: 'The study of the reception of wireless signals offers a wide field for research by experimentalists. The time is considered inopportune to license the operation of amateur sending-stations.'[11] Permittees — the holders of the provisional permits — were issued with an imposing four-page document. Particulars of the location of the receiving apparatus were recorded, as were the comprehensive conditions to be observed by the holder. One of these required the permittee to obtain the written sanction of the Radio Inspector for any changes made to the receiving apparatus and to supply 'descriptive diagrams'. When using valve receivers within 10 miles of a government station, permittees were required to continually listen for the 'Amateur Warning Signal — AAAAQRM', followed by a figure indicating minutes, and to immediately cease operations until the time had expired. The authorised apparatus was to be kept in 'secure custody' and no part could be removed without approval.[12]

The first permit holders did not merely apply. As L. S. Spackman of Auckland recalled, 'We had to pass an examination and we had to have references from the police.'[13] His brother, Bernie Spackman of Napier (later to become that city's first broadcaster) was ordered by the local constable to take down his first aerial for fear he might be spying.[14] Having obtained a permit, one could not easily purchase a receiving set. The permittees were able to get some parts, but to a large extent had to build their own sets. L. S. Spackman's first set included various pieces available in the home, from the eggbeater brace to the telephone receiver. Bill Huggins, grandson of the Timaru radio pioneer, recalls his own first set, built with his father's help:

> We made all the variable condensors, the plates out of zinc, and the coil holders out of red fibre and the panels out of ebonite salvaged from car battery jars. This was the only insulation that was available. The wire for winding the coils was obtained from the Patea tip . . . It was something completely new. I had no contact with anyone outside and had never seen a radio, only pictures.[15]

Like many of the early amateurs, Bill Huggins was unknowingly entering a lifelong career. His home-made set was successful, and he was able to receive the full range of available signals: 'I was able to pick up the ships all around the coast, and Awanui and VLW (Radio Wellington) and VLA (Radio Auckland) and VLC, the Chathams . . . The morse came through. I got plenty of signals. I couldn't read them of course.'[16]

The fact that Huggins and others could not read Morse code did not dampen their enthusiasm for listening to transmissions. The content of transmissions was, at first, not the point. That the set received signals from afar was a marvel in itself. They were intrigued by 'the extraordinary miraculousness of the whole thing'.[17]

Radio enthusiasts were not long content with a diet of Morse code signals. The few broadcasting advocates continued to urge an end to the ban on transmission. Senior among them was Strachan, now headmaster of Rangiora High School. He pointed out the significance of the developments in radio:

> Wireless telegraphy enables us to keep in touch with our inter-colonial steamers and is to all intents an extension of the land telegraph lines. But the wireless telephone has opened up almost immeasurable possibilities . . . Provided broadcasting stations are erected, a public speech, a musical concert, press news, market reports, etc. etc. may be heard simultaneously in any home in New Zealand no matter how remote, in any hall, in the saloon of the ferry steamer, and in the railway train. With proper arrangements as to wave length it is possible to adjust the receiving instrument so as to pick up at will any one of these. The era of the talking newspaper has arrived. Musical entertainments in any one of the centres may be enjoyed by the family in the loneliest farm house. Scholastic institutions may lecture to their students over a whole province.[18]

Strachan's was a rare voice speaking knowledgeably of the developments in radio. In 1921 he was joined by Robert Jack, professor of physics at the University of Otago. Jack contended the government was wrong to see radio enthusiasts in a negative light, as eavesdroppers on the messages of others. The new uses of radio, he contended, would be of great value to New Zealand. He stated to the *Otago Daily Times,*

> Wireless telephony will develop rapidly along its own special lines and will tend greatly to strengthen the bonds by which a civilised community is held together and formed into an organised whole. It will give ever-wider publicity to all news of public interest, to speeches and entertainments and will thus tend to bring country settlers into close contact with all the life of the town. Further than that, the life of each community will be broadened and educated by being brought into more effective touch with the life of the whole world. No country stands to benefit more than New Zealand by having the disadvantage of isolation removed.[19]

Late in 1921 these arguments bore fruit: the political decision was taken to allow radio transmissions from other than P&T Department and marine stations. Jack was the first to receive a transmitting permit.

THE BEGINNING OF BROADCASTING IN NEW ZEALAND

Jack was not, however, the first to broadcast in New Zealand. He was preceded by radio operators from at least two visiting ships. Their transmissions, intended as entertainment within the fraternity of ships' radio officers, were in fact broadcasts. All who could were welcome to listen. One broadcast came from the US merchant steamer *Eastern Planet*, berthed at Lyttelton. On Friday 20 August 1921 Ralph Lawrence used his own radio telephone set to play a jazz record and a march by Sousa, principally for the entertainment of the operator on the SS *Wairuna*, *en route* from Wellington. R. J. Orbell in Merivale, an engineering student at Canterbury College, was amongst those who heard the broadcast. Orbell was already well versed in radio. One of the quartet of schoolboys who had built and operated the Morse receiving station at Waitaki Boys' High School, he had continued his radio work during the war with the Royal Engineers. Orbell visited Lawrence the next morning and arranged for a further broadcast at 7pm that evening, after the ship had sailed for Timaru. A further concert was broadcast when the ship neared Timaru. The ship worked its way south, and at Port Chalmers its visitors included Professor Jack and his technician, Jack Sutherland.

The *Eastern Planet* was not the only ship to broadcast. Simpson, who had a receiving station in The Terrace, recalls a similar transmission in Wellington from the *France*: 'They were broadcasting music from a phonograph, with a vertical antenna strung up to the top of the mast.'[20]

ROBERT JACK'S OTAGO UNIVERSITY BROADCASTS

Robert Jack was appointed professor of physics at Otago University in 1914. Any experimentation into radio was impossible during World War I; Jack and Sutherland did not begin their research until after the war. D. G. Mitchell, an Otago wireless telegraphy enthusiast, reports Jack and Sutherland's unsuccessful attempt to make thermionic valves at this stage — 'They didn't have enough ginger in the engines to have a perfect vacuum to work the valves.'[21] Jack was finally able to develop his radio interests during a visit to Britain in 1920–21. Among his contacts in England was his younger brother, an electrical engineer with the British firm of Thompson Houston, which manufactured radio equipment. Jack returned to New Zealand with various pieces of equipment, much of it war surplus, including relatively high-voltage DC generators which were to supply the power to his broadcasting station. He also purchased radio valves — rare commodities in the early 1920s. He obtained Edison ES4 valves which Robert Nimmo, his student and later successor as professor of physics, described as very small and not meant as transmitting valves at all — 'But by running two valves in parallel and never having the anodes of the valves looking less than dull red hot, he achieved an appreciable power output.'

By May 1921 Jack had successfully transmitted voice and music within his laboratory. He sought official permission for longer range transmission, for broadcasting proper. His formal authorisation was dated 5 December 1921, but prior to this he received separate authorisations for individual broadcasts and, on 17 November 1921, transmitted the first in a series of concerts that included live voice and music along with gramophone recordings. He was required to submit all items to be broadcast to the P&T Department for approval: the broadcasts continued on these terms every Wednesday and Saturday from 8pm to 10pm until just before Christmas. He stopped for his summer holiday, then resumed broadcasts until the start of the university teaching year. Regular transmissions then ceased, though there were occasional further broadcasts in 1922.[22] Stan Hughes, who joined the physics department as a technician in 1924, described the equipment used as 'a one valve transmitter with grid modulation. The generator supplied 800 volts for the plate of the valve and their aerial current was in the order of ⅓ to ½ an amp. The total output was about 2 to 4 watts. The aerial was a single wire extending down to the Arts' building from the tower of the University.'[23]

Jack's broadcasts began with the sound of a buzzer lasting for up to five minutes to enable listeners to tune into the sound. He broadcast records supplied by a Dunedin music house, which also supplied the gramophone. To transmit this music Jack and Sutherland jammed the microphone into the gramophone through the louvres, watching the fluctuations in their aerial ammeter. They also broadcast live music. As Nimmo records, 'It was a case

Robert Jack, professor of physics at Otago University. HOCKEN LIBRARY

of putting on what seemed relatively suitable and what was readily enough available.' Among the singers who provided live music was the matron of Knox Theological College, Miss Manson — soon to become Mrs Jack. In August 1922 Jack relayed from the Allen Hall the whole of a concert given by the University of Otago Musical Society.

Jack was the leader for these first broadcasts. It was he who, with very limited financial and material resources, gathered the necessary equipment. But considerable credit goes also to his technicians, in particular to Jack Sutherland, who was a radio enthusiast before he joined the physics department. Nimmo makes the point clearly: 'Jack Sutherland . . . made the apparatus and . . . was there every time the transmission was going and for quite good reasons . . . [He] carried a good deal of the responsibility for the technical side of getting it to work.'[24]

The broadcasts were heard over a wide area of New Zealand. Naturally reception was mainly in Otago, Southland and Canterbury, but reports also came from much further away. At various times the broadcasts were heard over the Southern Alps on the West Coast, in Wellington, and further north in Hamilton and Auckland. The most distant reception reported was from a ship two days out from Melbourne. For L. S. Spackman, listening in Auckland, the signals 'were quite audible'. Their sound quality was 'pretty terrible but they sounded marvellous in those days of course'.[25] The quality of the reception varied considerably with distance. The Otago reception was best: Frank Bell in Shag Valley reported the 14 December 1921 broadcast as 'a triumph . . . When Mr Sutherland was speaking, without thinking that his words were being heard over half New Zealand, they were the plainest I have ever heard over any telephone, wired or wireless. The bells in the gramophone record were really extraordinary, and seemed to be ringing right in the room.'[26] The reports also varied according to the enthusiasm of the listener. Nimmo, who listened to the broadcasts on a crystal set, recalls, 'I was immensely impressed. I thought it was wonderful to be able to do this. But my father who was musical listened to it and he thought it was pretty poor. He felt it was not even up to the standards of the gramophone records that existed at that time.' Similarly a broadcast in April 1922, done especially for the diamond jubilee of the Canterbury Philosophical Institute, was reported in a telegram from J. A. Orbell, listening at Canterbury College, as 'Whole affair most successful. Flute heard by over 300 people in hall.'[27] The *Press*, in regard to the same broadcast, reported, 'At times a voice could be heard amidst what sounded like one long rattle of musketry, though the actual words of the speaker could not be distinguished.' Regarding the music, 'the never-ending blare prevented one from discovering what was the actual air being played.'[28]

The quality of the broadcasts was open to interpretation. What was indisputable was that they were broadcasts. Something new was taking place and, while it was to become an everyday event, it provoked an emotional response in the first listeners. Often they were awe-struck. Even Clive Drummond, by 1921 an experienced telegraphist and knowledgeable about the developments in broadcasting, more than fifty years later still vividly recalled his first reception of Jack's broadcast. At work in Radio Wellington, he found himself listening to the song 'Come into the Garden Maud': 'I can't tell you what I said, but I nearly went through the roof as you can imagine. All the sound you ever heard was morse and static, but to hear a voice — well!'[29]

Jack, a far-seeing enthusiast for radio, was capable of appreciating and publicising the new mode of communication. His broadcasts did much to awaken public interest. Among his achievements, besides the actual broadcasts, was the effect they (and he) had on the acceptance of broadcasting in New Zealand. Importantly, his work led to a developing press discussion. In the various centres, interest and pride were stirred by reports that a broadcast had been received locally. The widening public understanding of broadcasting sounded the death knell for the view that radio was exclusively a medium for secret point-to-point communication. During 1922 the government's attitude softened considerably. The bottle had been opened, and the broadcasting genie was abroad.

Jack's achievement was the modulation of a continuous wave with speech and music, and many of his students took aspects of radio research as the subject of MSc theses. However, he made it clear that he and the university were not interested in continuing as broadcasters. He was happy to hand over the development of broadcasting to others. He and the 572 holders of listeners' permits did not have long to wait: in 1922, broadcasting stations began transmissions in all four of the country's main cities.

BROADCASTING STATIONS IN 1922

In the week before his first broadcast, Jack was elected patron at the inaugural meeting of the Otago Radio Association. The public meeting, attended by some 150 enthusiasts, formed the association to 'set up a modern receiving station, keep a log of all messages received, check standard time, record atmospheric phenomena, investigate wireless telephony and direction finding, and provide technical information for members'.[30] Broadcasting was not mentioned but the association started a station which, as 4XD, is still transmitting. The first committee members included the earlier wireless telegraphy experimenters, James Passmore and Ted Meinung, while another, D. G. Mitchell, was secretary-treasurer. Members of the association began to build a transmitter, and applied for permission to broadcast. The completed transmitter, with a power output of 50 watts, was demonstrated to members, though only with an indoor aerial, on 4 October 1922. The first public demonstration was held on 15 November 1922; among the guests were Sir Robert Stout, the Chief Justice, and E. A. Shrimpton, by then the Chief Telegraph Engineer.

Further north a similar association, the Radio Society of Christchurch, was formed even earlier, making it the first organisation of its type in New Zealand. Its inaugural meeting was held on 15 February 1921 and R. J. Orbell, the convenor, was elected the first president. The Radio Society members were keen observers and supporters of Jack's Otago transmissions and were moved by his example to build and operate their own transmitter, Orbell being the principal constructor. They also were broadcasting intermittently by the end of 1922.

The Otago Radio Association and the Radio Society of Christchurch were groups particularly interested in the technology of radio. Their licence applications, after the publication in 1923 of the first broadcasting regulations, reflect this: rather than for broadcasting stations, they were for Grade I transmitting and receiving station licences. Members of the groups did, however, develop their interests in broadcasting; and the Otago Radio Association especially is a long-standing broadcaster. The groups are an example of a type of broadcasting later followed in other centres — community and association-owned stations run by unpaid members as a combination of hobby and community service.

The groups did much to publicise radio and develop a wider awareness of the medium. Lionel Slade, a foundation member of the Christchurch society, has recorded the disbelief their early efforts encountered.

> People used to tell us we were mugs . . . They used to just simply laugh at you . . . I used to go round to church bazaars and such like and give demonstrations . . . You would see people come up. They would go on the stage and look at the radio set and put their hands in their pockets and just casually wander round and look under the table and underneath the stage just to see if they could find a gramophone.[31]

These cooperative amateur associations were soon joined by other would-be broadcasters who coupled their broadcasting enthusiasm with a desire to make a profit from the new medium. In Dunedin two such broadcasters, Norman Arundel and Fred O'Neill, were associated with the group of enthusiasts around Professor Jack, but both left because of their commercial plans. Jack could not allow any commercial use of broadcasting through the university. Furthermore he personally disliked commercial development of what he regarded as his teachings.[32] Arundel and O'Neill obtained temporary transmitting permits in 1922, and both established stations. Arundel started the British Radio Supply Company. He constructed his own transmitter using four french R valves as oscillators and achieved a power output of 20 watts. This station began from premises in Moray Place in Dunedin, but soon moved to Princes Street after being offered the use of rooms in the Savoy Restaurant. O'Neill's station operated from the Moray Place premises of his business, the British Electric Engineering Company. He obtained a De Forest transmitter in Wellington, and his station was opened on 29 October 1922.

The commercial activity that both men associated with broadcasting was the sale of radio sets and related equipment. For them the commercial objective of broadcasting was to develop an audience to whom they could sell the necessary receiving equipment. Further north, in Wellington, two stations were already developing commercial broadcasting beyond this understanding.

The first Wellington station opened in February 1922 — New Zealand's second broadcasting station. It was owned by the International Electric Company, soon renamed the International Radio Company, which opened with transmissions on Monday and Friday evenings from 7.30pm to 9.00pm. The principal figure was Charles Forrest, the man from whom O'Neill purchased his transmitter. Forrest, then in his early twenties, joined in partnership with Hope Gibbons, a Wellington businessman, from whose newly erected multi-storey premises the broadcasts were made. A one-room station was built on the top floor.

Forrest began transmissions, which he properly refers to as 'research work', in 1921 from the O'Neill building at the corner of Tory Street and Courtenay Place. These first transmissions were contemporaneous with the Dunedin work of Professor Jack. Unlike Jack, Forrest did not have a temporary transmitting permit. He began without a licence, operating on a verbal permission from Shrimpton, the Chief Telegraph Engineer. Whenever his transmissions were 'causing too much trouble', which meant they were interfering with the work of VLW (Radio Wellington), he would be rung by Shrimpton or his assistant, Arthur Gibbs, and asked to stop transmitting. Forrest's first transmission was immediately on receipt of a De Forest OP3 transmitter from the US. It illustrates the looseness of the arrangement, along with the interest and goodwill of the P&T Department officials:

> Just to make sure it was working I put on gramophone records and adjusted the dials on the transmitter. Next thing I heard was a telephone call. 'This is Mr Shrimpton, Chief Telegraph Engineer. Mr Forrest have you received and are you operating this transmitter you were telling me about?'. . . I said, 'Well, why, why, no.' 'Well there is music in the air and there is nobody else in the country that can do it. It must be you.'. . . It seemed just a few minutes afterwards, Mr Shrimpton . . . walked in and said, 'Let's see this. Let's see how you operate it.' I mean I was really caught . . . But we became awfully friendly. There were no bad feelings.

The single room station in the Hope Gibbons building housed the studio along with the transmitter and the other equipment, most of which was constructed and installed by Forrest and by Hugh Simpson. The recorded music was played on an ordinary gramophone, with the sound picked up by a carbon microphone. The microphone often ceased functioning and needed the granules to be knocked apart. Usually the operators knew of this only when a listener rang to tell them their broadcasting had stopped.

Forrest was happy to broadcast whatever local talent he could find. The first person to sing from his station was Mrs Violet Giles, whose husband was a printer in the firm of Giles and Sons, housed in the same building. Forrest calls her 'the first person to sing over the radio in New Zealand', but this was after the earlier of Jack's transmissions which also included singers. As with Jack's transmissions, most of the music broadcast was not live but from records.[33]

The principal of the other Wellington station was Arthur McClay, the pre-war wireless telegraphist. His war service in the Medical Corps had ended with his spending nine months in an English military hospital with a lung infection, during which he read all he could on wireless telephony. He decided at that time to become a broadcaster. After the war, back at his parents' home in Newtown, he restarted his wireless telegraphy station, and various enthusiasts regularly gathered there. They included Gordon Harcourt, later McClay's broadcasting successor, and Clive Drummond, the Radio Wellington telegraphist.[34]

In 1921 McClay gained the Chief Telegraph Engineer's permission to start a broadcasting station. He entered into partnership with Bob Apperley and Hugh Simpson. Simpson and McClay became friends when Simpson moved next door to McClay's parents' home. Apperley, a young but senior technician at the telephone exchange, was introduced to McClay by Shrimpton. The three set to work to get a station ready and had to construct most of the necessary equipment. Simpson described their equipment as having 'three modulating valves and three transmitting valves with a total power of about 15 watts, as far as we knew. The antenna was strung between two large pine trees on Arthur McClay's premises.'[35]

The station started broadcasting in July 1922. The intention was to broadcast on Wednesday, Friday, Saturday and Sunday evenings and in addition to transmit any special items as the opportunity arose. The station premises were at the address of McClay's parents' home. The property included a two-room detached cottage which had housed Arthur's wireless telegraphy equipment and now was the station site. There were some recitations, but in general very few artists were used and live broadcasting was infrequent. On those occasions the lounge of the parents' home, with its carpet, drapes and piano, was used as the studio. As with Forrest's station the great majority of the broadcast material was from phonograph records with the sound picked up by a carbon microphone. The microphone originally was fitted with a funnel, but eventually this was replaced by a small horn over the phonograph. The broadcasters were to use a wavelength of 275 metres but had no accurate way of measuring their wavelength and were satisfied with an approximation. The important matter was to avoid interfering with ship or coastal station traffic.

They were all aware that what they were doing was new, and derived great satisfaction from the fact. A joyful enthusiasm boosted their endeavours. Drummond, who spoke of his time with the station as 'the days of thrills', describes the experience: 'We began broadcasting just for the fun of it . . . We thought we were Christmas when ships just a couple of miles out in the Wellington Harbour picked up our broadcasts'.[36] Drummond soon became the major public figure in the new station. The others, noting his pleasant voice and good vocabulary, readily accepted his offer to be the regular, though unpaid, announcer. Drummond had begun a new career.

In his recollections, McClay spoke of four broadcasts in particular. At the 1922 general election he was invited by the Post Office to broadcast the returns as they came in. This was done successfully and for the first time New Zealand election results were given over the radio. On another occasion the parliamentary Speaker, Charles Statham, invited them to present a programme which the members could listen to in their recreation room. For many of the parliamentarians and their wives and friends this was their first experience of radio. Another broadcast less accepted by authority took place in collaboration with the publicity manager of the James Smith department store in Wellington. A broadcast concert was arranged for a Friday evening from 7pm to 9pm outside the store, facing Cuba and Manners Streets. The store advertised the programme, McClay hired several Magnavox loudspeakers to amplify his broadcast, and the crowd that came to listen was so large that traffic had to be diverted. Drummond gleefully recalled, 'We were the first to create a traffic jam in Wellington.' Another innovative and well-received broadcast took place when the Presbyterian Bible Class hired the Wellington–Days Bay ferry, the *Duchess*, for a Saturday night dance and sing-along. The music was broadcast from the station and successfully received on the ferry.[37]

The commercial broadcasting that is a notable feature of the Wellington

The equipment — a single circuit receiver with a three stage amplifier and Magnavox loudspeaker — used by the McClay group to bring music via radio to the P&T Engineers' Ball in the New Century Hall, Wellington, 11 August 1922.

stations began with Charles Forrest. At first Forrest did not see commercial broadcasting as an end in itself and had similar views to his Dunedin colleagues: 'I was very keen for broadcasting to commence in New Zealand so that I could sell radio bits and pieces and parts so that people could buy them and listen to the music I was sending out.' It was in obtaining the records that made up the greater part of his broadcasts that Forrest entered into a business arrangement which introduced advertising to New Zealand broadcasting: 'A firm called Williams, Music Dealers . . . used to supply the records . . . provided I mentioned over the air that these records were being loaned [by their firm].' McClay followed a similar practice at his station: he obtained his records from Sid Vause, then manager of a Willis Street gramophone store, and later in charge of recorded material at 2YA. In return, the station acknowledged the loans and informed listeners where the records could be purchased. Nor did the advertising efforts stop with this. Although his plan was to build up a listening audience to which he could sell radio sets, Forrest rapidly found that he was faced with the problem of financing a broadcasting station. He sought further advertisers and successfully asked several Wellington firms to 'donate so much a week towards the running of the station . . . Each evening . . . I would announce that this music was being transmitted through the cooperation of the following firms and I would call out the list of the firms. They used to pay . . . about thirty shillings a week each.'

While some of the paid-for material might more appropriately be called sponsorship, much of it was clearly advertising. Nor was it unsophisticated. Forrest recalls one campaign, initiated by the music manager of Charles Begg and Company.

> He said to me, 'Would you do something for me. Would you broadcast over the air, "Yes, we have no bananas."' I said, 'Tell me, what does it mean?' He said, 'I can't tell you what it means now but I will eventually' . . . I was looking for monetary support for the station and I said 'Yes'. I carried on this for quite a long while. People would say, 'He's gone mad. What does he mean, "Yes, we have no bananas"?' And eventually Beggs came out with 'Yes, we have no bananas'. . . and it became a great hit.[38]

McClay had a similar commercial rationale to Forrest. He began a business, Federal Wireless, and by providing something to listen to, hoped to boost his sales of receiving sets and related equipment. But McClay hoped this business would be only temporary: his long-term aim was the development of a substantial broadcasting concern. He and his associates dreamed of establishing a nation-wide group of stations. Like Forrest, McClay sold advertising time — and it was this side of broadcasting that he wished to extend.[39]

Forrest remained with his station until the end of 1923. Throughout that time there was no official concern with his commercial broadcasting: 'Never at any time did the [P&T] Department tell us to cease advertising or mentioning these other names.'

A further station opened in 1922 when Charles Forrest sold a 5-watt De Forest transmitter to an Auckland firm, Radio Ltd. The transmitter was installed in an upstairs room in the firm's premises in the Strand Arcade in Queen Street, and it was from this site that Auckland's first broadcasting station began transmitting. The broadcasters involved were Douglas Shipherd and Bob Burrell. Shipherd in particular had long been interested in radio: he had made his first receiver in 1908. His friendship with Burrell started during the war. In peacetime the two joined with H. V. Coe in Radio Ltd, with the aim of selling radio parts and sets. Again their commercial motive was to stimulate interest so as to sell radio equipment. The station operated on Saturday afternoons. As Shipherd recalls, they 'put over programmes from gramophones, did a bit of talking, a bit of topical stuff out of the newspapers. About two or two and a half hours in all.'[40]

Thus by the end of 1922 there were seven stations — three in Dunedin, two in Wellington and one each in Christchurch and Auckland. None were broadcasting more than a few hours a week. In highly changed guises, five of these stations continue today. Four of them became the YA stations, and the station run by the Otago Radio Association eventually became 4XD.

It is distinctly possible that this is not a full coverage of broadcasting stations in 1922. There are other claimants. Roy Keith, who was soon to establish a station in Auckland, came from Hawera and claimed he was broadcasting there in June 1922.[41] Furthermore, temporary transmitting permits were issued in 1922 to, besides some of the individuals already mentioned, R. Bansgrove and C. Ward of Stratford, W. K. Bunce of Timaru, L. G. Freeman of Wanganui, and R. J. Patty and W. J. Sinclair of Gisborne.[42] Anyone who went to the considerable inconvenience of obtaining a temporary transmitting permit in 1922 can reasonably be assumed to have used the permit and transmitted. However, there was no distinction made in the early permits between broadcasters and amateur radio transmitters ('DXers'), who are involved in point-to-point, two-way communication. It was not until 1922 that it became possible in New Zealand to distinguish these two types of radio users — and the distinction was not officially made until the publication of the 1923 regulations. Even after that it was not clear, with many of the amateur radio transmitters at times operating a two-way, point-to-point communication, and at others broadcasting music and other material for whoever might be listening.

BROADCASTING POLICY

By late 1922 people were beginning to realise that broadcasting offered wide-ranging possibilities that would change the nature of public communication in New Zealand. J. E. Strachan, reporting to the Canterbury Progress League, noted, 'It is difficult for us in New Zealand to realize that we have entered upon a new era in the matter of communication.'[43] Few shared his appreciation of the new medium. Many officials were convinced that broadcasting would have a short life. One P&T Department official was quoted in late 1922 as expecting 'wireless listening' to 'have a bit of a run, like ping-pong, or put-and-take, and then die down'.[44] Businessmen were yet to be convinced that investment in a broadcasting station would be profitable. However, the topic of broadcasting was beginning to attract public attention. The first mention of the word 'broadcasting' in Parliament was in August 1922 when D. G. Sullivan, the member for Avon, asked the postmaster-general whether his department had prepared a policy with regard to the broadcasting system of wireless telephony. Coates replied that the government intended to relax the restrictions banning transmission. He also indicated that the government was proposing to develop regional broadcasting: 'It was proposed to cut the country up into areas and then issue licences in those areas.'[45] Returning to the topic the following week, he made it clear that 'The Government would not go in for broadcasting . . . A licence would be issued to firms interested, and they would make their own arrangements with the people in the districts concerned in the matter of broadcasting news between the city and the country.'[46] This announcement aroused considerable interest, but left many matters unresolved. Specific regulations were promised, and were awaited with interest. The government received a number of submissions in the next months.[47]

The long-awaited Radio Telegraph Regulations for Amateur, Experimental, and Broadcasting Stations were formally presented on 17 January 1923. Radio telephony was defined as within the meaning of radio telegraphy. The country was divided into four districts in the same manner as for the telegraph and the telephone: 1 Auckland, 2 Wellington, 3 Canterbury and 4 Otago. This division gave first numbers to the various station call-signs which are still used today.

Six types of station were differentiated. Three were varieties of amateur transmitting and receiving stations. There were two types of broadcasting station, toll and private — but what a toll station could do that a private station could not was not made clear, and the category was omitted in the 1925 rewriting of the regulations. The distinction between the two types of broadcasting stations did not imply that toll broadcasting stations could compete with the government communication services. Any transmission or reception calculated to cause a loss of revenue to the P&T Department was prohibited for all stations. The station licence applied for by members of the broadcasting audience was for 'an amateur station for reception only, including reception from broadcasting stations'. The annual fee was 5s. It was another regulatory splitting of the transmission and reception capabilities of radio. As well as paying a fee, all applicants, including those wanting a licence for reception only, had to be fourteen years of age or over and were required to supply various particulars including a character reference and evidence of British nationality. Every licensee also had to declare that they would not divulge to any unauthorised person any information relating to public correspondence that came to their knowledge.

The 1923 regulations signalled an acceptance by the government of both broadcasting and public access to radio. However, the government still regarded private ownership of radio with considerable trepidation; and it retained its previous concerns about wireless telegraphy — public access to private communications such as radio telegrams, and the consequences of disloyal people using radio, if only for reception.

Broadcasting stations were classified and controlled so as to minimise mutual interference and make the service as widely available as possible. To this end the regulations included a listing of the fifteen largest centres, with the wavelengths to be used in each. Stations in these centres were allowed a .5 kW power. Nineteen smaller centres were also listed, with their wavelengths, and their stations were allowed a .25 kW power. Some centres (such as Gisborne and Nelson, and Whangarei and Timaru) which were regarded as far enough apart not to cause interference problems, were allocated the same wavelength. If there was only one station in a centre, broadcasting hours were unlimited. Where there were more than one, the minister was to determine the hours of operation.

It was made clear that this listing of centres was liable to future amendment. It in no way indicated the actual path of development of broadcasting stations, which was one of powerful metropolitan stations whose broadcasting range would meet the needs of the entire country. But as of 1923 the regulations gave effect to the government's proposal to develop broadcasting on a regional basis with low-power stations in most population centres.

The regulations governing the technical matters of wavelengths, power output and hours of broadcasting were accepted happily enough. Coates had explained the rationale behind them to the House the previous September:

> . . . if you allow the wireless people to use any wavelength or any number of watts, the man with the largest station would jam out the smaller stations . . . it is necessary there should be fairly drastic regulations to control wavelengths, watts and so on. If we do not have such provision then in a short time amateur wireless in the Dominion will be impossible.[48]

Shrimpton much influenced the regulations. He had been sent by his department to the United States, England and Europe in 1919–20 to study the expanding field of telecommunications. Although this was just before the first burgeoning of broadcasting in the United States, Shrimpton saw the effects of the different countries' policies. He returned with a clear view of the best path for New Zealand. Harris recalled that his aim was 'to strike a happy medium between what was practically a "free for all" in the U.S.A. and the very reluctant, restrictive and almost impossible conditions imposed in England'.[49] On technical matters the regulations were a good initial guide for a policy which aimed to allow as many broadcasting stations as possible while guarding against the mutual interference of their signals.

The regulations, and Shrimpton's subsequent interpretation of them, were regarded favourably by most amateur radio transmitters. Shrimpton, who had a keen personal interest in amateur radio and was the first president of the New Zealand Association of Amateur Radio Transmitters (NZART), consistently acted to limit the maximum output of amateur stations and thus curb some of the more exuberant practitioners. But his personal interest was recognised as restraining the restrictive practices of officialdom. He was soon dubbed 'Dad' by the amateurs, and 'Talk it over with Dad' became the advice given to amateurs having difficulties with bureaucracy. From this point on the P&T Department was seen in a different light by the amateurs. For instance, the Gisborne operators R. J. Patty and Ivan O'Meara, who had been clandestine operators previously, now saw the department as not at all standing in their way.[50]

Curbs on what could be transmitted included a ban on material of a seditious, profane, obscene, libellous, or offensive nature. Also there were four particular regulations regarding broadcasting material. One required priority be given to the "broadcasting of religious services and kindred matter" on Sundays from 11am to 12.30pm and 6.30pm to 8pm. The second prohibited both direct and indirect advertising except for the provisions of the third which, in addition to requiring stations to repeat their call-signals three times at the commencement of each separate item, also allowed them to announce the full name and address of the station. The fourth, regulation 79, was the most detailed and gave rise to considerable difficulties of interpretation. It reads:

> Broadcasting stations shall not be used for the dissemination of propaganda of a controversial nature, but shall be restricted to matter of an educative or entertaining character, such as news, lectures, useful information, religious services, musical or elocutionary entertainment, and such other items of general interest as may be approved by the Minister from time to time. The licensee of a broadcasting station shall not transmit radio communications which, in the judgement of the Minister, do not come within the authority of this regulation or do not conduce to the public interest.

It was clear from these regulations that, although the government itself did not intend to become a broadcaster, it intended to be firmly in control of broadcasting. But much was still unclear, particularly the financial status of broadcasting. The regulations allowed private broadcasting stations, but broadcasters were faced with a problem. The 1922 stations with a commercial motive had been started to promote an audience which would buy radio sets. The Wellington experience had already shown that revenue from this was not enough to keep a station on the air, and what broadcasting there was had begun to be supported by advertising. This was now banned. It was not obvious how broadcasting stations were now to meet their costs. There was an annual licence fee of 5s per listener, but this money was revenue for the P&T Department; it was not collected to support stations.

1923 BROADCASTING STATIONS

It was some months into 1923 before the broadcasting regulations were implemented and call-signs given to the various stations. The first was issued in Auckland: on 2 May 1923 the call-sign 1YA was given to the station

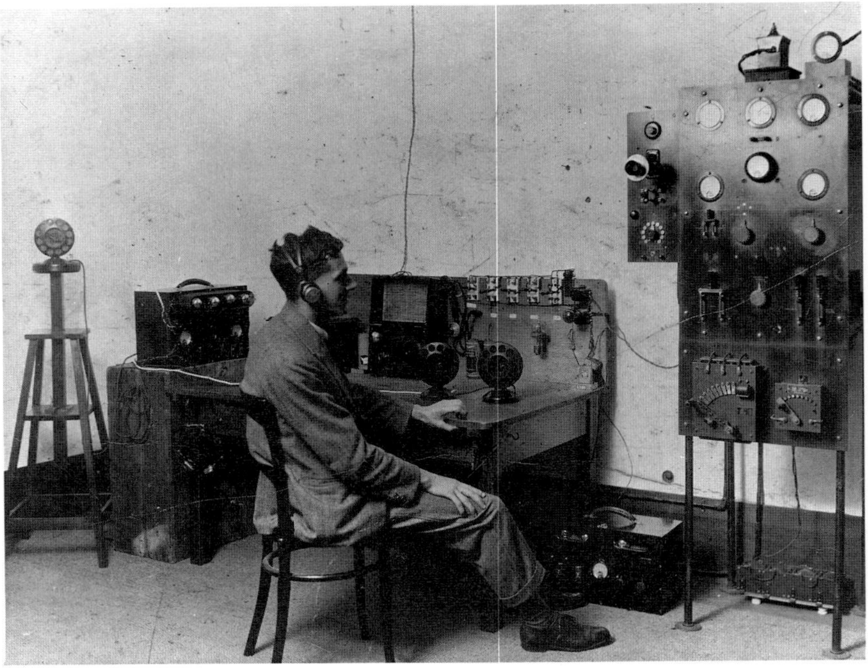

Bill Huggins at 1YA, Scots Hall, Symonds Street, 1924. ALEXANDER TURNBULL LIBRARY

of Radio Service Ltd. The station had by then been transferred from its original Queen Street site and was transmitting from Scots Hall in Symonds Street. It was operating with a 50-watt De Forest transmitter, again bought from Charles Forrest. The original operators had left the concern, and one of them, Douglas Shipherd, was soon involved in Auckland's second station. The operation of the station was under the control of John Bingham and Bill Huggins, with Jack Davies as manager. Bingham, the ex-Christchurch wireless telegraphy experimenter, and Huggins, now moved up from Patea, were beginning careers as broadcasting engineers.

A second Auckland station was licensed on 20 September 1923 and given the call-sign 1YB. The licence was issued in the name of Charles Pearson, and the station operated from his premises at 157 Karangahape Road. The operator was Douglas Shipherd. Pearson ran a music business, selling records and sheet music as well as radio sets and parts. As with the other Auckland station, the aim was to provide a broadcasting station so as to sell sets and parts to would-be listeners. Pearson was also able to publicise new music.

Station operation was not the only task — installation and demonstration of radio sets was a major part of the job. Bill Huggins describes what was involved: not just the radio receiver, but masts, aerial wire and batteries had to be transported to the purchaser. 'You would get the aerial up, get the lead in through the house . . . put the set on a table and battery on the floor, careful not to spill acid on the carpet, and connect up the batteries and loudspeaker . . . You would go back again as late as you possibly could, because there were no signals on the air until it was dark, and give a demonstration of the set.' Giving a demonstration was a 'very difficult thing', for tuning a radio set required considerable skill and was a major reason for the initially slow acceptance of radio by the public: 'You usually had at least five controls that you had to jiggle before you got any signal . . . you had three knobs that had to be synchronised. The volume was controlled by varying the brightness of the filaments . . . you had two coils to tune and at least two condensors to adjust. They were very ticklish to get right.'[51]

In Wellington the two stations — that of Charles Forrest and Hope Gibbons and that of Arthur McClay — continued into 1923. Charles Forrest and Hope Gibbons incorporated their business as Wellington Broadcasters Ltd in August 1923, and the station was given the call-sign 2YB. The incorporation was a protection against possible legal action by AWA, which claimed patent rights for much of the technology used in broadcasting. Forrest recalls, 'It appeared to me at the time that all Amalgamated Wireless wanted to do was to stop broadcasting.'[52] This was the first indication from AWA of its understanding of its patent rights. Nothing came from AWA's actions at this time, but two years later they were offset only by legislative action. Shortly after the formation of Wellington Broadcasters Ltd, Charles Forrest sold his interest in the firm to Hope Gibbons and left for Sydney, where he started station 2BL.

Arthur McClay also withdrew from broadcasting in 1923 when a new group, headed by Gordon Harcourt, announced its intention to start a station and to import and sell radio sets and parts. In November 1923 this group formed the Dominion Radio Company Ltd, which operated from the Dominion Newspaper Building alongside the Plimmer Steps. The transmitter, under the control of Ray Haggitt, was housed in a small cabin on the roof. The aerial, which was stretched between two masts about 150 feet high, was a four-wire flat-top type, on 12-foot spreaders. An area about 14 feet square was partitioned off in the paper-roll storeroom to make a studio.[53] Dominion Radio took over McClay's licence and continued on the same wavelength with the same recently allocated call-sign, 2YK.[54] Apart from changing to a more powerful 50-watt transmitter, 2YK continued in a similar vein under its new ownership — still transmitting four nights a week, with Clive Drummond still doing the lion's share of the announcing.

It was from this station that Drummond broadcast the result of the final test of the 1924 All Blacks' tour of Great Britain. This broadcast established a speed record for news transmission, and illustrated the growing power of broadcasting to spread information through New Zealand. Arrangements had been made with the P&T Department for a direct line to the Pacific Cable office in Auckland, and listeners were advised that 2YK would be

The transmitter and carbon microphone of 2YA, Nelson, in 1923.

on the air in the early hours of Sunday morning to transmit the result of the match. At 2.43am, 2¼ minutes after the final whistle, 2YK first broadcast the news of the 17–11 New Zealand victory. 2YK's reception of the news from London was of such low quality that Drummond was able to decipher the score but not which team had won. Fortunately his patriotic guess was correct.[55] The score was heard by few listeners but was read by many more, for it was chalked on railway engines and station notice-boards, on footpaths and walls of buildings in many towns. As the *Evening Post* noted, 'There were not many who by 7 o'clock [on Sunday morning] had not heard the result either as a result of wireless broadcasting or by word of mouth, and all were feeling elated at the victory.'[56]

This initial pairing of broadcasting and rugby is not incidental. By waiting up for a rugby test result, Drummond was illustrating that while broadcasting itself was a new and even revolutionary activity, there was no intention to link it with other than the established verities of New Zealand culture. Sport and broadcasting were to be combined to unite the population in a shared celebration of its representatives' exploits and to affirm a particular understanding of New Zealand as a nation.

In spite of the fact that they were the first to broadcast, the two Wellington stations were not the first in the Wellington region to be given a call-sign. The first licensed station in the Wellington region, and thus the one to receive the call-sign 2YA, was a Nelson station owned by Wilkins and Field Ltd and licensed in July 1923. It was a battery-powered station, for Nelson's electrical supply did not begin until October 1923. The transmitter was a small De Forest 5-watt set — a box about 10 inches square which, although another valve could be added in parallel, was essentially one valve modulated with a hand-held carbon microphone. The station ceased broadcasting before the end of the year; the call-sign lapsed, and was given in 1927 to Wellington city.

In Otago, the three existing concerns all obtained licences in 1923. Fred O'Neill's and Norman Arundel's stations were both licensed on 13 August 1923, and given the call-signs 4YA and 4YO respectively. The Otago Radio Association was also given a licence, as was the Radio Society of Christchurch, but these applied for and were given Grade I licences for transmitting and receiving rather than for broadcasting. The Otago association received the call-sign 4AB, and its Canterbury counterpart 3AC.

In the North Island a new broadcasting station, 2YM of Gisborne, was licensed on 27 October 1923 in the name of the Gisborne Radio Company — the collective title for the three long-standing Gisborne wireless amateurs, Ivan O'Meara, R. J. (Bob) Patty and P. R. (Percy) Stevens. All three also took out amateur transmitting and receiving licences. Percy Stevens was the main figure in the station. All three remained as friends and associates, but O'Meara and Patty continued to concentrate on point-to-point wireless telegraphy and telephony, and were famous for their long-distance communication achievements. Stevens took up broadcasting using a 25-watt transmitter of his own construction, first from the lounge of his home and later, after obtaining a licence and a call-sign, from a studio at the rear of his shop in Gladstone Road. There is some doubt as to just when Stevens began broadcasting. Stan Vincent, the 2YM announcer, claimed that Stevens had started transmitting some months before Professor Jack in Otago.[57] Judging from the Gisborne enthusiasts' long-standing interest in radio, their considerable technical ability and their lack of concern for the niceties of the regulations, the claim is quite possible.

Again these stations were not the only ones broadcasting. Various amateurs holding Grade I or II transmitting and receiving licences broadcast for the entertainment of whoever might be listening. One was Bernie Spackman of Napier, who had built several crystal sets for friends: 'They were not able to get any music. So I started broadcasting.' Spackman ran 'children's sessions from about half past five until soon after six'. He also borrowed records from a local music store and played them on a horn gramophone with the sound picked up by a microphone he had made 'out of a little brass button with a mike diaphragm about an inch across and a centre contact filled with carbon granules'.[58]

Morton Coutts, on 2AZ of Taihape, was also licensed as an amateur transmitting and receiving station, but he 'normally sent out a little music for the benefit of local hams on Mondays and Tuesdays and sometimes on a Saturday night'.[59] In December 1925 he was told by his local radio inspector that the inspector 'had received word from headquarters that someone was broadcasting in Taihape, and they were to stop immediately'. Coutts took the hint and ceased transmission, but after press discussion of the matter he received a telegram from the P&T Department secretary stating that no such orders had been issued and that he was free to continue broadcasting. He thus resumed transmissions, intending to carry on until a more permanent broadcaster was on the air.[60]

The broadcasting stations transmitting after the passage of the 1923 regulations were soon faced with the financial difficulties inherent in the prohibition on advertising. This regulation was often disregarded and stations did receive some income from this source. But radio advertising was illegal and, unlike newspaper advertising, was something to which neither advertisers nor the public were accustomed. Furthermore it exposed stations to the wrath of the P&T Department. These were not conditions under which advertising could flourish; that it did continue intermittently was an indication of the stations' financial plight. It was these changes that led McClay to abandon broadcasting. No longer able to advertise or to attract financial backing to keep his Wellington station going, let alone to implement his nation-wide plan, he withdrew, frustrated at the seemingly impossible situation for broadcasting.

Stations soon began to seek donations. They opened 'subscription lists' and invited the largesse of listeners. At times listeners were generous, especially when money was sought for a specific purpose. Charles Forrest recalls that when he asked listeners to each send in half-a-crown (2s6d) so as to raise £100 to buy a piano for 2YB, the station was inundated with more than sufficient coins.[61] But as a regular source of income to offset the continuing expenses of running a station, subscription lists were not successful.

Station owners felt they were carrying an inequitable share of the cost of broadcasting. Most stations were set up in order to build an audience, to whom station owners, who were also radio dealers, could sell sets. Station owners were allowed to announce the name and address of their station, which meant in effect broadcasting that they were the dealers who ran the station. This was the only gap in the general ban on advertising, and it gave station owners some advantage in the competition for sales. But the number of radio dealers grew rapidly, and although all were benefiting from the audience growth brought about by the stations, most were not carrying the cost of running one. In Wellington and Auckland various of the non-broadcasting retailers accepted that they should pay a levy to assist in defraying the local stations' expenses, but this cooperative movement never became complete.

During 1923 it became increasingly obvious to the various parties concerned that the state of broadcasting was not satisfactory. An audience was developing, but there was little growth in the number of stations or increase in power output and transmission hours from the existing stations. The most common views were either that the stations should be allowed to raise money by advertising, or that the licence fee should be increased, and the extra revenue paid to broadcasters. Professor Jack, for example, advocated that fees should be 10s for crystal sets and £1 for valve sets; the government should retain 5s and pay the balance to broadcasters.[62] With the regulations only months old, many argued that broadcasting required reorganisation.

A NEW ORDER FOR BROADCASTING

In 1923 Coates received representations from broadcasters and other interested parties. He concluded that changes were needed. Broadcasters and P&T Department officials agreed the power outputs of the existing stations were too weak to offer a reasonable signal to listeners. Various parties, including the different station owners, requested that the government arrange a new order for broadcasting; and a Wellington conference of Radio Club members also asked the government 'to do something to provide a national service'.[63]

Coates agreed that a national service would be the best way to develop broadcasting as a form of communication, especially for rural areas. 'Properly handled,' he contended, 'this form of transmitting news could be made use of in backblock districts — in fact in all isolated parts of New Zealand.' Coates understood that stations were hampered by a lack of revenue. He decided to move the organisation of broadcasting from a regional to a national basis, and to allocate to the national organisation a portion of the radio licence fees. These intentions were first disclosed in a speech Coates made in November 1923 at the opening of the new Dannevirke Post Office. They were given substance in the Post & Telegraph Amendment Bill of 1924 and in new radio regulations gazetted in March 1925.

The government won considerable support and goodwill from the broadcasters with a clause of the 1924 Act that showed the government's readiness to promote broadcasting. This clause was occasioned by a further action of AWA. Still claiming patent rights over broadcasting equipment, AWA approached all the New Zealand stations and threatened to sue them individually. Mindful of their own parlous financial position, stations began to capitulate. All had ceased transmission in October 1924, when they were saved by government intervention. A new clause was added to the bill already before the House and passed through all stages in the House of Representatives in one day. The clause deemed all broadcasters to be agents of the P&T Department. Thus the liability for any breach of AWA's patent rights was shifted from the individual to the consolidated fund. Broadcasting continued, but broadcasters, while safe from the threats from AWA, were now even more under the wing of the P&T Department. The negotiations with AWA took four years to settle: the government paid AWA £10,852 on 10 November 1928, and agreed to pay a proportion (3s) of all future receiving licence fees to AWA. This money was to come from the P&T Department's portion of the fee, not from the broadcasters'.[64]

The government ultimately failed to win the broadcasters' support, however. Initially, although some broadcasters preferred the right to advertise to an income from licence fees, there was general acceptance of the government's intentions, including Coates's proposal to combine the existing stations. 'A complete system of broadcasting' could be guaranteed, Coates considered, 'if we could amalgamate the whole of the wireless and broadcasting interests by having a company sufficiently strong to be able to supply useful information, such as market reports, amusements, important speeches, weather forecasts, and other matters of public interest.' But added to this amalgamation were arrangements which were not approved by the existing broadcasters. Full details were never given, but it was obvious the postmaster-general was envisaging something different from the usual commercial company:

> The proposed company consists of all interests — listeners-in, wireless dealers, existing broadcasters, and so forth. The company is to have sufficient capital to be able to guarantee a service, and the company will be entitled to a proportion of the listening-in fees to provide for the items and news. It will be a company properly formed, but under the direction of a Board, and on that Board will be representatives of all the interests concerned, including officers of the State.[65]

The postmaster-general's proposals raised many questions, but few of these were addressed publicly. It is not known how the government reconciled its view that it should not be a broadcaster with the presence of its officers on the proposed board. Nor were the powers of the Board made clear — in particular whether the various station owners would lose control of their stations to the Board. What was obvious was that the government, while admitting that broadcasting should be undertaken by private enterprise, wanted to maintain administrative control over broadcasting policy and standards. Coates argued that this control was the aspect which gave New Zealand a better service than Great Britain with the BBC, 'where there is no control'.[66] The BBC to which he referred was the British Broadcasting Company — not the British Broadcasting Corporation, which did not start operations until 1 January 1927.

The existing broadcasters were thus being required to relinquish their autonomy in return for a licence fee income. Already a disparate group for whom amalgamation was difficult, they were asked to accept direction not from a cooperative of broadcasters, but from a board on which they had no automatic majority.

Although the broadcasters favoured the institution of a new order for broadcasting, they also appreciated the technical difficulties involved in its development. The government's proposal was for powerful stations in the four main centres giving a coverage over the entire country. Coates had been informed 'on reliable authority' that 'the man with a one-valve set' would 'be able to listen to the whole four stations in New Zealand'. This was an enormous leap forward from the isolated, low-power stations that were then transmitting — small-scale affairs which it would be an exaggeration to describe as regionally based. Town or city stations, they concentrated on the immediately surrounding population. It was understandable that the government, with a rural-based ruling political party, should insist on a national system. But broadcasters realised the scale of the demand and viewed it with unease.

Acceptance of the proposals was rendered even less likely by the further political decision that any agreement was for an initial period of only five years. An agreement was renewable but for no more than five years at any one time. The financial implications were obvious to all. Would-be broadcasters faced the difficult task of raising capital. The novelty of broadcasting made investors reluctant to commit the necessary funds to the venture. Coates and others were convinced that radio was a useful innovation, and a few perceptive individuals foresaw something of the later growth of broadcasting, but the general public, and particularly investors, had yet to be enlightened. As of 31 March 1924 only 2900 listeners were licensed in New Zealand. Although broadcasting was to experience a rapid growth, investors were not prepared to make the leap of faith required. This problem was exacerbated by the government's restriction of any agreement with a broadcasting company to a five-year term. Although there was a possibility of renewal, this was not automatic. Even those who foresaw an immediate growth for broadcasting regarded a five-year term as too short to repay the necessary investment. The postmaster-general's own opinion was that 'Personally, as a business man, I should not be inclined to rush after an investment of my money in any undertaking of this kind for five years.'[67]

Thus Coates and Sir James Allen, the man who became postmaster-general in May 1925 when Coates became prime minister, had considerable difficulty in forming a national organisation. This was not for lack of interest. The principal Wellington broadcasters, Gordon Harcourt and Hope Gibbons, were ready to increase their broadcasting commitments. They pooled their resources, with 2YB and 2YK merging in November 1924. The Dominion Radio Company, 2YK, were the name and call-sign that continued. This Wellington syndicate withdrew from negotiations after failing to persuade the government to agree to a ten-year contract. They were unwilling to risk their capital and effort on only the possibility of renewal. Nor was this the only party: the various negotiations were confidential and out of the public gaze, but other groups were reported as interested, including a Christchurch syndicate based around the Radio Society of Christchurch. AWA, represented by its managing director, Ernest Fisk, also entered into at least preliminary negotiations. None of these discussions were fruitful, and Coates's hoped-for amalgamation of all the existing stations into one concern did not take place.

While the various negotiations were under way, new regulations were gazetted in March 1925. Those that applied to broadcasting stations and the sale of radio apparatus were separate from the regulations for receiving, amateur transmitting and receiving, and experimental stations. In line with the intention of providing an income for a national broadcasting system, the annual fee for a receiving station licence was raised to £1.10s. The bulk of the broadcasters' income would come from this listeners' fee. Broadcasters were to receive two thirds of this money. The P&T Department was to receive the other 10s — twice the amount it had been receiving from each licence holder. In practice, however, this division of funds did not eventuate. As part of the negotiations by which a new broadcasting service was formed, this regulation was changed so that the broadcaster received £1.5s per licence fee and the department received the remaining 5s, as before. There were two important changes in the particulars required of licensees: a character reference was no longer requested, and licensees did not have to be of British nationality.

The broadcasting regulations indicated that all current licences under the 1923 regulations would expire on 31 March 1925 and would not be renewed. Four stations, one in each radio district, were issued with wavelengths and were each required to have a 500-watt power output. This was a considerably higher output than the existing stations were generating; the belief was that this would allow almost all New Zealand listeners to receive all four stations. Although these four stations received particular attention in the regulations, there was no understanding that they would be the only stations. Other stations could be established and allotted a wavelength. However any new stations, unless they were part of the national organisation, were not entitled to a share of the licence fee income.

The regulations were considerably stricter with regard to the nature and construction of broadcasting equipment. In addition to the new power requirements, all stations had to be able to accurately measure their power output and their wavelengths. Antennae were to be properly secured. Procedures were established to deal with harmonics and other spurious transmissions. Studios were to be upgraded. One-room operations were no longer sufficient: 'The studio or room housing the microphone apparatus shall be separate from the room containing the transmitting-apparatus, reasonably free from interior reverberation, and so designed as to conduce to maximum efficiency of operation.'

Stations were required to broadcast for a minimum of twelve hours per week, in addition to Sunday broadcasts. They were permitted to have one silent day per week other than Sundays. They were required to broadcast, free of charge, 'any announcements by the New Zealand Government or by any Department thereof'.

The regulations concerning what could and could not be broadcast were as in 1923, but with three changes.[68] The requirements for religious programming were not renewed. The ban on direct and indirect advertising remained, but the regulation now concluded with the phrase, 'unless specially authorised by the Minister'. Many broadcasters saw a ray of hope in this. The other major change concerned the use of gramophone music: 'mechanically operated musical instruments' were permitted but were 'strictly limited to a comparatively small part of the time occupied by the station in transmitting programmes'. In practice, this regulation was interpreted as allowing a maximum of 25 per cent of transmission time for recorded music. The reasoning behind the regulation was not made public, neither was there any newspaper discussion of the point. This reflected the general lack of concern with broadcasting matters. The effect was a considerable increase in the amount of live music transmitted, and thus in the amount of New Zealand programming. Later broadcasters have argued that the aim of the regulation was to foster local talent.[69] This is not likely: broadcasting was not yet considered a cradle for local talent. There are more probable explanations. Record companies were quick to show concern that broadcasting might adversely affect sales, and the regulation may have been at their request. It could also have been a deliberate attempt to raise the quality of broadcasting. The electric pick-up was not developed until 1926, and was not available to New Zealand stations until 1928. There was a recognition at the time that recorded music transmitted via a microphone close to the gramophone left much to be desired.

BROADCASTING IN 1925

The changes to the administration of broadcasting in New Zealand that were ushered in by the legislative and regulatory activity of 1924 and 1925 were indicative of a growing concern to determine the nature of the new medium. Broadcasting was accepted. The changes in regulation allowing all citizens, rather than only those of 'good character' and British nationality, to be radio listeners are evidence of this acceptance and of the allaying of the government's fear of dire consequences if all were allowed such access. The New Zealand government only reluctantly conceded the right to broadcast, but having done so, at first accepted the American rather than the British example by allowing any citizen to establish a broadcasting station. Within a year of Robert Jack's first broadcasts, broadcasting was being developed regionally as a commercial private enterprise, with sponsorship and advertising as the funding basis for the new stations. This situation was shortlived. Regional broadcasting was repudiated and, although private enterprise broadcasting remained the government's choice, the prohibition of radio advertising and the insistence that broadcasting be coordinated nationally by the one enterprise funded from the licence fee, placed the new broadcaster in a close relationship with the state. In theory, all citizens retained the right to broadcast. In practice, there was to be one preferred broadcaster, private enterprise in name but funded in such a manner as to occupy a government-determined monopoly position.

Would-be broadcasters continued to view the prospect with suspicion and, as 1925 continued with no formation of a new broadcasting system, public criticism grew. This focused particularly on the sixfold increase in the listeners' licence fee. The fee was higher than Coates's original suggestion for an increase to £1 or £1.1s, with half going to the broadcasters and half to the department. Even at this level the fee had been criticised, mainly for the large share going to the department and for the fact that there was no reduction in fee for crystal set owners.[70] With the increase in fee to £1.10s and with no evident increase in service, the reduction in licence renewals was so marked that the P&T Department refrained from publishing the number of licence holders in its 1925 report.

From 1 April 1925, the government resorted to a temporary subsidy programme for some of the existing stations. Four stations were selected, one from each of the main centres, to receive £15 a week towards their costs. The four stations were Radio Service's 1YA in Auckland, Dominion Radio's 2YK in Wellington, J. I. Smail's 3AQ in Christchurch, closely associated with the Radio Society there, and Fred O'Neill's 4YA in Dunedin. The subsidies were originally to continue for four months, until the end of July. They were extended for a further month, by which time the government had finally reached an agreement — with William Goodfellow and Ambrose Reeves Harris, who began the Radio Broadcasting Company and became the new national broadcasters.

3
The Radio Broadcasting Company

The Radio Broadcasting Company began operations in 1925, and remained the dominant New Zealand broadcaster until the end of 1931. The company succeeded small, separate stations which had been started to assist the sale of radio sets rather than to develop broadcasting in its own right. In this chapter we shall follow the first years of the company's activities when it focused principally on technical and engineering matters to do with station establishment and operation. While it was concerned only to a lesser degree with programming (and received much contemporary criticism for the lapse) what is notable in retrospect is the speed with which it developed an enduring broadcasting style. By the middle of its term the RBC had formed most of the distinct programming departments, and was beginning to coordinate its broadcasts over all its stations.

GOODFELLOW, HARRIS AND THE FOUNDING OF THE RBC

In August 1925 an agreement was reached between the government and the Radio Broadcasting Company to begin a national broadcasting system. The principal figures in the RBC were William Goodfellow (chairman of directors) and Ambrose Reeves Harris (managing director). Goodfellow had had no previous interest in radio. His background was in the dairy industry, particularly in the development of the New Zealand Cooperative Dairy Company (NZCDC) — an achievement for which he was later knighted. Harris had the necessary technical expertise. Born in Halcombe, he worked for some years before World War I with Thomas Edison in the United States on various of the famous early experiments in the transmission of sound. On his return to New Zealand in 1913, he founded A. R. Harris Co. Ltd in Christchurch — a business which represented Edison's interests.

The Radio Broadcasting Company had its beginnings in the second half of 1923 when Harris was contracted to install electrical equipment in the NZCDC's dried-milk factories in the Waikato region. During their many discussions, he persuaded Goodfellow of broadcasting's potential: 'I realised

William Goodfellow, chairman of directors of the Radio Broadcasting Company. RADIO NZ

Ambrose Reeves Harris, managing director of the Radio Broadcasting Company. *LISTENER*

broadcasting could give the country people a marvellous service and could greatly strengthen the cooperative movement in the South Auckland province.'[1] Goodfellow, who addressed up to eighty meetings of dairy farmers each winter, realised that radio could relieve him of this arduous task.

> When I heard of the radio, I at once visualised the possibility of greatly reducing the number of meetings, yet keeping dairy farmers and the general public continuously informed, not only about the company and its activities, but also about improved farming methods; and at the same time providing entertainment.[2]

Goodfellow arranged for Harris to go to Sydney to enquire into the type and cost of a station able to broadcast in an 80-mile radius from Hamilton, thus covering the NZCDC supply area. With the station cost estimated at around £2500, Goodfellow obtained approval from his fellow directors to develop a proposal to establish a cooperative, non-profit broadcasting station, owned and operated by the NZCDC, primarily for the benefit of the company and its suppliers. Included in the proposal was an intention to import valve radio sets in bulk from the United States and to sell them cheaply and on easy terms to the suppliers, 'so practically every farmer would have a set'.

The proposal seemed to fit well with the government's announced plans for regional broadcasting, and Goodfellow and Harris went to the postmaster-general with some confidence. Coates informed them 'he would agree to give us the required licence to have the exclusive rights to broadcast in Hamilton and within an eighty mile radius'.[3] The agreement was withdrawn, however, when government policy changed to favour national rather than regional coverage.

In 1925, further discussions were held between the government and Goodfellow and Harris. There is some disagreement as to how the discussions began. Goodfellow recalled that the government, after failing to establish a broadcasting scheme, approached Harris through its P&T Department officials and asked for a meeting with him and Goodfellow.[4] Coates stated that the initiative came from Goodfellow and Harris.[5] After a series of meetings, an agreement was reached.

There was a political sympathy between Goodfellow and Harris and the government: both men were members of the Reform Party.[6] It was later contended by the government's opponents that Goodfellow and Harris's political sympathies were of major importance in their obtaining the broadcasting contract. At a dominion conference of Listeners' Leagues it was argued that several others were prepared to take the government's contract but were not given the chance to do so, and that the Wellington syndicate was not allowed to submit fresh proposals.[7] However, while the political affiliations of Goodfellow and Harris were no doubt welcome to the government, the agreement was no rushed affair and came only after considerable expression of dissatisfaction with the slow progress.

Eventually a contract was signed. As with the other would-be broadcasters, the greatest obstacle was the five-year term. The terms finally agreed were that if the minister were willing to renew the contract for a further five years he would give six months' notice of this, at which time the terms and conditions would be renegotiated. If the minister were unwilling to renew, the Crown would buy out the concern at an agreed price or, failing agreement, at a price settled by arbitration. Goodfellow, however, considered he had reached an agreement with Coates and Parr on the matter of renewal.

> I pointed out to Mr Coates that five years was too short a contract period and that what I visualised would happen was that the company would do all the pioneering donkey work and then the P&T Department would say 'Thank you very much' and would take over the concern at valuation in terms of the agreement . . . He assured me the Department had no such intention but I told him that we wanted something in writing. He finally agreed to ask Sir James Parr, the Postmaster-General, to give us a letter of intent which satisfied us that, if the service was satisfactory, renewal would be assured.

Six years later Goodfellow and Harris were to realise that this assurance from a previous government was of no value; but in 1925, satisfied with it, they moved on to negotiate the other details of an agreement with the government.

The contract did not involve the NZCDC: it was a personal contract with Goodfellow and Harris, who became directors of the Radio Broadcasting Company of New Zealand Ltd. A third director, Arnaud McKellar, a Christchurch accountant, was appointed in 1927. The RBC's contract, which was for five years from 30 August 1925, was to purchase the four stations then receiving the £15 per week subsidy and continue their broadcasting. The four stations were to be upgraded in line with the 1925 regulations; the Auckland and Christchurch stations within six months, and the Wellington and Dunedin stations at a later date. This timing was the outcome of the government's political desire for four stations giving a nationwide coverage, and of Goodfellow and Harris's protests that the income would be too small for some years to upgrade four stations.

The RBC was to receive 90 per cent of the fees from radio-dealers' licences and 25s from every receiving licence fee.[8] The share of the receiving licence fee was by far the greater part of its revenue. The agreement covered the use the company could make of this revenue. This was restricted to the purchase, maintenance and operation of broadcasting apparatus, the provision of suitable material for broadcasting and the payment of interest on shares and debentures.

The agreement also dealt with the relationship between Harris's own electrical engineering and importing company and the RBC. His company was to build the stations and supply equipment to the RBC at a commission rate of 10 per cent. This agreement, which amounted to a considerable business relationship with the RBC, in effect excluded the AWA from New Zealand for the duration of the RBC.

The RBC was incorporated with an authorised capital of £20,000 divided equally into ordinary and preference shares. The postmaster-general, acknowledging Goodfellow and Harris's expenses in founding the company, authorised the issue to them of 2000 £1 shares ranking as fully paid. By agreement with the government, the RBC's dividends were limited to 7.5 per cent per year, cumulative and free of income tax. This limitation was, however, also an effective guarantee that annual dividends would not be less than 7.5 per cent.

The government had earlier accepted liability for the use in broadcasting of patented inventions. This protection was now extended to the RBC. This was, however, the extent of the government's liability. The company took on all the usual business risks, as well as the unknown dangers that broadcasting might bring.

The board legislated for in the 1924 Act appeared in the agreement as 'an Advisory Board . . . consisting of the Postmaster-General, the Secretary of the Department, the Chief Engineer of the Department, one representative of the company, and one elected representative of the listeners-in'.[9] Its powers were not specified publicly, but the clear majority of state officials was a further indication of the extent of the government's control.

STARTING THE FOUR STATIONS

Goodfellow and Harris moved promptly to implement the various decisions made. Even before the company was incorporated, they had ordered the 500-watt transmitters required in Auckland and Christchurch. They reached agreement quickly with the station owners at Dunedin and Christchurch and shortly after with Radio Services, the Auckland owners. Only in Wellington did they encounter considerable problems.

With Harris as managing director based in Christchurch, the head office of the RBC was also set up there. Harris was on friendly terms with the members of the Radio Society of Christchurch, and the society agreed to broadcast on behalf of the company until the installation of the 500-watt transmitter. The existing subsidy was thus transferred to 3AC from 3AQ, which ceased transmissions. This was a change of call-sign rather than transmitter. Smail's transmitter was used until March 1926, when the Radio Society's new 50-watt transmitter became operational. While a new studio was being built, the station was moved to temporary premises in what had been a dance hall in Hereford Street, and the call-sign was changed to 3YA. Lionel Slade, a Radio Society foundation member and, in 1925, the 3AC organiser, joined the RBC as superintendent of stations.

In Auckland, as well as purchase of the station, the negotiations also involved employment of personnel. When 1YA became part of the RBC on 20 November 1925, John Bingham and Bill Huggins joined the company. Bingham was appointed chief engineer and moved to the head office in Christchurch. Huggins and Drayton T. (Tom) Venables were the full 1YA staff. The entire concern at Scots Hall was housed in one 20 by 15-foot room. It was, as Huggins recalled,

> nothing very elaborate. We had scrim for acoustic damping on the walls, coconut matting on the floor, and a wind-up gramophone. The transmitter was in those days quite modern. It wasn't very big. We used to have to close down about every hour for the generators to cool. When we wanted to broadcast the microphone used to be hung up in front of the gramophone.[10]

Huggins later joined Bingham in Christchurch, where their first job was to build a transmitter for 4YA. In Dunedin, the purchase of 4YA was completed in October 1925 when Fred O'Neill accepted £500, the price determined under the arbitration of E. A. Shrimpton.[11] O'Neill agreed to continue running 4YA until the company's new transmitter was built and installed. Although Dunedin was not originally scheduled to be the first to get a new transmitter, it rose on the priority list as radio plans there became connected with the planning for the New Zealand and South Seas Exhibition. AWA asked the government for a licence to broadcast from the Exhibition. The government refused, but referred the matter to the RBC. The RBC took the hint, and a transmitter under construction in Christchurch was rapidly completed and installed at the Exhibition in time for the opening ceremony on 17 November 1925. The station was located in a building at the back of the Festival Hall, with its antenna supported by two 100-foot wooden masts. A small studio was attached to the transmitter room, but most of the broadcasts originated from other parts of the Exhibition, with lines run to the Band Rotunda, the Festival Hall and other sites.

O'Neill's 4YA ceased broadcasting at this point, and the company took over in Dunedin as VLDN. It was a mixed blessing for Otago listeners. In the haste to complete and install it, the transmitter had not been fully tested. Nor was the Exhibition site a good one for radio transmission and, although there was a gradual improvement, it was not until the end of the Exhibition and the transmitter's move to O'Neill's premises at 219 Moray Place that the problems diminished. At the conclusion of the Exhibition the Dunedin call-sign reverted to 4YA. Even though transmission was more satisfactory from the changed site, the station, which had a power output of 370 watts, was not easily heard in Southland and was not regarded as any great improvement to Otago broadcasting. Although Dunedin was the birthplace of New Zealand broadcasting, Otago radio did not receive much support from its citizens. Over the next few years the number of Otago receiving licences fell to below that of the other main centres.

All the station purchases presented difficulties to both sides. The RBC considered it was at a disadvantage, since it had no choice but to purchase. It also had little room to manoeuvre, and was obliged to accept the valuations determined by the chief telegraph engineer.[12] The station owners were under no such obligation. In Wellington the disagreement continued when the chief telegraph engineer's valuation of 2YK was close to the amount offered by the company.[13] The owners' refusal to accept this price led to silence on the air — the last thing Goodfellow and Harris wanted. The break in transmissions was almost averted when the P&T Department decided to run the station itself, until the RBC could take over. From the beginning of August the department became the broadcaster of the station's usual Tuesday, Thursday, Friday and Sunday sessions. There was little change in the actual broadcasts: the sessions were at the same time, and the announcer was still Clive Drummond. Drummond's position had never been full-time with 2YK. He remained a P&T Department employee and, while the department ran the station, he was relieved of his normal duties and became a full-time, paid broadcaster.[14] This arrangement continued for several weeks. 2YK ceased transmissions when, on 14 September 1925, the Dominion Radio Company requested that the keys to the station be returned the next day.[15] The directors felt that the temporary arrangements were not aiding their negotiations over the price to be paid for their business. Apart from reporting the election results, 2YK remained off the air until almost the end of the year. Although the fault was by no means entirely that of the RBC, the considerable criticism that followed was directed

The George Court's premises, Karangahape Road, in 1926 with the 1YA transmitter building and masts on the roof.
RADIO NZ

largely at the company. The whole affair earned the company no friends and did nothing to promote the cause of broadcasting.

In Christchurch and Auckland the reputation of the company was not helped by the failure to install the 500-watt transmitters in either city within the stipulated six months. The fault was not that of the company; the delays were caused by English Western Electric, the manufacturers. But again criticism was directed at the company. Aucklanders felt particularly aggrieved. The new transmitter was not installed on schedule, and Auckland had become the most distant outpost from head office. Furthermore, on taking over from the previous owners, Goodfellow and Harris had cut the 1YA transmission hours back to the minimum required in their agreement. While this minimum was above what most stations had been offering, Aucklanders had been enjoying longer transmission hours from 1YA.

The government conceded the delays were not the fault of the RBC and extended the six-month period to the end of June. Even this new deadline was not met. It was August and September respectively in 1926 before the 1YA and 3YA transmitters were finally operational, and listeners in those cities began to receive broadcasts of the promised strength. The company was thus more than a year into its five-year term before stations of the required strength were on the air in two cities. These stations were both situated in new premises which were well up to the standards required in the 1925 regulations. The 1YA studios were in France Street: the main studio was a room that measured 40 feet by 24 feet. The transmitter was erected some 400 yards away, connected by underground cables, in newly constructed quarters on the roof of George Court & Sons premises in Karangahape Road. The antenna was a prominent landmark — four wires linking two 104-foot steel towers placed on diagonally opposite corners of the roof. In Christchurch the new station and studios were located at the A. R. Harris Company premises in Gloucester Street. 3YA was the only station at which the studios and transmitting rooms were adjacent to one another. The 3YA antenna, another prominent landmark, was another four-wire flat-top type, connecting two 154-foot steel towers.

In Wellington it was almost another year before a new transmitter was installed and a new station opened, on 16 July 1927. 2YA's new transmitter was no less than 5000 watts — ten times the original power contemplated, making it, the prime minister stated at the opening, 'the second most powerful station in the British Empire. The Great British station at Daventry alone is more powerful.'[16] The change in plan indicated a growing, though still too optimistic, appreciation of the difficulties presented by the New Zealand landscape to a successful nationwide radio coverage. It also added a further financial relationship between the RBC and the government which dogged both with controversy.

The government's original advice was that most New Zealand listeners would obtain good reception from four 500-watt stations. Doubts were soon cast on this. E. J. Howard, who at the time considered he was the only MP to own a radio receiver, first spoke against this optimism in the House in September 1925, well before any of the stations were operating.[17] By July 1926 there had been considerable public discussion of the matter and Howard was able to argue that, 'The consensus of opinion was that the output of the proposed broadcasting stations would be hopelessly inadequate.'[18] The postmaster-general, William Nosworthy, continued to maintain that the plan for 500-watt stations was the best, but with the opening of the new stations it was obvious that the envisaged clear reception over all New Zealand was not being achieved. 1YA gave good reception, but did not reach over anything like the entire country. Nor did 3YA, though it gave the better reception at a

The prime minister, Gordon Coates, speaking in the main studio at the opening of 2YA, 16 July 1927. *LISTENER*

distance — a matter which Harris put down to the nature of the earth in Christchurch.

The RBC had first considered a 5-kW station at Wellington when it originally took up its contract, but had rejected the idea. The eventual decision to opt for a 5-kW station for Wellington was taken by the government. Goodfellow stated that the government wanted the more powerful transmitter for military purposes, and the P&T Department, in its annual report, also saw the station as valuable 'for defence and national purposes'. Another reason was that the government was extending its broadcasting plans to include the Pacific. It was keen to allow residents of other Pacific countries access to a New Zealand station rather than have them listen exclusively to Australia.[19] However, the change was also an acknowledgement of the faults in the original plan. Wellington's central position and the greater station strength were seen as finally allowing something closer to a nationwide coverage.

The premises chosen for the new Wellington station were in the old Wellesley Club building in Featherston Street, and the site for the transmitter was on the summit of a prominent Wellington landmark. Harris and Bingham subscribed to the conventional view that height was the major consideration for a transmitter site: they chose Mount Victoria, which gave a clear path north and south. The assembly and installation of the transmitter was carried

The transmitter building and masts for 2YA, built in 1927 on Mt Victoria. NZBC

out by Bingham and Huggins. While the design was acceptable for its time, this was shortly before the introduction of crystal control, and the self-oscillating transmitter had continuing problems with frequency stability. The water cooling system also presented difficulties. Huggins described it as 'a pig of a transmitter' which could never be left unattended.[20] The antenna was a single wire connecting two 154-foot steel towers. The Featherston Street premises were the largest and most lavish that the RBC made. There were two studios — the grand and the auxiliary (which was used mainly for solo and news items), with a control room between the two. There were three lounges — the grand, the smoking and the ladies', along with various offices and other rooms. As with the 1YA and 3YA and later the 4YA studios, the decorating and furnishing of the 2YA studios were the responsibility of Mrs J. M. Bingham, the wife of the chief engineer.[21]

Even the 5-kW 2YA transmitter gave less than the hoped-for nationwide coverage, with the performance figures claimed by the factory proving over-optimistic. The early broadcasters overestimated the coverage potential of their transmitters, and underestimated the difficulties presented to their plans by the New Zealand landscape. Although they are little more than 400 kilometres from east to west at their widest point, the two main islands of New Zealand stretch some 1400 kilometres from north to south and include mountainous terrain that made substantial areas largely impervious to the RBC's broadcasts. Many areas such as Nelson, which was relatively close to 2YA's transmitter, were enclosed by hills that thwarted clear reception. Other areas, such as the South Island's West Coast, were so sheltered by the Southern Alps from the RBC's signals as to be, for all practical purposes, part of Australia's radio audience, for it was from that country that clear reception was received at night. The *Radio Record* argued that 2YA was 'able to cover almost the whole of both North and South Islands in daylight'[22] and even better at night. This contention was fiercely debated in the coming years; reception was neither even nor adequate over the entire country. Many argued that their area was particularly badly served; certainly, along with Nelson and the West Coast, Hawke's Bay, Poverty Bay, Taranaki and Southland had problems with reception from the RBC stations.

The decision to install a 5-kW station led to further financial ties between the government and the RBC, and to an extension of the company's licence. It was agreed that the transmitter would cost the RBC £15,000 more than a 500-watt station. The government loaned this sum to the company for a five-year term; it secured the loan by holding a mortgage over the 2YA station. The company's contract was extended to run for five years from 1 January 1927, which meant a total term of six years and five months. The arrangement received much criticism, and the government came under attack in the House. The loan's security was considered questionable, and Opposition members made it the main focus of the P&T Department supply debate in 1927.[23] The loan also opened the government to a barrage of accusations of political favouritism towards Goodfellow and Harris. Many Opposition members, including Sir Joseph Ward and the Labour parliamentarians, held the view that broadcasting should be a state monopoly. Even those members who accepted the desirability of private enterprise in general began to question the wisdom of the government's policy. In 1927 broadcasting became a matter of divisive political debate. The actions of the company and the government that had given it a mandate were opened to parliamentary scrutiny. Members of future governments suggested that, given the opportunity, they would change the administration of broadcasting.

In many respects the increasing political discussion was not of the actions of the RBC, but of broadcasting itself — a sign that it was becoming an influence in New Zealand life. Broadcasting was a rapidly growing innovation world-wide. The particular form it took in any one country could advance or retard its growth, but public acceptance of the medium was gaining momentum. Inevitably it attracted political and public attention that went beyond the details of its local administration.

EARLY RBC PROGRAMMING

Broadcasting was a new venture when the RBC took over. Although the company employed some people who had acted as announcers at earlier stations, there was no established understanding of what was involved in programming. The company's senior personnel had no experience in broadcasting — a craft that had yet to develop a tradition and expertise. Neither were they familiar with the related arts and entertainment areas.

For some time, the mere fact that broadcasting was happening kept the audience satisfied. As Clive Drummond recalls, 'It was the novelty. The interest throughout the country was tremendous. We could do comparatively simple broadcasts that the whole of New Zealand would sit up and notice.'[24] The broadcasters themselves were enthused by the pioneering nature of their activity. A. L. (Arch) Curry, an announcer at 4YA, remembers 'the busy early days of trial and error of handling an instrument which was creating something where nothing had been before. Every occasion was a first occasion. Programmes, outside broadcasts, relays, commentaries.'[25]

Generally the stations were on the air two hours a night, from 8pm to 10pm. Music became the mainstay of programming. Regulations permitted only a small proportion of recorded music. For some time there was a willing flood of singers, musicians and other performers keen to be on the air: according to Clive Drummond, in the early months of 2YA, 'We used to use a lot of studio artists and of course it was all gratuitous. They used to come down and give their services and loved doing it. It was a novelty of course.'[26] John Gordon, a long-time broadcaster who retired from announcing in 1980, writes, 'I started acting on radio at the old 1YA in 1925 . . . Of course we never got paid. We were not members of the staff, we were just performing. Occasionally we would get the odd seven and sixpence for a performance, but not often . . . I wouldn't have done it if I hadn't enjoyed it.'[27] At first the number of willing artists was such that programme times were often extended. They were bustling nights for the small staff, as Mrs Drake, the 4YA accompanist, remembers:

> I was the only pianist to do all the work then — play in a trio, accompany all the instrumental and vocal work and monologues etc. I can tell you those were very busy nights with flesh and blood artists giving the whole programme from 8 till well after 10pm . . . this entailed much hurrying between the two studios so there should be no gap between items.[28]

Even with two hours' programming each night, the programme organisers found that broadcasting had a voracious appetite. At first the RBC assumed that artists would continue to be eager to be on the air. However, the novelty wore off, and the continual need to persuade people to turn up at the station and perform soon defined the daily duties of programme organisers. Drayton Venables recalls that the task of recruiting sufficient unpaid artists to fill a night's programme was difficult even before 1YA had moved to France Street; the Scots Hall programmes, in breach of the regulations, consisted 'mainly of recordings with occasional player piano solos . . . I'd be going around the whole of the day trying to persuade people to come at night and they'd all promise and probably out of twenty that would promise you'd get two. That meant you'd have to fill in with recordings and player pianos, do solos yourself and elocute and that sort of thing.'[29] Many of the early programming staff were also musicians; their skill was a major factor in their appointment. Announcers were regularly called on to improvise, often for substantial periods. Venables recalls spinning out a Mark Twain story, which he found on the paper wrapping from Huggins's fish and chips, for a full twenty minutes. Such incidents often worked well as humorous items, but programming was in danger of being characterised by studio broadcasts chosen for availability rather than quality, augmented by the improvisatory talents of the announcers betraying more than an occasional hint of desperation.

A small amount of money was available to the programme organisers to pay artists. In its first year, to 31 July 1926, the company received £5262 from licence fees: it spent £5000 on capital development, and £319 was paid to artists. In the next year, with the more powerful transmitters coming on air, licence-fee income increased more than fourfold and almost £4000 was paid to artists.[30] Not until 1928 was there a general policy that artists should be paid. Until then the programme organisers were given funding according to the degree with which their case was presented and accepted. Dudley Wrathall at 1YA was given three guineas a week for talent. When 2YA began, Colin Trim received £10 a week for the same purpose. This funding was not provided to all four stations — Dunedin received nothing.[31] As those funds that were allocated were to provide for around ten hours of live programmes per week, the reliance on volunteers was scarcely eased. Nor did the payment necessarily improve the artistic standards. It was, for instance, a frequent practice for music teachers, who were paid to perform, to bring along various students who would fill out the evening's programme. Dudley Wrathall, who was himself a singer and knew most of the Auckland music teachers, encouraged this when he was spending his weekly three guineas: 'I'd say, "Look. Would you come along and broadcast two or three songs. I'll pay you a guinea if you'd like to fill in the rest of the programme with some of your artists."'[32] It was the same with regard to broadcast plays. John Gordon writes of those performed by J. F. Montague and his pupils: Montague 'ran a speech training school in a small room in Swanson Street, and his "pupils" were his clients there, whom he brought to the studios for the experience and to practise what he had taught.'[33]

The need for programming led to the development of 'relays' — broadcasts taken from another venue and relayed over a standard telephone line to the studio. The practice began so as to broadcast dance bands from cabarets and orchestras from the silent picture theatres. Dudley Wrathall explains the way the system worked in Auckland:

> At that time the Empress, the Rialto, the Majestic and one or two other picture theaters had an orchestra. They would open at eight o'clock with an overture and then at nine o'clock they played in the interval with a selection of some kind.

> So we put relay lines in and at eight o'clock we would go over to a theatre and take the overture of six or eight minutes. That was a very valuable six or eight minutes to the programme. Remember three guineas a week. We'd fill in each night with the overture from the various theatres. Then at nine o'clock we'd take the selection of another eight or ten minutes . . . We also did on Saturday nights fairly long relays from the Clik Clak Cabaret in Newmarket and the Dixieland Cabaret in Upper Queen Street.[34]

There were soon complaints about the programmes and their reliance on amateur talent. The *Radio Record*, in its opening issue, argued that this criticism was misplaced: 'There is ample evidence to suggest that a proportion of these newspaper criticisms emanate from those who have been disappointed either in the failure to be included in the programmes through inadequate talent or artistry, or because of unsuitability of the proposed contribution.'[35] This did not explain away the increasing dissatisfaction, and the *Radio Record* itself printed in its next issue a classic complaint:

> I will give you my opinion on your amateur jazz and foxtrot music. It is so hideously ugly and so badly played that I wonder you can bear to endure to announce it. There must be something wrong with people who can make such a dreadful noise, even to themselves, but to bring it before the public is a painful nuisance which ought to be suppressed. I do hope you will take my hint which is well meant.[36]

In June 1927 the P&T Department secretary, George McNamara, raised the question of the standard of broadcasting and suggested that 'now that an income of over £25,000 is assured', the RBC was in a position to do something to improve matters.[37] In a lengthy and uncompromising reply Harris began by arguing that McNamara's letter came 'just at the time when we have completed laying the main foundation work of broadcasting in New Zealand and are just about to build up . . . the only thing that interests listeners, namely the programmes'. He gave examples of staff planning under consideration: 'Proposals for the appointment of a sporting editor and a superintendent of stations are being put before the directors this week . . . The appointment of a Programme Superintendent who will cooperate with the programme organisers at the various stations in raising the standard of programmes generally is also under consideration.' He also mentioned the imminent *Radio Record*: 'The publication of a magazine which will contain advance programmes and keep the public well informed about the activities of the Company is . . . under immediate consideration and preparations are now advanced.' But Harris also argued that the company's financial position was difficult, necessitating a 'down to bedrock' approach to fees for artists. Far from being content with an income of £25,000, Harris estimated the annual costs of the company running the four stations 'on a proper basis' to be at least £55,000 — a figure which would require more than 40,000 fee-paying listeners.[38]

Harris's argument was informed by considerable financial caution. Like the other parties who sought the government's broadcasting contract, the RBC directors experienced difficulty in raising money. Goodfellow considered the venture was always undercapitalised and that they were never able to spend enough in the first years to develop stations and build an audience.[39] The five-year term of the contract also increased the financial caution. While in 1927 they were still confident of a renewal, the directors were unable to encourage similar optimism in investors, and they never had the freedom of looking at their venture as a long-term development.

RBC STAFF

Harris had informed the P&T Department that he was in no position to move rapidly on the matter of programming. But it was at this stage that the RBC began to increase its staff. The engineering and technical side of broadcasting had been the first to receive staffing attention, with the appointment of Jack Bingham as chief engineer, and the appointment of full-time technicians at each station. Station managers were also taken on at each station. None remained long. The resignation of one of them, 1YA's J. M. Prentice, received parliamentary attention.[40] Prentice had been recruited from Sydney, where he was a well-known personality. He was chosen for his experience, and was to have been supervising announcer for all the stations, inaugurating new ideas and practices. Disagreement and his departure were rapid. The station managers felt considerable dissatisfaction with the lack of resources. The longest-lasting, 4YA's J. W. Webb, resigned in 1928.

With the quick turnover of station managers, the critical early appointments for programming were those of the announcers and the programme organisers. These were jack-of-all-trades positions. That of programme organiser was described by Bill Huggins, the 1YA technician in 1926, as 'office boy, announcer, artist and manager as well'.[41] To these crucial positions the RBC fortunately appointed individuals who became lifelong broadcasters and broadcasting executives. Drayton Venables was 1YA's first programme organiser. Culford Bell succeeded Prentice and was appointed announcer in 1927. Colin Trim was 2YA's programme organiser and, in June 1927, John Ball was appointed as editor–announcer. His staff included Clive Drummond as part-time announcer. The manner of Drummond's permanent appointment to the RBC highlights the way in which announcing staff were becoming public personalities. Promoted in the Post Office, he was transferred in 1928 to Paeroa as postmaster. By this time he was into his sixth year as a broadcaster from 2YK and 2YA. His departure, which was broadcast in the 2YA children's session, aroused concern throughout the 2YA reception area. Within a few weeks he was back in Wellington as 2YA's permanent announcer.

3YA's staff began with the Rev. Clyde Carr as announcer. He was appointed in May 1927 after all seventy-three original applicants for the position had been turned down. The 4YA announcer was A. L. Curry, who had been promoted after a few months as junior announcer at 2YA. Other appointments included that of S. J. (Sid) Hayden as a musician at 1YA in 1926, and of Dudley Wrathall in 1927: both rapidly became important figures in the company. Carr was the only one from this group not to make broadcasting a long-term career; he moved to a political career instead. He successfully contested the Timaru seat in the 1928 election and was the first broadcaster to become an MP.

Three years into its contract, the company was employing fifty-six full-time staff — seventeen at head office and eleven, twelve, nine and seven at 1YA, 2YA, 3YA and 4YA respectively. The largest grouping, fifteen in all, were the administrative staff. The engineering grouping was the next largest. There were twelve operators in total at the four stations, headed by Bingham as chief engineer. The head office staff included a director of music and a music librarian, along with a principal of the children's department. Each station had a station manager, a programme organiser, an announcer, a studio pianist and a children's session organiser. In addition to the permanent staff each station was also employing people on contract to do a variety of things, including sports broadcasts, religious sessions, or acting as 'uncles and aunts' on the children's sessions.

In 1927, as it built up its staff, the company also brought sports broadcasts, children's programmes and religious broadcasts under systematic organisation and head office control. The other major development of 1927 was the start of a weekly magazine.

THE *RADIO RECORD*

The *New Zealand Radio Record* was founded to publicise the programmes and to act as the official organ of the company. It received an annual subsidy of £2000 from the RBC. The *Radio Record*'s first issue, on 22 July 1927, was timed to coincide with the opening of the new 2YA station. The magazine had been planned by Goodfellow from the time he originally contracted with the government, and the deed incorporating the Radio Publishing Company of New Zealand Limited, the publishers of the *Radio Record*, was signed on 22 August 1925, the same day as the incorporation of the RBC itself. A. J. (Arthur) Heighway was appointed as managing editor at the same time. A journalist of long standing who had been employed by Goodfellow to write a book on the activities of the NZCDC in time for the Wembley World Exhibition of 1924–25, Heighway also edited the *Dairy Exporter* for many years.

The *Radio Record* began with an issue price of 3d, though its first two issues were given away in an effort to build up circulation. It was printed on Tuesdays so as to allow effective distribution throughout the country before the weekend, with full programmes for the coming week. The programmes were copyrighted. Only the *Radio Record* could print a full week's programmes in advance, but newspapers were allowed to publish individual daily programmes on the day of performance. The magazine was a substantial publication from the beginning. Originally all of its articles dealt in some way with radio matters, but this changed somewhat from the end of January 1930, when the magazine became the *New Zealand Radio Record and Electric Home Journal*. This change came after the general reticulation of electric power through the country — a matter of great significance for radio. The new format led to various articles on the uses of electricity in the home, a cooking section and an array of advertisements for the growing range of electrical appliances. The *Radio Record* was not a magazine of general comment; it was the RBC's mouthpiece, advocate and publicist.

SPORTS BROADCASTING

Sports broadcasts were an early specialised development. Harris argued that the RBC was the first service to broadcast direct descriptions of sporting events.[42] Sports broadcasts are of considerable significance. They became a highly popular part of programming in which many of the technical innovations associated with international and outside — as opposed to inside or studio — broadcasts made their first appearance. Furthermore, it was in regard to sports programmes that many discussions on the uses and abuses of broadcasting were most public.

The first sports broadcast took place in Nelson on 28 April 1923. The short-lived 2YA station there, before it received a licence or a call-sign, obtained permission to give a commentary on the race for the Australasian professional sculling championship between Hannan, the New Zealand champion, and Felton, his Australian counterpart. The race on Nelson harbour was broadcast by relay from a launch following the scullers.[43]

Christchurch soon became the centre of sports broadcasting. The head office was there, and so was Alan Allardyce — a man described by his broadcasting colleague, Arch Curry, as 'the most outstanding broadcasting personality of those days'.[44] Allardyce began his sport commentaries before the start of the RBC stations. 'I went round to the amateur radio society in Christchurch and suggested that a football match . . . might help them to get people a bit interested in radio. They were very pleased to accept my offer and I got permission from the Rugby Union and we managed to get the gear on Lancaster Park and did the match on the 29th of May 1926.'[45] The match was the annual charity match between Christchurch and High School Old Boys.

A week later Allardyce obtained permission from the Canterbury Park Trotting Club to broadcast commentaries on the trots at Addington. His next broadcast was a boxing commentary. This was followed by broadcasts of two hockey matches between teams from Canterbury and the Indian Army:

Allardyce was commentator for the first match and Sidney Holland, later prime minister, for the second. With the start of 3YA Harris employed Allardyce to arrange and conduct sports commentaries. '3YA engaged me to see if I could get permission to do the Grand National meeting in August [1926]. I approached the Jockey Club and the Trotting Club and got permission in both cases. The only restriction that the Jockey Club placed on me was that I had to be out of the stand because I'd interfere with the people.' Searching for a place from which to broadcast, he settled on a haystack outside the back straight, three furlongs from the judges' box, with an extension from the private phone line of a neighbouring owner-trainer, Jack McCombe.

Allardyce was soon in a better position at race meetings. The clubs realised he was not upsetting their patrons and accommodated him, usually in the stand, and provided him with the necessary telephone line. Allardyce's personal contacts helped him obtain permission to broadcast, but he felt the sporting authorities did not take the matter seriously: 'I fortunately knew some members of the committees of the Canterbury Rugby Union and the Trotting Club . . . and they were quite willing to give it a go at that time . . . They thought it was just a joke.'

Allardyce's early broadcasts were made without his receiving any payment. This was a temporary arrangement, however, and he soon became a full-time sports editor for the RBC. He was responsible to the company as a whole rather than to 3YA, and in the spring of 1926 he went to Auckland to broadcast the trotting and galloping meetings from 1YA.

A big boost to the popularity of sports broadcasting and of Allardyce personally came with his commentary on 9 July 1927 of the Ranfurly Shield match between Hawke's Bay and Wairarapa — 'a magnificent record of a magnificent game magnificently described'.[46] For broadcasting, the significance of this commentary was that it was the first from 2YA, not yet officially opened, and, with 2YA's power, was heard in many parts of the country. It was a test programme — it required the P&T Department to have a landline available at a greater distance from a station than had been attempted before. The *Radio Record* later reported, 'Crowds in every town and township stood around loudspeakers and listened to a very graphic description of a great event. No one who heard will ever forget.'[47] Stations were not yet capable of networking — of relaying the same programme — and no earlier broadcast had been heard other than in the local reception area. This was the first approach to a national broadcast.

Allardyce selected the first cadre of commentators, which included Gordon Hutter, Bill Meredith, Charlie Lampburgh and Whang McKenzie. Rugby became a regular feature on Saturdays in winter, with club and representative matches being broadcast; and broadcasts of race meetings were also common. Other sports which were given commentaries included cricket — but only after some persuasion from Allardyce: 'They couldn't imagine how you could

Alan Allardyce giving the commentary of the Christchurch vs High School Old Boys match on 29 May 1926 — the first rugby match broadcast. *LISTENER*

make cricket interesting. But they gave it a go.' The *Radio Record* gave a listing of the number and duration of sports broadcasts for the year ending June 1928, which indicated they had averaged over eleven hours a week.[48] This impressive tally brought its own problems — the very growth of sports broadcasting led to pressures to stop it.

CHILDREN'S PROGRAMMES

Children's programmes rapidly became a popular feature. They began when the transmission hours were extended and covered the late afternoon or early evening hours — times considered acceptable for children's listening. Such programmes had developed before the RBC began, with musical items and other renditions by children, and stories and general announcing by a variety of adult presenters called 'aunts' and 'uncles', many of whom were contracted specially for the children's sessions. Most of the regular announcers also took part, regularly or intermittently. Culford Bell of 1YA, for instance, was 'Tom Thumb', and Clive Drummond of 2YA was 'Uncle Jasper'. Some programme ideas were quickly copied by other stations. E. J. Bell, the city librarian in Christchurch, was also 'Uncle Jack' on 3AC and later 3YA. He introduced the idea of reading birthday greetings, which produced a continuing stream of letters from children and became a regular part of nearly all children's programmes.

In 1927 the RBC started a separate children's department with Anna Rose Hall as its inaugural principal. Children's session organisers were appointed at each station, and the *Radio Record* announcement of Hall's appointment was accompanied by a lengthy statement of policy from Harris regarding children's programmes. This statement, 'bearing in mind the plasticity of the childish brain', was largely concerned to assure listeners that 'no item will be broadcast that may prove a possible danger or detriment to the children's future health . . . Stories . . . will not be of a gloomy hair-raising type.' Fear of giving offence to parents and educators alike meant that the initial RBC children's programmes were vapid, but they became more robust and entertaining during the 1920s. Harris also indicated that the content and timing of the children's programmes would cater to 'our juvenile audience (whose ages vary from two to sixteen years of age) as well as to a large number of adults, who find the children's session vastly entertaining . . . In this way the "tiny folk" will be sent off to bed in good time by the wise parent who realises that "early to bed" is the best means of ensuring their continued good health.'[49] Harris's aim was to have the programmes under the supervision of adult coordinators, but performed as much as possible by the children themselves. This is another instance of the general RBC policy of using unpaid talent. As with adult artists, this practice was not greeted with universal satisfaction. Child performers, and no doubt their families, were pleased with their individual appearances, but many people argued that this was not good broadcasting, and that children seldom wished to be entertained by their peers.[50] Another view, however, was that by relying so much on local contributions, the children's programmes developed lasting relationships in their local communities, with individual children and with the various school, church, cultural and leisure organisations.

Some young performers, in particular those involved in children's orchestras, bands and choirs, were able to continue their broadcasting and gain from the experience. These choirs and orchestras were a feature of all stations, but it was in Auckland under the stewardship of Rev. Thomas (Uncle Tom) Garland that they became most accomplished and best known. Garland was an important figure in New Zealand broadcasting because of his training of choirs and his later association with private radio. He organised many of the 1YA religious broadcasts and he had his own weekly children's session, which he opened with his theme song:

> As I was walking round the corner, I heard somebody say,
> There goes our good friend Uncle Tom, he's off to 1YA;
> He sings and talks to little girls and boys on Wednesday,
> So hurry home and listen in, Hip, Hip, Hip, Hip, Hooray.

Garland's association with children's choirs lasted thirty-five years; it became an Auckland institution and a regular broadcasting feature.[51] The children's programmes, in spite of criticism of the amateur talent, gained a devoted audience and were among the first programmes to illustrate the popularity of broadcasting personalities. The Rev. Ernest (Uncle Ernest) Weeks of The Terrace Congregational Church presented the 2YA children's programmes on Fridays and Sundays until December 1928, when he left New Zealand to live in England. At his farewell, which was broadcast, he was presented with a large leather-bound album containing the autographed names and addresses of 3000 children who had listened to his programmes. Another of the 2YA presenters was Gwen Shepherd, 'Aunt Gwen' from 1928 to 1930. When she married and left to live in Sydney, her marriage service was broadcast, and attracted a crowd of more than 2000 on the street outside St Paul's.

RELIGIOUS BROADCASTS

Religious broadcasting began with the opening, on 29 October 1922, of Fred O'Neill's Dunedin station. The opening, which was on a Sunday, began with a hymn and included a sermon. The presiding cleric was the Rev. C. J. Bush King. After the passage of the 1923 regulations, Sunday broadcasting had to give priority to religious programming. Although the requirement was not continued in the 1925 regulations, this did not lead to less religious programming. Actual church services were sometimes relayed, although there was some concern over the propriety of such broadcasts. These fears were generally allayed, however: the *Evening Post* noted after the first broadcast of a service from The Terrace Congregational Church, 'It must have been reassuring to many dubious listeners to find that the service lost nothing of reverence and impressiveness by being transferred from the church to the home.'[52]

In 1927 an attempt was made to coordinate the programmes for religious broadcasts. Until that time the RBC had been trying, with little success, to rotate broadcasts from the various denominations. The original idea, as part of the view that all stations would be heard all over the country, was to coordinate the broadcasts so that members of the major denominations would be able to hear a broadcast from their church from one of the stations most weeks. This range of reception was not obtained. The churches were not particularly cooperative in this scheme. The company argued that it tried to give all churches broadcasting opportunities, but that the churches were unwilling to abide by any roster that it devised. In particular the company had difficulties with the Church of England which, as the largest denomination in New Zealand, had been unhappy with the broadcasting opportunities offered. Matters came to a head when the Anglicans announced that they intended to apply for their own broadcasting licence.[53]

At this stage Harris arranged for a conference with representatives of the various churches. He proposed that the churches, not the RBC, be responsible for deciding when churches should broadcast. A full agreement was not reached at the conference, but the principle was accepted. It was decided that preference should be given to the 'more prominent denominations', of which there were eight: Church of England, Presbyterian, Roman Catholic, Methodist, Baptist, Salvation Army, Church of Christ and Congregational; and that 'Every Sunday regular evening services or special studio services be broadcast, the studio services to be at the regular church hour.' The conference also agreed that a schedule be devised for church broadcasts. These decisions were not unanimous, however. The Baptists wanted the studio services to be at a different time from the regular church hour, and the Catholics decided not to be part of the broadcasts at all. It was also decided to continue the discussions by setting up a committee with a representative from each church along with company representatives. Churches opposed to broadcasting could join at a later date if their views changed.[54]

This was the start of the church service committees, which were set up for each station and which coordinated religious programming among the denominations for many years. Each church was concerned that no other church receive an inequitable share of broadcasting time. This concern led to difficulties for the committees and the company. At times, churches made independent arrangements with private broadcasting stations. For instance, after the start of 2ZW in Wellington in 1931, the Anglican Church arranged for regular broadcasts from that station. The 2YA church services committee reallocated the Anglican broadcasts from 2YA to another church — a move which led to complaints and a query from the P&T Department secretary.[55] But generally the rostering by the committees was amicable, and was welcomed by the RBC.

SCHOOL BROADCASTS

The idea of broadcasting to schools was suggested in New Zealand even before the RBC began operations. In speaking against the suggestion the minister of education raised as objections matters that were close to the eventual method used:

> It is difficult to see how broadcasting of messages or communications from the large centres to outback schools could be carried out . . . There would need to be a number of central sending stations established or utilized, wireless apparatus installed in a large number of outback schools, and permanent arrangements made for various people to speak to the children from the centres. The whole arrangement would be very complex and expensive.[56]

In spite of the minister's objections, broadcasting was recognised as having educational value. In reply to the chief telegraph engineer, who asked what use would be made of an educational broadcasting scheme, the assistant director of education saw it as useful, 'firstly for the improvement in the instruction given children in small backblocks schools and the children at present catered for by the Correspondence School; secondly [for] the general broadcasting of lessons on Music and Musical Appreciation by the Director of Musical Training.'[57] Educational broadcasting, like broadcasting in general, was useful as a way of bringing information to the isolated. Many New Zealand schools were in isolated places, and with the Correspondence School (for pupils otherwise unable to attend school) having opened in 1923, it was an obvious and long-lasting application. The assistant director's second reason indicated an early appreciation that radio could be used to broadcast to all school pupils, urban and rural, material beyond the expertise of their local teachers.

The P&T Department promised full cooperation, but suggested matters stand until the RBC stations were operational.[58] Once 2YA was established the RBC offered the Department of Education regular broadcasts of up to one hour every school morning.[59] A first test was carried out on 13 December 1927, when 2YA broadcast musical items and short addresses by various officials between 2pm and 3pm. Many schools from Northland to Southland were alerted to test the reception. Their reports varied: 21 per cent reported excellent reception, and 10 per cent found it less than satisfactory. This was enough to encourage the Department to consider the matter further. T. B. Strong, the director of education, was of the opinion that,

> Given successful reception, the country schools could be greatly assisted through judiciously selected programmes. I believe that wireless could be used in conjunction with the 'School Journal' with very beneficial results. Suitable literature published in our journals could be dealt with by experts, and children taught by wireless to appreciate more deeply the delights of good music, art and literature.[60]

The following year the RBC proposed that each of its four stations broadcast a lecture or lesson four days each week from 2.30pm to 3pm. Nothing came of this proposal until 1931. The delay was caused by the director of the Department of Education, who advised his minister that the department would proceed 'if the Government is prepared to assist in the equipment of schools to the extent of a £ for £ subsidy'.[61] However, the director also advised against such a subsidy because of the expense involved: 'The more remote the school the more expensive the machine required and unless the school were connected with the electric current, batteries would have to be used and these teachers in remote localities would experience difficulty in keeping in order.'[62] Apart from the cost, the department favoured the use of broadcasting in schools: 'The opinion was definitely formed that good use could be made of it . . . the only real difficulty in the way is that of expense.'[63]

New Zealand was comparatively slow to begin a broadcast to schools programme. The Department of Education had received considerable information on the BBC and various United States initiatives in educational broadcasting before it introduced such broadcasts into New Zealand.[64] On 14 April 1931 the first regular school broadcasts began. They were for one hour a week, on Tuesdays from 2pm to 3pm, and were made from the one station, 2YA. The *Radio Record* regarded it as 'strange that so far more has not been made of it [radio] in schools', but welcomed the new programme as 'likely to be an important milestone in the history of education in New Zealand'.[65] The *Radio Record* was correct and the new broadcasts, which followed a policy of making available the services of experts in particular branches of teaching, became a feature of New Zealand schooling.

MUSICAL PROGRAMMING CHANGES

The development of these specialised areas of programming was accompanied by changes in the major area of programming — musical entertainment. In 1927 the RBC addressed the complaints about the nature of radio's offerings. These included the type as well as the quality of performances. The RBC could exhort but was in no position to insist its volunteer performers vary and extend their repertoire. Frequent repetition of material not entirely appreciated on the first hearing became a regular feature. Added to these difficulties was the fact that any one programme might contain a considerable variety of musical styles, which did not always appeal to all listeners.

The first changes were to the pattern of evening programmes. The problems of radio programming compared to musical, theatrical or operatic entertainment were discussed in the *Radio Record*: 'People go to a specific hall . . . and they choose the entertainment that their tastes dictate . . . Radio is asked to produce a programme which will satisfy all tastes. It is an impossible demand.' The *Radio Record* suggested there were 'two methods by which an ideal programme might be reached', but it rapidly discounted the first:

> [A] mixed programme of items to suit all tastes. This would be comparable, say, to inflicting upon an average individual, in one evening, five minutes of a scientific address, five minutes of a religious service, five minutes at a boxing tourney, five minutes at a Shakespearean recital, five minutes at a classical revue, five minutes at a musical masterpiece, five minutes at a jazz cabaret, and five minutes at a Salvation Army repentance form. Contemplate the experience! Only an experienced newspaper reporter would survive! . . . The alternative is the provision of special programmes for each type, or allied types, at one session; and for another type at a later session . . . Under this arrangement of programmes a single type or limited types of listener can be catered for at one session, and other types at

> later sessions at one station, and alternate with a different type of programme each night at each of the company's stations.[66]

This proposal was introduced in the second half of 1927. By the end of the year, Harris had developed it into a plan to 'synchronise through New Zealand the class of programmes to be performed each night by each station.' Each station was given 'skeleton programmes showing the nights to which these various classifications have been allotted for your station, and also giving an outline of the performers who should be used, and the approximate time they should perform.' The object was to enable the devotee of, say, vaudeville to be able to follow it on as many nights as possible. Harris made it clear that suggested additions to the programme were wanted from the stations. 'You are specifically asked to suggest anything original which will tend to create further variety.' He stressed that any such ideas had to be submitted through head office.[67]

This policy of coordinating musical programmes over all four stations was known as complementary programming. It was a major innovation in policy and a significant feature in the development of broadcasting as a centrally controlled national system. Introduced by Harris, it remained a feature of New Zealand broadcasting for twenty-six years. Complementary programming was immediately available to the majority of listeners. Their ability to tune in to many stations varied with their location and the type of set they owned. Those with crystal sets were generally able to hear only their local station, while those with valve sets could usually also contact at least some of the more distant stations. The number of listeners who could take advantage of the choice offered by complementary programming grew as the quality of radio sets and radio transmission improved.

These changes in musical programming required the RBC to have more control than previously over what was broadcast. This could not be done with voluntary performers. A significant effect of the programming changes was the company's realisation that musicians must be paid. This policy change was introduced by the *Radio Record* with reference to a 'policy of improvement' for programming, which was described as 'the employment of the best possible professional talent on a permanent basis, to be supplemented by others as proof of merit is given'.[68] Thus the RBC finally began employing musical groups, not as full-time employees, but on a relatively long-term basis. Within six months it was describing itself, with justification, as 'the largest employer of musical and artistic talent in the Dominion'.[69] The general trend of the company's policy was both to employ existing groups and to establish new orchestras for each of the four stations. The basis of choice moved from availability to ability. The quality of programmes improved, and stations were able to plan their programmes. By paying musicians, the company could demand that they extend their repertoire. The first of the permanent orchestras was formed at 2YA in May 1928. By August it had twelve

"This correspondence is now closed"

A Christmas cartoon in the *Listener* of 20 December 1940 indicates the 1928 Bellingham, Harris debate over musical quality would long continue. LISTENER

members, and was described as 'the largest and most efficient orchestra permanently employed in any broadcasting organisation in Australasia'.[70] This was followed in February 1928 by an eleven-member orchestra employed at 3YA, and in September 1928 by the start of a permanent orchestra at 1YA.

A full survey of the groups and the individual musicians employed by the RBC and its successors would be a study in its own right. Payment was made not only to those musicians who formed the permanent orchestras, but also to other individuals and groups who broadcast less frequently. It was an employment pattern of considerable importance for local talent. Like many other countries in the late 1920s, New Zealand was entering a period of economic depression that continued well into the 1930s. Employment opportunities generally were shrinking and musicians and entertainers found the change of policy in broadcasting a welcome development. The timing was particularly opportune for musicians, for the new positions in the radio orchestras were presented just as the orchestras of cinema theatres were being lost with the advent of talking pictures. Unpaid amateur performers still appeared, though less frequently, on the RBC stations, and the non-RBC stations continued to rely mainly on volunteers.

The broadcasting of music was under the control of the director of music — a new position announced in September 1927. The first appointee was W. J. Bellingham, FSM, of Christchurch. This appointment signalled the start of a continuing tension between Bellingham and Harris concerning the nature of the music that should be broadcast. The personal conflict between the two men ended in June 1929 when Bellingham, after twenty-one months with the RBC, resigned and moved to England. Essentially it was not a conflict between two individuals, but the first appearance in New Zealand of a clash between 'popularity' and 'quality', of different understandings of the purpose of broadcasting. Bellingham had begun his appointment by accepting that various types of music could and should be broadcast. What was wanted were admirable examples within each type: 'I believe in a varied programme with a major proportion of standard works. The essential feature is that what is done, whether it be foxtrot or symphony, must be well done.'[71] But Bellingham had been chosen specifically because of his musical background and he regarded himself as having a duty to 'music' as well as to the RBC. As the custodian of musical standards within the company, Bellingham particularly wanted to broadcast a substantial proportion of high-quality music — by which he meant the established repertoire of classical works. As managing director, Harris's overriding concern was to have programmes that would encourage people to buy radio sets, pay the licence fee and listen. He regarded the emphasis on classical works as acting against this, and he made moves to counter it. In his memo introducing complementary programming, Harris asked his station directors to give 'a very wide reading' to the terms 'classical' and 'operatic'. He encouraged them 'not to make the programmes too monotonous by adhering too closely to strictly classical and strictly operatic music'.[72]

Over the following year Bellingham became increasingly concerned with what he regarded as the RBC's concentration on low-quality material so as to attract an audience. He regarded the practice as contrary to the obligations of a broadcaster and, in the long term, counterproductive to the company's own interests. He argued his case to Harris in a lengthy report and questioned, 'whether it is not the duty of such a powerful influence as radio to present at least a definite proportion of the very highest forms of entertainment in order to develop the culture of the country'.[73]

For Harris the flaw in Bellingham's argument was the assertion that the broadcasting of popular as opposed to quality music would gradually lower the size of the audience. The growth and the appreciation of a new form of popular music was considerable in the 1920s. Although doubts were cast by established opinion on its quality, there was no doubt about its growing popularity. An individual popular song, unlike a work from the classical repertoire, had a limited life-span — but this did not mean that broadcasting the genre was in any way 'causing satiation', as Bellingham suggested. On the contrary, what was developing was an apparently inexhaustible supply of new material seeking its fleeting weeks of popularity amongst a listening public eager for more.

In Bellingham's view, the duty of broadcasting was to give people not necessarily what they wanted but what was good for them. This argument, supported by prevailing educated opinion, ran counter to the growing democratic ethos of the twentieth century. The debate went beyond just Bellingham and Harris, and it outlasted both men. The élitist views Bellingham was advocating were becoming known in England as Reithian views, after Sir John Reith, the former general manager of the now-dissolved British Broadcasting Company. Reith had been knighted in 1927 and appointed the first director-general of the British Broadcasting Corporation. In the understanding of Reith (and of Bellingham), broadcasting presented a wonderful opportunity to educate and enlighten the people in the ways of quality. The danger to be avoided was that of pandering to uneducated taste. The broadcaster stood at the boundary of the people's taste and had the power and the responsibility to improve it. At its extreme, the opposite view was one that denied any such responsibility, and instead considered it was the duty of broadcasting to cater to public taste. In New Zealand, Harris's quarrel with Bellingham was that of the pragmatic broadcaster attempting to build an audience and regarding élitist broadcasting as counter to that attempt. Harris had no previous connections with or allegiances to the established and accepted forms of cultural appreciation. Nor was he proposing to use radio for the revolutionary aim of supplanting those forms. But radio broadcasting was developing as a new and popular, as opposed to élite, type of entertainment. Its largest component, music, was increasingly of a type which concerned, even angered, devotees of established culture. Harris, like the majority of his growing audience, accepted the change.

NEWS AND PUBLIC DISCUSSIONS

Considerable debate and tension centred also around the different understandings of what were acceptable uses for broadcasting in the matter of public discussion and the spread of news. In this case, however, the difference in views was between the RBC and its controllers in the P&T Department. Inevitably the debate entered the political arena.

That radio should be a medium for presenting information as well as entertainment to its listeners was accepted from the beginning by the RBC. It was, after all, Goodfellow's reason for his first radio proposal — a station for the Waikato dairy farmers. But a news and commentary service such as was offered by the newspapers was not considered. Goodfellow and Harris would probably have eventually moved in this direction, as far as the regulations permitted, but this was not one of their early priorities. Hoping for a renewal of their five-year contract, they were in no position to make powerful enemies. They had no desire to offend the Newspaper Proprietors' Association by starting an independent news service. The one exception was sports reporting: the RBC began a commentary service, and continued the reporting of results even in the face of opposition.

From the beginning, however, the RBC did broadcast talks on a variety of subjects, and also started a news programme once the stations began transmitting for more than the original 8pm to 10pm hours. By the time the *Radio Record* began in July 1927, the stations had a half-hour news programme from 7.15pm to 7.45pm, later lengthened to a full hour from

7pm to 8pm. The four stations were all involved in the preparation of news and talks programmes, but the principal figure was John Ball, an ex-editor of the *Wanganui Chronicle*, appointed in June 1927 as editor-announcer at 2YA. The news was obtained from newspapers. The RBC was allowed to use items already published in that day's newspapers, provided they were not broadcast before 7pm — in other words, not before the newspaper subscribers had read the news. Radio was not competing with the press: news made up only a small part of the news programme. The bulk of it was reports on matters such as the stock and wool sales, along with talks. The talks were generally regarded as having an educative value and were referred to as 'lecturettes'. These were welcomed by listeners, many of whom were not impressed with the freshness of the RBC's news: 'Education lectures . . . are a welcome change from the sixty minutes of news which the average person reads in his daily paper and gets again on the air between seven and eight o'clock.'[74]

As well as being reliant on the newspapers for its news, the RBC was also dependent on government handouts. From its start, 2YA broadcast regular talks on imperial and foreign affairs. Compiled by Ball, these were semi-official broadcasts taken from material supplied by the Imperial Affairs branch of the Prime Minister's Department. The Prime Minister's Department also supplied the RBC with the daily British Official Wireless news despatches — a further source for the company's news broadcasts.

The RBC would have been hampered in any move it might make to become a source of independent news or commentary by the regulation which prevented it from broadcasting 'propaganda of a controversial nature'. Just how the company should interpret this phrase was unclear, and Harris raised the matter officially with the P&T Department secretary after complaints from the United Victuallers' Association that individuals had been permitted to broadcast their views 'on the controversial subject of prohibition'.[75] The RBC asked for guidance on the definition of controversy in general. Harris argued that, strictly interpreted, the regulation prohibited most spoken broadcasting, and certainly the best talks — for it was precisely the controversial matter that was the most liked. In his reply the secretary noted:

> For each of your stations there should be available a brief concise instruction to lecturers, stipulating that certain subjects of controversy should be avoided. It should be pointed out for instance that violent or definite remarks for or against prohibition or other subjects of the kind cannot be allowed. Further, nothing should be said having for an object the inflaming of political feeling . . . While not wishing to tie the Broadcasting Company down to any hard and fast rules in the matter, I certainly look to it to exercise a wise discretion and to see that there is a due observance of the spirit of the regulation on the broadcasting of controversial matter.[76]

Harris made it clear that he did not consider the secretary had offered any real help; but he did decide to give lecturers 'a card summarising the regulations concerning this matter'.[77] The RBC thus offered its station directors and speakers a summary of the regulations which it had been unable to clarify. This was merely an attempt to place the onus elsewhere. Nor was the company in practice able to ignore the matter, and leave the individual decisions to its speakers. The responsibility was clearly the company's. While 'news, lectures [and] useful information' could be broadcast, practices were developed whereby such material was always considered in terms of the ban on 'propaganda of a controversial nature'. The ban on controversy was the higher priority, and overrode the right to broadcast. Furthermore, the nature of the RBC's contract was such that the company considered it could not probe at the barriers of the regulation. Timidity was the result.

The RBC's news and talks broadcasts were consistently criticised for what was regarded as political bias in favour of the government. Opposition MPs argued that their views were regarded as 'propaganda of a controversial nature' and were not broadcast, but the same rationale was not used to stop broadcasts from members of the government. There were many sides to this criticism. There were accusations that the regulations were flouted by the government, as in E. J. Howard's statement that 'during the election the Prime Minister's speeches were broadcasted throughout the Dominion, despite the fact that the regulations prohibited such a thing . . . Apparently the Prime Minister in broadcasting his speeches must have obtained permission from himself, as Minister in charge'.[78] There were charges that political favouritism was a result of the relationship between the government and the RBC: John A. Lee held that the Reform Party had placed 'one of the best organs for propaganda purposes in the hands of private individuals who were well known to be supporters of the Reform party'.[79] It was argued that the news services used by the RBC presented only a government view. For instance, H. E. Holland, by then leader of the Opposition, criticised a broadcast prepared by Imperial Affairs on the military base in Singapore, and did not regard as relevant the prime minister's rejoinder that the statement 'represented by far the opinions of the great majority of the members in Parliament'.[80]

BROADCASTING'S GROWING PUBLIC PROMINENCE

The parliamentary discussion of the RBC's performance was intensified by a growing understanding that broadcasting increased the public's awareness of the speaker. Opposition members would not accept that a minister's broadcast could be non-political; they argued that equal opportunity should be given to all to express their opinions. M. J. Savage, later prime minister, argued that controversy should be aired, rather than suppressed:

> It was admitted by all shades of opinion that the wireless service was one of the finest achievements of the present age from an educational viewpoint, and

> he would be sorry to hear anyone from the Opposition side of the House saying that any censorship should be applied to the utterances of any responsible person or any person representative of a large section of the community.[81]

The postmaster-general, reminded by Savage that 'the Government party might soon find itself back in the political wilderness', replied 'Yes' when asked whether he intended 'to continue the existing policy of censorship'.[82] It remained to be seen whether Savage, when he emerged from the wilderness, would implement his own views.

The increase in the public prominence of broadcasting was linked, of course, with the increasing sales of radio receiving sets. Far from being a passing fad or a medium for isolated rural dwellers, broadcasting was beginning to play a prominent and permanent role in New Zealand life. Whether or not it continued as a private enterprise, it was becoming a public activity which was taking up a growing amount of the leisure time of a growing proportion of the population, and acting as an increasingly influential carrier of public information. The nature and future course of broadcasting were matters of substantial public discussion. By this time most newspapers had regular weekly columns devoted to radio broadcasting, and the topic was considered important in parliamentary debates. The discussion did not abate during the remaining years of the RBC, and eventually led to the demise of the RBC itself.

4
Disputes and Growth

RADIO OWNERSHIP AND THE BROADCASTING AUDIENCE

Towards the end of the 1920s attitudes towards broadcasting changed. The medium regarded by many earlier in the decade as a joke soon demanded serious consideration. In this chapter we examine this change. It followed from the growth in the number of broadcasting stations and, particularly, in the listening audience. During the term of the RBC the audience growth was in some measure a result of the efforts of the RBC itself. The provision of regular programmes and their transmission from stations with a power output considerably higher than previously increased the public desire for radio. But the company was by no means solely responsible for radio's growth. It was joined by other non-RBC stations which attracted their own following. More importantly, the RBC years coincided with the start of a world-wide growth in radio ownership. New Zealand has a record of early and widespread use of new means of communication and the ownership rates for radio paralleled the country's past record.[1] In the 1920s radio listening was a minority but growing activity, with the 4000 licensed listeners of 1925 becoming 54,000 during 1930. In the 1930s the radio set became a normal household possession.

The RBC began its term with a drop in the total of radio licences. This reflected the increase in the licence fee from 5s to 30s at that time, and the slowness of the company to start new stations and begin regular transmissions. The next two years' growth more than made up for this decline and indicated there was a ready audience that had only been waiting for an incentive to purchase a set. By 1928 this first burst in listening numbers had returned to a slower growth which continued well into the 1930s. The RBC experienced both the first slow years and the frustration of losing its contract as the number of listeners increased.

The early lack of radio equipment for sale meant that listening was restricted largely to enthusiasts who could construct their own sets. However, New Zealand radio dealers soon attended to the potential market.[2] Knowledge of the workings of the crystal set was widespread and a large number were in use in the 1920s and beyond. Articles on how to make your own crystal set were regular features of the time.[3] It was a simple technology that was in use well before the start of broadcasting. By the time broadcasting started, crystal sets were available for sale in New Zealand: the earliest known advertisement for a New Zealand-made set appeared in the *New Zealand Wireless and Broadcasting News* on 10 February 1923. For many, the crystal set, although easily bettered for quality of reception, never lost its pre-eminence as the instrument of wonder. Eddie Hegan, himself later a broadcaster, explains, 'There was just a little box with a set of earphones and you fiddled a little bit of wire against the crystal and out of the air from nowhere came voices, a disembodied voice. This was sheer magic. It was absolutely wonderful. I couldn't believe it.'[4]

The disadvantages of the crystal set were its general inability to receive with clarity anything other than a local station, and the need to regularly search for the sensitive position on the crystal on which to put the 'cat's whisker' — a piece of fine wire. Its advantages were its ready availability and low cost: it was the most widespread type of receiving equipment in the early years of broadcasting. In 1924 valves became generally available and valve sets went on sale in New Zealand. These were expensive to buy and use; the first sets were inefficient and used a lot of power. Like crystal sets, they were listened to on headphones, which meant that they could be heard by only one person at a time. Loudspeakers were first used in 1925. Originally they took the form of a headpiece fitted to a horn, but by the end of the decade, cone speakers, with their higher quality of sound reproduction, were in general use.

There were many inconveniences associated with the early sets. They required some skill and experience to tune correctly. Newspaper wireless columns of the time were full of advice on how to do this, and on the problems of interference from 'howling valves' emanating from incorrectly tuned sets. The term 'wireless' was a misnomer for these cumbersome early sets: they were distinguished by the number of wires requiring attention. But the major inconvenience of the early radio sets was their reliance on battery power. Until electric power was made widely available in New Zealand in the 1920s, either wet-cell batteries, such as a 6-volt car battery, or dry-cell B batteries were needed. These were expensive and required regular recharging or replacement.

The great boost to the spread of radio was the reticulation of electric power and the consequent manufacture of electric-powered sets. Apart from their convenience, the electric sets were superior to their battery-operated counterparts in tone and volume, as well as in their ability to pick up distant stations. They soon took over the market. The battery-operated sets were usually housed in small metal boxes. The electric sets were placed in wooden boxes which, by the end of the decade, were enlarged as radio sets became elaborate pieces of furniture, set in more and more splendid cabinets. This trend continued during the 1930s, but small, relatively inexpensive sets were

A radio display at the 1928 Wellington Winter Show.
ALEXANDER TURNBULL LIBRARY

also manufactured from the start of that decade, for a population hard-hit by the depression.

Portable sets were being manufactured towards the end of the 1920s. Advertisements declared, 'Few indeed are the camping parties, fishing parties and even touring parties that now go far afield without their portable set.'[5] But sets were cumbersome. The Airzone Portable, a much promoted 4-valve model, for example, was 'only 27 lbs in weight'. It is doubtful if portable sets were taken far from a motor car. However, the introduction of portable sets was an indication that radio was not solely a living-room fixture.

The spread of radio beyond the home was in part due to the efforts of broadcasters. Allardyce, for instance, remarked with some justification that after he started his commentaries of horse racing, 'It was no time before the pubs were getting them in of course.'[6] However, radio broadcasting was predominantly a home-based medium. Its growth seemed unimpeded by economic difficulties; it is notable that during the depression years there was a boom in sales of sets. Whether the boom would have been greater if economic times had been better is not possible to determine. Advocates of radio argued that the hard times led people to decide to purchase a radio: 'With less money available for entertainment . . . [radio] gives a definitely better return for every pound spent than any other expenditure on amusement.'[7]

LISTENER REPRESENTATION

As the radio audience grew it was gradually recognised as a new type of collectivity. This presented particular difficulties in selecting a listeners' representative for the Advisory Committee, as required in the agreement between the government and the RBC. The listeners' representative was elected by a postal ballot of members of the radio societies. He was Gordon McBride Salt, chairman of the Auckland Radio Association and member of the physics department of Auckland University College. While all were careful to cast no aspersion on Salt, there was a general recognition that he was in no sense fully representative. The Radio Societies Conference that elected him recognised that 'it was neither a constituted Dominion organization nor did it constitutionally represent listeners'. The *Evening Post* noted, when discussing the conference: 'It is by no means certain that the societies can in all cases be considered fairly representative. In the Wellington district, for example, there are about 1600 licence holders and a very large number of them reside in Wellington itself. But the local society has less than forty members.'[8] A further conference held in Wellington two months later to form a central society failed to do so and acknowledged its inability to organise listeners.[9]

The problem had to do with the nature of the audience for the new medium. Meetings of listeners were called and were often well attended. There were many instances when dissatisfaction with radio was shown by large public gatherings. But there were questions as to how representative such meetings were. The early radio societies were groups united in their enthusiasm for experimentation with radio. They were followed by the amateur radio transmitter and receiver associations. The growing number of listeners differed from these groups in two important ways. As an audience they were increasingly interested, not in radio itself, but in the listening fare offered. Also their numbers were expanding beyond that of any other voluntary organisation. In theory, and eventually in practice, the audience approximated the entire population. Listeners were not amenable to representation by the traditional means of organisation in voluntary societies.

The Advisory Committee appointed when the RBC began its contract was an ineffective body. Few meetings were held. At a Dominion Conference of Listeners' Leagues held in September 1927, the chairman stated that the Advisory Committee had not functioned at all for more than a year. Those meetings that were held were not followed by any public statement. Salt, the listeners' representative, was described as a keen battler for the rights of listeners and was usually at odds with the rest of the committee.[10] After Salt publicly expressed his contrary views, the postmaster-general made it clear that the committee was merely an advisory body with no legal status.[11]

Initially, officials of the RBC and the P&T Department did not favour any form of public contribution to broadcasting decision-making. They were required to accept it but ensured such public contribution did not threaten their own control. Goodfellow and Harris saw no need for the Advisory Committee. The company had a clear idea of its priorities and was embarking on a programme of station development. There was little desire to pay attention to a committee which might not agree with the RBC's priorities. A. T. Markman, the P&T Department secretary, was not keen to have regular committee meetings either. Under the minister, he was the official responsible for the actions of the RBC. The day-to-day control was his, and he was unwilling to share this responsibility with an advisory committee that included members who were not public servants.

Public discontent with the nature and administration of broadcasting was frequently voiced in the press. Dissatisfaction was greatest in Auckland, where a Listeners' League was formed. Six hundred and fifty people attended the first general meeting on 2 June 1927, and 1600 became members in the first six weeks. The 1YA station director reported to Harris that the League was expressly started to attack the RBC. He saw it as cooperating with the Auckland newspapers: the city's three radio writers were all on the League executive. At its inaugural meeting the League 'urged the Government to assume immediate and complete control of broadcasting on the basis similar to that now operating in Great Britain'.[12] The allusion was to the BBC, now a corporation. That corporation control represented 'complete control' by the government was later disputed, for corporation control came to be seen as an acceptable midway point between two extremes — private enterprise and state control. Criticism was also made in Parliament and at a 1927 Dominion Conference of Listeners' Leagues.[13] At this point the government suggested the formation of a new advisory board on which listeners would have three representatives.

J. M. McLean of the P&T Department head office, in a 'private note to give you information that could not very well be stated in an official communication', assured Harris that 'the Company need have no apprehension as to the powers of the proposed Board'. The listeners' representatives 'will not have any more power than they have at present. The Company will still be supreme in controlling the stations subject, of course, to the existing control by the Government.'[14]

Harris was not impressed; he objected to the nature of the listeners' representation and to the formation of a national advisory board: 'No country has yet been able to properly organise its listeners, and if such is the case in New Zealand, what is going to be the value of the so-called representatives' decisions?' He argued instead for local listener involvement in programme development — an alternative that had been discussed within the company: 'The majority of listeners are mostly interested in the programmes of the local station . . . The time and place of any conference is at the station of broadcast and at the time the programmes are made up.'[15] Harris's view prevailed and was made public the following March with the announcement in the *Radio*

Record that the RBC had decided that 'the time has now arrived when it might establish some form of public relationship with its customers'.[16]

The RBC plan was to establish relationships with various community organisations as a way of developing its programmes and also to establish a system of 'official listeners' so as to gain information on the quality of reception. The official listeners programme involved the appointment of honorary listeners in each of twenty-two localities. They reported on the reception of radio transmissions and investigated reports of interference.[17]

Relationships with community organisations were established through a system of committees. The RBC saw broadcasting as touching 'upon work already being undertaken in the more usual way by well-defined public bodies or societies. There are the churches, musical and dramatic societies, child welfare leagues, educational bodies, sports clubs and radio and listeners' societies, all organised, and inevitably interested in what radio is doing towards the furtherance of their respective objects.' The committees were to 'consist of representatives of the bodies concerned and representatives of the company. For instance, the sports committee shall consist of representatives from football clubs, cricket clubs, tennis clubs, athletic clubs and other sporting clubs and the company's representatives.' The committees were to recommend items to the company, consider applications and suggestions, and propose resolutions of competing claims.[18]

The church service committees were already in operation. After a trial in Christchurch, they had been started for the other stations. With their name changed to 'religious advisory committees', they remained part of New Zealand broadcasting well beyond the years of the RBC. Advisory committees for children's sessions were also established at the four station centres. The most general committees were the musical and dramatic committees, which were also set up, one for each station, with representatives from each of the various local musical and dramatic societies. The local station contributed two members to each committee and Harris was an *ex officio* member of every committee. Primary production advisory committees were added to the scheme in 1929 after a conference of agricultural and pastoral interests convened by the RBC. By the end of 1929 there were fourteen separate committees. There were, however, no sporting committees: in a reconsideration which indicated there was a limit to its readiness to allow others to determine programme content, the RBC decided that it should judge what sports were to be broadcast.

There were various reasons for the introduction of the committees. The recently established church committees were successfully coordinating the programme of church broadcasts, and stopping much of the previous criticism of the company. It was felt that an extension of this system to other areas would have a similar effect. Also, the company needed to obtain regular and reliable information on the quality of reception in the many centres. A system of listener involvement was an obvious way to approach the matter.

Furthermore, there was a growing appreciation from Goodfellow and Harris that broadcasting was a quite different venture from those with which they had previously been connected. Public relations and listener and community involvement had not been part of their original planning, because it was against their understanding of business practice. They were starting a private business which, within the confines of the regulations, they should run as they saw fit. They gradually moved from this attitude and sought an involvement with community organisations. One could argue that this was a desire to appease criticism and forestall the formation of a different and, to the RBC, less acceptable form of listener involvement. However, when the P&T Department suggested a new form of advisory board which essentially left the situation unchanged, the company instead opted for a scheme of considerable community involvement. In the RBC, more quickly than in the department, there was an acceptance that the new world of broadcasting demanded new practices.

B STATIONS

The RBC's inauguration of a system of local committees was also an attempt to combat the growing competition from other broadcasting stations for listeners. While the RBC had the enormous advantage of being the national and government-contracted broadcaster, the new world of broadcasting was not one in which it held sole sway. The 1925 radio regulations allowed for stations other than those of the RBC. The RBC took over one station in each of the four centres when it began its contract, but three other stations continued broadcasting. All outlasted the company. The oldest was 4AB, the station of the Otago Radio Association. The others were 2YM Gisborne and 1YB Auckland.

A policy decision not to allow the establishment of any more private broadcasting stations was made in 1933. Up to that time a further thirty-six stations, including one at Apia in Western Samoa, were licensed.[19] The RBC's stations were increasingly outnumbered by a growing group of B stations. The term 'B stations' was used to signify any non-RBC station, although 'Z stations' would have been more appropriate: from 1926, 'Z' was part of all non-RBC station call-signs. The term 'B stations' was not the official term in New Zealand, but was adopted by all commentators and was used in parliamentary debates. The term originated in Australia with the July 1924 regulations there which divided stations into A stations, financed by licence fees from listeners, and B stations, financed by the sale of advertising time. The difference in New Zealand was that while the YA stations — those controlled by the RBC — were funded from listeners' licence fees, the B stations were not allowed to advertise and had no obvious means of financial support. They were protected by the government's action regarding broadcasting patent rights, but otherwise had neither shelter nor income.

The B stations outnumbered the four YA stations, but their power output was considerably less than that of the RBC, and they never obtained the coverage that the company achieved. All concentrated their efforts in their local area. In 1928, for instance, the most powerful of the seven non-company stations operating was 3ZC, with a power output of 50 watts; four of the others were on 20 to 25 watts, and the two Otago stations were on 10 watts. Some of the later stations had more powerful transmitters, but none equalled those of the company. The company, however, was vulnerable and was challenged in the main cities and in provincial areas.

Early in 1925, the year the RBC began, one other station was also started. This was 2YF, Palmerston North. It began as 2GI, the call-sign of the newly formed Palmerston North Radio Club. The signal transmitter had an output of 5 watts and, at the start, the power was supplied by torch batteries. Two hundred of these were connected to give 300 volts.[20] At the end of the year, when the station was licensed to broadcast and the call-sign changed, the output was lifted to 25 watts and a 1500-volt motor generator was installed to supply the power. As with most of the B stations, transmission hours were few — Wednesday and Sunday evenings at the start. The station was housed by the local firm of Collinson and Cunningham Ltd who provided a room in their building for a studio and equipped it with a piano and with heavy drapes for the ceiling and walls to reduce echo.

Ken Collins, who had a broadcasting career which continued to the early years of television, was then a high school pupil and member of the Radio Club. He has written of the station's broadcasts, which relied heavily on gramophone records but gradually added children's programmes, local concerts, weather reports, local news items and sports results. The Palmerston North station, like many of the early B stations, was sustained by an enthusiastic club membership and by local support, mostly unpaid. Collins, one of the station operators and announcers, received no payment but did not find his radio interests entirely unprofitable. He earned £70 a year in his last two years at High School from the *Manawatu Evening Standard* for writing a weekly 'Radio Notes' column.[21]

A second Auckland B station, along with 1YB, was set up in 1926 when Roy Keith's 1YD, soon renamed 1YQ, began transmissions in January. Keith, who had only recently moved to Auckland from Hawera, first broadcast from his home in Mount Albert. Keith's Radio Shack, as the station was known, had a chequered history and was often in strife with P&T Department officials, usually for breaches of the advertising regulation. Keith's transmissions were published in 1929 as from 8pm to 10pm on Mondays and Thursdays and 4pm to 6pm on Sundays. 1YB was transmitting for three hours a week, from 7pm to 10pm on Mondays.[22]

Later in 1926 in a general change from 'Y' to 'Z' in the call-signs of the B stations, 1YQ became 1ZQ and 1YB became 1ZB. In Palmerston North, 2YF became 2ZF, in Gisborne 2YM became 2ZM and in Dunedin 4AB, the station of the Otago Radio Association, became 4ZB.

Three more B stations were established in 1927: 3ZC in Christchurch and 4ZM and 4ZL in Dunedin. They were joined the following year by two more, 2ZK in Wanganui and 4ZO in Dunedin. Thus, by the time of the general election of 1928, there were ten B stations in addition to the four stations of the RBC. Three of the B stations were in provincial centres but the other seven were in the cities. Of the RBC stations, only 2YA was operating without a B station in the same city. Dunedin had four stations in addition to 4YA.

To a large extent the B stations operated when the YA stations were off the air. 1ZB, for instance, operated on Mondays, the 1YA silent day, and 4ZB on Wednesdays, the 4YA silent day. 4ZM transmitted from 3pm to 5pm on Sundays and 10pm to 11. 30pm on Mondays, again hours when 4YA was silent. 3ZC transmitted from Monday to Friday but for four of those days only from noon to 2pm and 5pm to 5. 45pm, hours when 3YA was silent. Its evening transmissions were on Tuesdays, 3YA's silent day. This could be seen as a system of cooperation with the YA stations and, when it suited them, B station owners stated that cooperation was their intention.

During 1929 a further seven B stations, the largest number yet in any one year, were licensed and began broadcasting. All were in provincial areas. One of them, 2ZG, closed before the end of the year, another, 2ZQ, in 1930 and two more, 2ZE and 4ZI, in 1932.

Without a share of the licence fee and lacking permission to advertise, the B stations were in an impecunious situation. They were small-scale affairs run by part-time enthusiasts, many of whom were eager for a greater involvement in broadcasting. Otherwise they were a disparate group, more isolated than coordinated. The provincial B stations were essentially an attempt to bring radio to areas neglected by the company. The city B stations were united in their dissatisfaction with the company stations but otherwise were varied. Some were the hobbies of enthusiasts. Some were set up in an effort to give their owners a competitive advantage in the radio (or related) trade. The RBC saw them all as challenges to its position and as attempts to reorder New Zealand broadcasting. In spite of the short life of many of the B stations, the RBC was unhappy with their influx. When presented with a request for assistance from the would-be promoters of one of the new provincial stations, the RBC seized the opportunity.

The station in question was that planned for New Plymouth, 2YB. The major figure in its establishment was E. J. L. (Ted) Payne, who had been a radio enthusiast since the wireless telegraphy days. He called the inaugural meeting of the North Taranaki Radio Society and suggested starting a station in New Plymouth in cooperation with the RBC. The reception in New Plymouth from 2YA was not good and the society, which at its peak had a membership of 200, was well supported by a community keen to have its own station. Payne approached Harris and suggested that the 50-watt transmitter that had been used at 1YA be sold to the society. Harris did not dispute the nature of New Plymouth's radio reception and regarded Payne's suggestion as an opportunity for cooperation rather than competition with non-company stations. Thus the RBC assisted the establishment and running of the new station. The station broadcast its own local programmes and also rebroadcast material from 2YA. In heralding the station the *Radio Record* described the enterprise as 'the forerunner of a campaign of development on similar lines in other centres'.[23]

The question of suitable premises for the station was solved when the owner of Colliers Music House offered the first floor of his building for a studio. The premises were rent free and, as Payne recalled, only one condition was laid down: 'You must bring in the name of my building.' Collier's offer was generous and of great benefit to the society but, like others before and after him, he was keen to make sales from his association with a radio station.[24]

The station began transmissions on 27 April 1929. The licence was in the name of the North Taranaki Radio Society, not the RBC as the company had wished. However, as was evident in its call-sign, which included the letter 'Y', denoting an RBC station, the station was regarded officially as being in a special relationship with the RBC. The equipment was put together in Christchurch by Bingham and Huggins and installed by them in New Plymouth. It was made mainly from the old 1YA equipment but included parts from other stations the RBC had taken over. Programmes were broadcast from 8pm to 10pm on Monday, Wednesday and Sunday evenings, with sports relays on Saturdays. Payne was station director and usually the announcer. The opening night's programme, however, was announced by John Ball of 2YA.

2YB was the only B station to operate in conjunction with the RBC, though the company attempted to institute similar schemes elsewhere. On 11 April 1929 Harris addressed a public meeting held in Invercargill to consider establishing a station. He suggested not an independent station, but one connected by landline to 4YA and able to transmit either 4YA or local programmes.[25] This was not to be, for the minister would not allow any further 2YB-type development. Within a year Invercargill had two B stations — 4ZI and 4ZP. Neither was connected to the RBC.

It was not only the RBC which attempted to institute relay stations in association with the company. In Napier and Hastings in Hawke's Bay, and in Hamilton, prospective B-station proprietors asked for an association with the company in a relay scheme. The RBC was interested but unable to obtain a decision from the minister. By late 1929 the company had stopped attempting to develop relay stations in provincial centres and was stating that the matter 'was purely one of Government policy, quite beyond the province of the company under the terms of its licence'.[26]

A significant aspect of the RBC's difficulty was the reluctance of the P&T Department to provide the necessary landlines to enable relay connections between stations. Landlines were standard telephone lines, owned and operated by the department. The quality of their sound transmission was open to question. But more importantly their availability, in the years of rapid growth of telephone use, was never sufficient for the company.[27] The P&T Department was reluctant to allow other organisations permanent access to landlines. The department's lack of cooperation was not directed solely at the RBC. Other organisations, such as the Newspaper Proprietors' Association, had similar experiences. Departmental actions consistently ensured that the department obtained and retained ownership and control of telecommunications.

The P&T Department's rights were made clear to Goodfellow and Harris with the advent of shortwave radio, which allowed international transmission and had applications both in broadcasting and point-to-point communication. The RBC was eager to construct a set. Not only would it be useful for international broadcasts, but Goodfellow and Harris were enthusiastic about the commercial possibilities. The company's solicitor was in no doubt that the RBC had no authority to operate a shortwave service and would be in breach of several of the 1925 regulations were it to do so. Goodfellow and Harris persisted in their enquiries until they discovered the department had ordered a shortwave radio set.

The New Zealand Shortwave Station was a 5-kW transmitter, purchased by the department from AWA and installed at the government radio station on Tinakori Hill. It opened in December 1930 and was in contact with the 20-kW AWA station at Pennant Hills in Sydney, Australia. New Zealand was thus in contact with the international telephone system. The service was available for broadcasting whenever the department allowed, but the RBC had failed in its attempt to own and operate a set. The use of shortwave radio for commercial applications and broadcasting was monopolised by the department.

The RBC felt it did not rank highly in the P&T Department's priorities. The department was, in effect, the company's master and publicly received only praise from the RBC's organ, the *Radio Record*. Away from the public gaze the company's records indicate a growing exasperation with the difficulty experienced in obtaining landlines and with what was regarded as lack of cooperation from the P&T Department.

Not until 1931 was the RBC able to link with further B stations. Lines were not available on a permanent basis, but when they were available, they were used to connect RBC and B stations in significant acts of cooperation. By May, 2ZF Palmerston North and 2ZD Masterton were rebroadcasting via landlines the 2YA Wednesday night programmes. By July they, plus 1ZH Hamilton and 4ZI Invercargill, were regularly receiving programmes from YA stations. Further applications for lines to Greymouth were turned down by the P&T Department because no line was available, and from Napier because the lines were 'inconducive'.[28]

CHANGES IN BROADCASTING POLICY

The RBC's difficulties with the P&T Department reflected a change in the political attitude towards broadcasting in general and the RBC in particular. From 1928 the policy for broadcasting changed to one favouring state control. The general election of 1928 saw the defeat of the Coates ministry and, after a sixteen-year absence, the return of Sir Joseph Ward as prime minister. Ward had previously expressed his opinion that broadcasting should be operated by the state. This remained his view during his final term as prime minister. Ward returned to office with an inexperienced United Party and was forced to include in his Cabinet four ministers, including J. B. Donald, the new postmaster-general, who had not even served in Parliament before. Ward had considerable personal interest in the postmaster-general portfolio and had held the position for all but three years of the 1891–1912 Liberal reign. Donald's initial term as postmaster-general lasted one year and he was succeeded for the remaining five months of the Ward ministry by Ward himself. However Ward was not well, and Donald was generally considered to be doing the day-to-day duties of postmaster-general during these months. After Ward's death and George Forbes's rise to prime minister, Donald was again appointed postmaster-general and held the post until the formation of the Coalition Government in September 1931.

State control of broadcasting had always been the favoured option of the Labour Opposition. After the 1928 election this became the wish of Ward's colleagues as well as his opponents. Forbes was also not convinced that private enterprise was the appropriate form of administration for broadcasting, though he did not express his view as directly as his predecessor. The changing political attitude with respect to the need for a state control of broadcasting was not a complete turnaround in government policy, for even the supporters of private enterprise, such as Coates, had never questioned the need to link private enterprise with considerable political control of broadcasting. Although the RBC's contract remained current throughout the term of office of the United Party, the government and department view moved towards an understanding that the contract would not be renewed and that a new organisation for broadcasting would be found.

It was in this changed political climate that both the B stations and the RBC presented proposals for the future of broadcasting. The B stations' proposals raised some difficulties in their preparation, as they required coordination among the stations. The first attempt to coordinate the B stations came in late 1928 from three of the Dunedin station owners, who suggested the formation of 'an association to make an effort to get some monetary return for the time and trouble incurred in the running of their stations'. Their suggestion was for either a grant from the government, or the right to devote a proportion of transmission time (they suggested 10 per cent) to advertising — or both of these options. While the B stations' financial status was the major concern, the writers also saw other reasons for cooperation: 'There are many other questions, such as the high price of land-lines, etc., that could be overcome if the owners of the eight stations, class B, get together and pull in harmony, but for one station to attempt to do this is like a flea biting an elephant.'[29]

A conference of B-station representatives was held in Wellington in January 1929. L. E. Strachan, station director of 3ZC and the conference representative, released details of the proposal agreed on at the conference. The basis of the proposal was a request that the government permit radio advertising. B stations were 'to utilise up to 10 per cent of their total hours on the air for publicity purposes'. Both subsidies from the government and the idea that B stations should receive a portion of the licence fee revenue were explicitly excluded. Indirect rather than direct advertising was contemplated: 'The scheme proposed was based on that which was the basis of the radio industry in America, namely, that programmes shall be provided by business firms, the only advertising being a statement at the beginning and end of the programme that: "This programme has been supplied by Messrs — , manufacturers of — ."'

Various aspects of the proposal were unclear. Just what counted towards the up to 10 per cent advertising time when what was proposed was only one sentence at the start and end of a programme was uncertain. There was also an apparent contradiction in the suggested hours of transmission. Strachan stated that B stations would 'utilise the present silent hours and keep off the air during the hours when the YA stations are broadcasting'. Yet he also indicated that, if the scheme were approved, 3ZC, at least, would form 'a company to provide a continuous programme from early morning until midnight'.[30] The following month the minister announced he was 'not disposed to accede to the request'.[31] Thus ended the first attempt by the combined B stations to introduce advertising.

After the 1928 election the RBC presented its own proposal for radio's future. It was a plan to extend broadcasting into the provincial regions. In its essentials the proposal had been considered and discussed publicly soon after the RBC's contract was signed. The *Evening Post*, for example, reported the company as holding that 'the most economical plan would be to establish relay stations of 50 watts antenna power at places where experience shows relay stations are desirable. These stations will receive their programmes by the land line from the nearest main station.'[32] Although the postmaster-general had been informed by his department officials that the four city YA stations would give an adequate national coverage, the RBC never believed this. It had always considered a chain of secondary stations would be needed. The *Radio Record* reported, 'It is an open secret to those familiar with the conception and plans of the original broadcasting service of New Zealand that the provision of relay stations at suitable provincial centres was contemplated as an integral part of the original scheme.'[33]

This scheme was further detailed in the later months of 1928, and presented to P&T Department officials at the end of January 1929.[34] The plan was for a series of stations in the main provincial centres, connected by landline with the four major stations. No proposal was made as to the power output of these stations. The stations were now envisaged as being able to operate at different times both as independent stations transmitting local programmes and as relays transmitting YA programmes. The scheme was submitted to the department, as a tentative proposal for it required government approval and considerable cooperation from the department. The company was keen to learn of the official reaction and of the department's view of the power required for the provincial stations. The plan was dependent on the existence of the necessary landlines and the readiness of the department to make them available. This was one of two rocks on which it was to founder.

The other difficulty and the major point for consideration was the time needed to implement the scheme. Harris contended it would need 'to be spread over five years both for technical and economical reasons; technically because the landlines would not be available, and economically because it would take that time to develop public patronage to the extent required for commercial working'.[35] A five-year period would, of course, have extended beyond the company's contract period and thus assumed there would be a renewal. The proposal also assumed the government would both continue a policy of unified control and inhibit the growth of independent B stations.

The scheme awaited an official response which, to Harris and Goodfellow's mounting frustration, was not forthcoming. In the changed political climate George McNamara, the P&T Department secretary, indicated 'there was not a ghost of a chance for a renewal' but Harris clung to a view that this was only McNamara's personal opinion.[36] The department argued that any delay was not of its making, stating that it would place no obstacle in the way of the RBC establishing relay stations.[37] However by July the company had been advised that the matter would not be considered by the government until the following year.[38] The delay, of course, came not from the nature of the scheme itself, but because it required a decision on the renewal of the contract. Harris and Goodfellow were well aware that the acceptance of their scheme was inextricably connected with the matter of renewal, and that the political climate was not to their advantage.

Meanwhile the P&T Department prepared for a change in broadcasting organisation. In September 1929 McNamara returned from an overseas trip during which he had studied, among other matters, broadcasting control. Two new models, other than those of state control and private ownership, now became part of political consideration. One was the British system of corporation control introduced at the start of 1927. The other, a version of which McNamara favoured, was the Australian system. In 1928 Australia established a national broadcasting service with the A stations owned and operated by the government, but with programmes provided by independent contractors. The Australian B stations continued as before, with advertising as their source of income. McNamara argued for a similar state relationship to that in Australia. He wanted his department to take control of the RBC stations without being responsible for programming. McNamara's preference for the Australian system did not include its B stations — he was not keen to allow New Zealand's B stations to continue independently and flourish with an advertising revenue.

At the end of the year George Kent, the RBC's solicitor, discussed the topic of contract renewal with the minister and secretary. He advised that the minister was in favour of a renewal upon suitable terms and would be taking the matter to Cabinet early in the new year. But he ended with a note of caution: 'The question of broadcasting has, however, become a political as well as a business one, and it appears to be difficult to forecast with certainty what Cabinet is likely to do having regard to the present political situation.'[39]

The continuing uncertainty increased the reluctance of the RBC directors to plan and, particularly, invest for the future of broadcasting. While confident that the radio audience in New Zealand would grow appreciably, they were less confident that this would be a rapid growth. With only a five-year term, they adopted a conservative economic approach to their business. With the increasing political uncertainties their caution grew. It was most noticeable in the drawn-out 1927–29 decision, when new premises were constructed for 4YA, not to increase the power of that station to 1000 watts.

The Dunedin station costs exceeded the licence revenue from the Otago–Southland region. Harris complained that 'on a population percentage basis the figures show that the people of Otago have responded to just half the extent that each of the other districts has done'.[40] However, Harris failed to acknowledge that Otago's limited audience was a reflection of the RBC's performance in the area. 4YA was the Cinderella of the company's stations, receiving less company backing than the northern stations. After the dismal start of VLDN at the 1925–26 Exhibition, the station, reverting to the 4YA call-sign, had been reassembled at the pre-RBC O'Neill premises. The site was an improvement, but the transmitter still did not achieve the power output specified in the contract and was rated at 370 watts. Nor were the studios up to the standard of 1YA and 3YA, let alone those in Wellington.

A substantial reason for the low licence-holding in the Otago region was the bad reception in Southland, administratively part of the Otago region. Harris was aware that 4YA was not delivering the range originally envisaged and considered, 'a 1000 watt plant will ultimately be required for the territory'.[41] In 1928 new premises were found for 4YA. Arrangements were made with the Evening Star Company for space in its new building in Stuart Street, and for the right to erect towers on the roof to support an aerial system similar to those at 1YA and 3YA. The new building was intended to be a long-term site for the Dunedin station. It was the opportune time to address the question of the new transmitter. The RBC, however, decided against its purchase, and the old plant was transferred to the Stuart Street premises. The power output of 4YA was raised to 500 watts but the reception difficulties, especially in Southland, remained unresolved.

The reluctance of the RBC directors to invest in broadcasting reflected their suspicion that broadcasting would not remain under their private control. For them it was the most serious of the changes in attitude in the second half of the 1920s that accompanied the growth of broadcasting. But it was by no means the only difficulty faced by both the RBC and the directors of the various B stations. The growth of broadcasting altered the thoughts and actions towards the medium of most who were associated with it. In particular, organisations and individuals began to insist on payment for services rendered to broadcasting. Discontent surfaced first with regard to the largest component of programming, the broadcasting of music: the RBC was asked to pay for its use of copyright material. This demand affected the broadcasting of both live and recorded music.

MUSIC, OWNERSHIP AND BROADCASTING

The use of copyright material became a topic of concern soon after the start of the RBC when, in January 1926, the Australasian Performing Rights Association Ltd (APRA) was registered as a company in New South Wales. The parent Performing Rights Association was formed earlier in England. APRA, which claimed to control, in Australia and New Zealand, 98 per cent of the world's copyright music, presented the RBC with a claim for payment of copyright dues. The relevant legislation in New Zealand was the 1913 Copyright Act, substantially a reproduction of the English 1911 Copyright Act, which naturally contained no reference to broadcasting. The law, however, was clear on the point that the author or composer of a copyright work, or his assignee, had an absolute right to permit or prohibit its performance in public. Thus the company had no alternative but to come to an accord with APRA and, just before the new 1YA station opened, settled on an agreement for one year.

Goodfellow accepted that composers had entitlements that had to be met, and the RBC had no real quarrel with the terms of the agreement.[42] But the company saw that it would be in a weak position if and when APRA increased its demands. It asked the government to legislate on the matter, to determine a maximum amount that should be paid to copyright owners. There was some doubt as to whether this was permitted under the 1913 Act and the terms of the 1908 Berlin International Convention governing the law of copyright, to which New Zealand was a signatory. Therefore the government awaited the outcome of the forthcoming April 1928 International Conference held in

Rome, which considered the whole question of copyright. The government was sympathetic to the RBC's problem, and New Zealand's delegate to the conference, along with his Australian and Norwegian counterparts, argued successfully for the insertion of a clause in the convention which gave each country the right to control, within its own national area, the exercise of the principle of compensation.[43] New Zealand was thus free to legislate what it considered fair compensation for the copyright of broadcast material.

By this time the matter had become urgent. At the end of the initial one-year agreement, APRA raised its demands. No agreement was reached. From February 1928 the RBC ceased the periodical payments due and APRA sought an injunction against the company. The hearing was scheduled for 10 October 1928. With this pending, the government finally introduced legislation.

The Copyright (Temporary) Amendment Bill was introduced on 2 October 1928 and given its final reading in the legislative council on 8 October. The bill established that a maximum of 7.5 per cent of receiving licence fees should be paid in copyright claims. This money came from the RBC's portion of the fees, not from the P&T Department's share. The actual sum paid was determined by a tribunal appointed under the Commissions of Inquiry Act 1908. The Act was retrospective. It was applicable from 1 October 1927 and was to terminate on 31 August 1929. There was no parliamentary opposition to the measure — even E. J. Howard, the government's usually redoubtable opponent on broadcasting matters, welcomed the bill. The tribunal set up under the Act decided late in 1929 that the amount paid should be a flat rate of 6 per cent of the company's share of licence revenue. This was the equivalent of 1/6d per receiving licence fee. APRA, the only claimant, received this full amount. After the Act lapsed, the same rate of payment continued by agreement between the parties.

The situation was settled between APRA and the RBC, but the position of the stations not owned by the company remained unclear. Were any claims APRA had in respect of the B stations settled by the fact that 6 per cent of receiving licence fees had been paid? Or was this a payment from the RBC, covering only its stations, and thus leaving the other stations still in dispute with APRA? APRA stated its position in late 1930: it asserted to B stations that they were separately liable to pay a fee to cover copyright claims. The B stations disputed this, arguing that as all New Zealand radio set owners were required to pay a licence fee, the payment to APRA of 6 per cent of that fee covered all New Zealand radio stations. The fact that the 6 per cent was paid by the RBC, the sole recipient of the income from licence fees, was irrelevant. Harris supported the B stations, but APRA threatened to seek injunctions against the B stations unless they made separate payments.[44] The APRA demand was widely reported as being that each B station pay 4d for each side of a gramophone record. The B stations regarded this sum as enough to drive them off the air, and the demand was ignored.

There was little progress in the negotiations until APRA notified all B stations it would be enforcing the matter after September 1930. It then became apparent that APRA was more interested in asserting its rights than in making money from the impecunious B stations.[45] APRA's concern was to have B stations acknowledge a separate liability so that, if and when they were allowed to advertise, APRA could claim a percentage of their income. APRA thus announced that a collective payment of £200 would be accepted on behalf of the B stations for the year. This was around ten guineas a station — by no means an avaricious demand — and 2ZF and 4ZB, at least, paid.[46] Most B stations were reluctant to accept liability. As the 3ZC proprietor stated, 'If we admitted their claim we do not know what they would ask in the future.'[47] Again, various B stations ceased transmission. 2ZU Dannevirke closed permanently in February 1931, but most stopped only temporarily. No legal action followed, and the position remained inconclusive.

APRA was not the only non-broadcasting organisation with a financial interest in music. The gramophone record companies became increasingly ambivalent towards broadcasting. At first local music shops had willingly loaned records to broadcasting stations, and many continued the practice, particularly to the B stations. To play records was to advertise them to prospective purchasers. This, along with the extent of broadcasting stations' purchases of records, induced the record companies to give them substantial discounts. By 1930 the record companies had changed their view and all the New Zealand stations, both B and RBC, were faced with an ongoing dispute: the record companies argued that there was 'indiscriminate broadcasting' of songs by stations and sales were being reduced rather than increased.[48] In the hard times of the depression it took considerable economic blindness to blame broadcasters for the drop in gramophone record sales, but the companies considered so and moved to combat it.

The RBC received letters from all the record distributors in early May 1930. Each offered the same stringent conditions for future record purchases by the RBC.[49] Goodfellow and Harris broke the companies' solidarity and made secret arrangements with HMV and Charles Begg & Company.[50] The RBC also obtained records on loan from some retailers, and in return, announced at the beginning and end of a session who had supplied the records.[51] The RBC was little hindered, but the unresolved matter left a tension between it and the record companies.

In November 1930, record distributors made a collective decision not to loan records to stations. A record could not be broadcast unless it was owned by the broadcaster. This decision affected the RBC least, for it had been buying records since it began and had built up a large collection. B stations had fewer resources and relied on records borrowed from retailers. They combated the measure as best they could. Many appealed for records from listeners, with some success. 2ZF Palmerston North, was loaned 2663 records and nearly 100 piano rolls within 24 hours of its broadcast appeal.[52] 2ZQ

Masterton closed down and 2ZI Hastings was sold to a music store. It was reported of the Dunedin B stations that 'only one station will continue to operate and this is owned by a gramophone agency'.[53] Many stations ceased transmissions temporarily, but there was no large-scale closure. The distributors' solidarity was not long-lasting — B stations were able to make various arrangements with distributors. But again broadcasting stations had an uncertain relationship with the suppliers of records.

PERFORMANCE DISPUTES

The difficulties between broadcasters and the suppliers of recorded music were replicated in the relationship with live performers, who still provided material for the majority of broadcasting time. Again the debate centred on the payment for performances offered by broadcasters. The RBC performers were increasingly dissatisfied with what they considered low payment for their services. Once it had introduced its policy of payment for performances, the company was faced with demands for raises in the level of payment. Furthermore, demands for payment came not just from artists but from organisations such as sporting bodies, which had at first allowed free broadcasting of their activities. The matter of payment was not the basic debate: underlying it was a growing belief that broadcasting was detrimental to the success of the older means of public presentation.

In attempting to persuade performers to broadcast, the company was regularly faced with the anxiety, on the part of performers and particularly their managers, that broadcasting would lower attendances at concert performances. The RBC had experienced this difficulty from the beginning: many of the overseas artists present at the Dunedin Exhibition of 1925–26 had been approached by the VLDN management, but few had been permitted to broadcast. The problem resurfaced in the first week of the new policy of improvement. The baritone Barry Coney was in Wellington under engagement to the Royal Wellington Choral Union to sing in Verdi's *Aida* on 5 November. He had also agreed to broadcast from 2YA on 4 November, but at the last moment the Choral Union refused to allow this. The *Radio Record* presented the case for broadcasting:

> How would the fact of Mr Coney being broadcast on November 4 affect the financial success of the Union's performance on November 5? . . . Why not even contemplate the broadcasting of a full performance? The joy thus given over the air to so many new thousands of listeners would create a new circle of opera lovers which would desire the further joys of sight and feeling, and would fill the cash box rather than empty it.[54]

In February 1928, the committee in charge of the band contest held in Christchurch that month refused to allow the proceedings to be broadcast as it gave the RBC 'something for nothing'. Again the company's argument was that its actions would represent no loss to the performers who were being broadcast. On the contrary, they would gain from the publicity. At this stage the matter was at an impasse. Organisations such as the band contest committee would not allow their performances to be broadcast without payment, and the company's policy was that it would not pay organisations to relay them, but would do so without charging them.[55]

With the requirement that live broadcasting make up the majority of its broadcasting time, the RBC was dependent on its performers, and soon changed its payment policy. The change took place during the company's negotiations with the Auckland City Council regarding broadcasts of the Auckland Municipal Band. The band gave free Sunday concerts, which were originally broadcast free of charge. This arrangement was considered no longer acceptable to the council after the RBC's general introduction of payment to artists. In March 1928 the company agreed to pay the council £300 a year for band concert broadcasting rights. This arrangement did not last: later in the year the broadcasts ceased when the council demanded £750 a year. There was also considerable debate regarding just what concerts should be included. Although the council lowered its demand to £500, the concerts remained off the air.[56]

Over the years, the dispute over the price to be paid for the broadcasting of various forms of musical entertainment continued to see-saw. It showed a fundamental disagreement that applied also in other areas, such as religious and particularly sports broadcasting. The dispute reflected two opposing views of the nature of broadcasting. For the RBC and *Radio Record*, broadcasting was a new medium which would greatly change the world. Harris and Heighway had their own interests in mind, but they were genuine in their opinion that to broadcast an event would raise public awareness and interest in future events of the same kind. The opposing view was that there was already an established audience for concerts and sporting events. Broadcasting an event meant that at least part of that stable audience would listen to the broadcast rather than attend the event, thus reducing attendance numbers. Broadcasting was regarded as parasitical on established forms of entertainment, not as something which would in any significant degree change the existing order. There was no immediate solution to this disagreement, and it bedeviled broadcasting authorities long beyond the tenure of the RBC.

THE BANNING OF RACE COMMENTARIES

It was over the commentaries of horse races that the clash of attitudes to broadcasting received a first intense public discussion. Allardyce organised the commentaries, 'and we . . . never paid a penny to any of the organisations for permission to broadcast'.[57] Harris was no doubt pleased with this non-financial arrangement, but it backfired. With the sporting bodies receiving no financial reward from the broadcasts, the RBC was dependent on their continuing generosity. The very success of sports broadcasts, which Harris and Allardyce were so keen to achieve, soon led to a change in attitude.

A decision to ban racing commentaries was taken at the Racing Conference in July 1928. An identical decision had been made earlier in the month at the Trotting Conference. The matter had been under discussion for a year, informally at first, and then gradually more officially. The joint committee of the Racing and Trotting Conferences received a report from A. Ward, the chief racecourse inspector, which included this:

> It cannot be doubted that broadcasting as at present carried out does stimulate interest away from racecourses on race days, and is resulting in an increase of business for bookmakers. Receiving sets are to be found in clubs, hotels and other places where news from courses is received by wireless, and bookmakers are taking full advantage of the fact. If racing clubs were accused of providing a stimulus to bookmakers and their clients to indulge in illegal betting, it would be awkward to refute.

There was no doubt that the conference would stop all race commentaries, in the light of Ward's report and the endorsement of it by Sir George Clifford, president of the Racing Conference. The final resolution was to ban live race commentaries as 'such broadcasting (a) promotes and facilitates illegal betting, and (b) creates and maintains an attention to the happenings at race meetings during the progress thereof by many persons not present, which results in disadvantage to the business community.'[58]

That broadcasting was encouraging illegal betting was probably correct. It was certainly bringing racing to the attention of a large number of non-racecourse attenders and legal off-course betting, through the Totalisator Agency Board, the TAB, was still twenty-one years away. However it could be argued that the conference's apparent regard for the business community and support for the law against illegal betting was largely an attempt to protect its own revenue. The *Radio Record* hinted at this, without openly stating it.[59] Supported by parliamentary privilege, R. W. Smith, the member for Waimarino, was more straightforward:

> The claim that the broadcasting may lead to an increase in gambling is sheer hypocrisy, because, after all, the racing clubs look to the gambling — if one can call it gambling — through the totalizator as their chief source of revenue . . . One of the chief reasons for the Racing Conference making this attempt is that it wishes to force the public to attend the race meetings.[60]

The prime minister regarded the matter as a private business arrangement and said he had no desire to enter the affair. He stopped short of firmly stating he would not: 'The matter is purely a domestic one between the two parties, and I am not prepared to interfere with either unless there is a public demand that there should be such interference.'[61] This was a frank recognition that, while broadcasting in New Zealand was a private business, it operated under a public scrutiny where the usual private rights did not inevitably apply. The racing clubs took some years to recognise this. They had voluntarily allowed the RBC to broadcast races. They had no contractual or moral obligations to the company and were entitled to decide that broadcasting was not in their interests and cancel the arrangements. However the ban was not permanent, though it remained in force for the duration of the RBC's term. The racing organisations eventually accepted that their right to refuse permission for race commentaries was not the only consideration. In the new public domain of broadcasting, their Canute-like attitude was overwhelmed by the tide of public expectation for radio.

The Racing Conference allowed the RBC the usual press facilities, on the condition that results would not be broadcast until the evening. The company refused to accept this condition. While acknowledging that it no longer had the right to broadcast commentaries of races, it argued that results were news items and could be broadcast as soon as available. In particular, it was not prepared to wait until the evening newspapers had published the race results first. Thus the RBC made its own, at times novel arrangements to obtain and transmit race results. Usually company employees entered a race course by the normal method, paying for admission. They then left separately, taking each newly available race result to a nearby house which was used as a temporary studio. Sometimes the arrangements were more public and more chaotic. Covering the Auckland trots in September 1928, Gordon Hutter in one day worked from a step-ladder on the footpath outside the course, from three different motor lorries, a fruit delivery van and a vegetable-vendor's cart. Pressure of traffic made it necessary to move the observation point six times for the eight races.[62]

Although the efforts of its own staff were considerable, the main source of racing results for many years for the RBC and successive organisations were the country's bookmakers. They had a particular interest in having race results known as widely and quickly as possible. The matter was not publicised, but the origin of the mysterious telephone calls that gave rapid and accurate information of results throughout New Zealand was an open secret within the radio stations.[63]

RUGBY BROADCASTS

Rugby had long been the dominant sport in New Zealand, and match commentaries soon became popular broadcasting fixtures. The broadcasting of rugby in particular, and the place of broadcasting in New Zealand society in general, received substantial public discussion when rugby authorities, following the lead of their racing counterparts, questioned whether match commentaries should be continued. The New Zealand Rugby Union (NZRU) and the various provincial unions were increasingly concerned that broadcasts kept people from attending matches, and that gate receipts suffered. Allardyce felt that other matters had intervened: 'You have got to remember at the time things were starting to get bad. The depression was starting to creep up on us and they [the rugby unions] didn't know what was happening.'[64]

The unions tried various strategies as they grappled with the problem of broadcasting. The Canterbury Rugby Union first allowed commentaries, then banned them after deciding they were lowering match attendance figures. In 1929 it allowed broadcasts on fine days only. In 1930 it again allowed commentaries — on condition that all matches be broadcast. This condition the RBC would not accept.[65] A similar impasse had been reached with the Wellington Rugby Union which, in 1929, decided to allow a commentary of the Wellington v. Canterbury game 'on condition that other matches under the jurisdiction of the Union are broadcast for the remainder of the season'.[66] The unions considered, correctly, that unless all matches were broadcast, the company would choose to broadcast what it considered the best match of the week. Thus people were faced with the choice, not of attending or listening to a match, but of attending a local club match or listening to a provincial encounter. With this choice, the unions feared even more for their gate receipts.

The unions were pressed from both sides. They wanted to ensure good attendances at matches, and believed that broadcasting adversely affected this. They also received numerous complaints from the public, and realised that broadcasts were reaching many people who would not have been able to attend the match anyway. The chairman of the Wellington Rugby Union reported on some of the complaints, and read out a note from a man in Eketahuna: 'Let us have the best. Wireless today has become part of the life of the backblocks.'[67] Complaints such as these were coming not just from the general public, but also from the unions' constituency — rugby club members who were, after all, the keen rugby followers. It was not a matter of rugby organisations in opposition to broadcasting. Union members were divided amongst themselves.

The matter came to a head with the tour in the winter of 1930 by the British Rugby team. The NZRU made an original decision that the Saturday matches, including the four tests, would not be broadcast. The chairman stated, 'If we are going to broadcast the four tests, we are going to kill club matches throughout New Zealand on the days of the tests. The broadcasting of other matches on Saturdays will also have a great effect on club football.'[68] This decision was announced in February, leaving time for considerable public and press discussion before the tour. Previous arguments were

The *Dominion* cartoonist, A. S. Paterson, gives his view of the 1930 possible non-broadcasting of the rugby tests. Reproduced in the *Radio Record* 11 April 1930.

repeated: the company was not willing to pay for broadcasting rights; and the NZRU said that, while it would not suffer a gate loss at the British team's matches, considerable loss would be experienced by smaller unions. This NZRU argument was undermined when various of its constituent unions asked for the matches to be broadcast. The NZRU reconsidered its decision, and stated it would allow the broadcasts if the RBC would make a donation to equip hospitals with radio receiving sets. The *Radio Record* called this suggestion 'selling a "dummy" to the public'.[69] Nor was general press opinion favourable: the *Christchurch Times* held that the union had abandoned whatever case it had when it made the receiving sets for hospitals suggestion.[70]

During the debate in the autumn of 1930 the broadcasts began to be seen as more than a private matter between the NZRU and the RBC. The growth in the broadcasting audience, especially for an event such as a test match, meant that the NZRU decision would affect a substantial proportion of the population. Eventually this fact was accepted by the NZRU. As one of its committee members stated, 'We cannot afford to have a psychological knockback on this matter. I don't think we are concerned with the Broadcasting Company, but with the people as a whole.'[71] The decision was changed to allow the broadcasting of all matches by the touring team under the NZRU's jurisdiction. H. H. Sterling, who moved the successful motion allowing the broadcast commentaries, referred to 'a radical change brought about by broadcasting. It would not be keeping pace with the times if the union did not . . . place the people in the position of being able to get details of matches per medium of the new order.'[72] However, the NZRU did not go so far as to recommend the same action to provincial unions.

In 1930 the four tests, the game against the Maori side and six against provincial sides were broadcast. Individual unions still had their various attitudes and made different decisions on the broadcasting of local matches. In 1930 the Canterbury Union, in particular, held out against broadcasts and only two matches, the second test and the Britain v. Canterbury game, were broadcast from Lancaster Park. Until the fourth test, when public protest finally prevailed, the Canterbury Union also objected to 3YA's broadcasting commentaries on the British matches being held in other centres. This was in spite of the relays having to pass through Christchurch on the way to Dunedin, where the Otago Union had requested the rebroadcasts from 4YA.

The issues between the rugby officials and the RBC were in abeyance rather than settled. Debates on permission to broadcast resurfaced, and the matter of payment was raised. The NZRU, while acknowledging the new radio order and recognising that its own special constituency (rugby players) was generally in favour of the broadcasts, found the RBC's penurious attitude towards payment less and less acceptable as the number of listeners (and the RBC's income) grew.

Again in 1931 there were no broadcasts of Canterbury rugby. The union would allow broadcasting only on payment of £500 for the rights by the RBC. The company refused, and instead, in Canterbury that winter, rugby league matches were broadcast. However, the first payments for rugby broadcasts were made by the RBC in 1931. The Wellington Union asked for payment for the right to broadcast matches under its control. The company's reply was its usual one: that 'under no circumstances will it pay for the right to broadcast public news — and it regards the dissemination of reports of athletic and sporting contests as being in this category'. The RBC considered it should 'be put upon no worse a footing than the newspapers, which not only pay nothing for the privilege of publishing such reports, but also frequently charge for the insertion of advertisements relating to the holding of such contests'. This comparison with the newspapers ignored the crucial distinction that the broadcast report was a live running commentary, an innovation possible only via broadcasting. But the reply also included a major concession — an offer to make a payment to the injured players fund, 'on the understanding that your Union will confer on it the right to broadcast all matches played during the 1931 season under your Union's jurisdiction and control'.[73] While this was as much a sidestep of the issue as the NZRU's previous suggestion of donations to hospitals, it was a recognition from within the company that payment could no longer be avoided. Both the Wellington and Auckland unions were paid £50, in Auckland to the ground fund and in Wellington to the injured players fund.[74] Payment to sporting bodies had begun, and would be dealt with more directly by the company's successor.

BROADCASTING AND THE PRESS

Among the organisations that were influenced by and took an interest in broadcasting were the country's newspapers. Before the advent of radio, they monopolised the spread of public information. The New Zealand press showed an early interest in broadcasting. The activities of the first stations were well reported and by the time the RBC began transmissions the major dailies had regular, usually weekly, columns dealing with radio matters. Once the company's programmes were of a kind that could be published in advance, the newspapers began printing that day's programme, which was all they were allowed under the *Radio Record*'s copyright. Coupled with this interest was an early recognition by the press of the possibilities of competition from broadcasting,[75] both in news gathering and presentation, and in the sale of advertising.

The RBC and the press soon settled into a wary alliance. The possibilities for organising broadcasting were originally seen as two: private enterprise or state control. State control was unacceptable to the majority of the New Zealand newspapers. There was a philosophical acceptance of the RBC's existence as a private company. The fear of competition was alleviated with the regulatory ban of radio advertising and with the RBC's lack of entry, with the exception of sports reporting, into the news domain in any way which seriously competed with the press. Generally the company was happy to acknowledge that news reporting could not be its forte.[76]

The topic of radio advertising was raised when the B stations sought this right in 1928. The P&T Department's view was that, if advertising were permitted, it should be exercised by the RBC rather than the B stations. The department was keen for the company to extend its hours and saw advertising as the revenue source to allow such an increase.[77] Harris, however, made it clear the RBC would make no attempt to advertise. His view was that New Zealand needed to give a fair trial to the licence fee system for at least the five years of the RBC's contract. He also reported a discussion with A. G. Henderson, manager of the *Lyttelton Times* and chairman of the Newspaper Proprietors' Association. Henderson was naturally not in favour of B stations moving into an advertising preserve which the NPA regarded as its own. With the RBC itself repudiating radio advertising, Harris and Henderson agreed to cooperate against the B stations' desire to advertise.[78]

The *Radio Record* immediately began a campaign against radio advertising. It particularly noted that such advertising would be an unkind attack on a supportive press: 'We would not welcome the task of endeavouring to convince the newspaper Press in this country that the sale of time over the air by the radio company or companies would be a fair return for the hospitality accorded radio programmes by the Press in general.'[79] The RBC and NPA alliance was against B stations' advertising, rather than against B stations *per se*. In general the B stations were much appreciated by their communities. The

New Zealand press was a collection of locally based newpapers that were not prepared to oppose their readers by mounting a full campaign against the B stations.

Towards the end of 1930 the RBC was told that 'the hospitality accorded radio programmes by the Press' was no longer available. It was a time of economic depression, and in the quest for a continuing advertising income, the newspapers turned toward the company. The demands came from the Auckland newspapers, with both writing essentially the same message to the RBC. The argument, as phrased by the managing editor of the *Auckland Star*, was that 'in no other form of amusement is the cost of announcing the programmes to the public borne by the newspaper alone, and we think that your company should now accept the responsibility for this service.'[80] In reply Harris conceded the *Star*'s point but argued the RBC was unable to afford to begin payment.[81]

John Ball was sent to Auckland to interview the papers' managements and found that matters could continue as they were for 1931 if the RBC, in its turn, would give an assurance that it would begin to pay if and when its licence was renewed. When he conveyed the information, Goodfellow and Harris accepted the deal. They saw it as part of a pattern of cooperation between the newspapers and the RBC with, in particular, the press supporting an extension of the company's licence and the RBC continuing to affirm the principle of no advertising on the air.[82]

From early 1930 there was little about the RBC to which the press could take exception. Not only was its private enterprise format in accord with the majority newspaper view, but it was essentially non-competitive with the press in the news and advertising fields. Furthermore the company's concession with regard to payment for programme publicity meant it would be a major newspaper advertiser in the future. As it neared the end of its contract and sought a renewal, the company was confident of substantial newspaper support. While such confidence was not misplaced, newspaper opinion was of little consequence in view of the understanding that broadcasting should be under increased state control.

5
The Extension of Broadcasting and the Fall of the RBC

PROGRAMMING

The political acceptability and the future of the RBC were of interest to New Zealand audiences. The majority, however, were more concerned with the programmes that were offered. The RBC's contract required it to transmit at least twelve hours of programmes from each station weekly. The government regarded this as a minimum to be exceeded as the company became established and its income increased. Broadcasts began with two hours on each of five evenings per week, from 8 to 10pm. At first broadcasting was not for the full two hours as stations had to cease transmissions for 5 to 10 minutes at least once an evening to allow the transmitters to cool. Each station had a silent day. Stations very early introduced a 6.30 to 7.15pm children's session once or twice a week. Saturday afternoon sports broadcasts were also added but these were irregular. Sunday programmes were from the beginning a common, though not ever-present feature. They usually began at 6.30pm and featured a church service and a studio recital with the close-down at around 9 to 9.30pm.

A 7.15 to 7.45pm news and reports session was added in 1927, though not every night and not from all stations. Afternoon sessions from 3 to 4.30pm were also gradually introduced and by late 1927 were regular parts of the programme. By this time the children's sessions had been increased to a full hour from 6 to 7pm, and were on the air most evenings. During 1928 the news and reports session became a full-hour session from 7pm to 8pm and was broadcast from all stations. The close-down time remained at around 9 to 9.30pm on Sundays and 10pm on other nights except for Saturdays when transmission often continued to 11pm. In early 1929 the *Radio Record* was able to state the company was 'voluntarily giving an average service exceeding 30 hours per week from each of the four stations'.[1] The next addition to the programme came in August 1929 with the introduction of a daily 6–7pm dance music session. At this time the children's sessions were changed to 5–6pm daily.

With these additional hours of transmissions there was a gradual introduction of what were to become regular broadcasting services. Time-pips were added to the RBC programme from the start of 1928, when they were transmitted four times daily from 2YA. This service, provided by the Dominion Observatory, had begun on radio twice a week from 1 October 1920 on VLW, the government Morse station in Wellington.

Weather forecasts were broadcast at 9pm nightly from 2YA once that station began, with 3YA doing the broadcasts on 2YA's silent night. It was a sparse service, but developed considerably over the years. At first it was considered a service for country listeners rather than listeners generally, and any special additions were made in this light. These included frost warnings in fruit-growing areas, wind warnings during fruit harvesting times and, in November 1929, a temporary introduction of an afternoon forecast for sheep farmers during the shearing season.[2]

The requirement that 75 per cent of broadcasting hours be from other than 'mechanically operated musical instruments' meant the RBC relied heavily on live broadcasting by local talent. This reliance increased with the extension in broadcasting hours. The expectations of performers were high and the musicians, in particular, were worked hard. The 2YA orchestra, for example, in its first ten weeks broadcast for an average of three hours daily.[3] Harris regarded the dependence on local talent as a continuing problem. The company, and the four stations' musical and dramatic committees, found it difficult to provide a varied and quality programme from the available talent. The 1YA committee particularly, under its energetic chairman Karl Atkinson, encouraged new talent and urged existing performers to increase their repertoires. For all stations the need to meet the conflicting programme demands of quality and quantity posed constant difficulties.

Studio live performances were presented as concerts. This included dressing for the occasion. Eddie Hegan, in the late 1920s 'The Radio Rascal' of 3YA, who did everything — singing, telling gags, vocal impersonations, and tap dancing (which oddly enough he found very popular with listeners) — considered the YA stations to be 'stiff with formality. Although we never worked to a live audience all performers were expected to dress formally; formally then meant a boiled shirt and dinner jacket for the men, and for the ladies an evening gown. Even the duty announcer wore a dinner jacket.'[4] Formal attire was the custom of the time. As John Gordon has noted, in the 1920s society 'returned very much to the social conventions of the Edwardian age. We always wore gloves — both men and women — to dances and balls. We always wore evening dress for the theatre.'[5] The various station managers were merely keeping to convention by insisting their staff and performers dress formally. It was also important to the success of the performances. For some years to come there was only live broadcasting; formal dress added to the sense of occasion and promoted the necessary intensity for these live performances.

The public were allowed into many stations; at 2YA the lounge was

The Wanganui Maori Party pictured in the 2YA studio at the time of their 1928 broadcasts. Their guest of honour, Gordon Coates, the prime minister, is seated in the centre. RADIO NZ SOUND ARCHIVES

crowded most nights with people watching the studio through the large double plate-glass observation windows.[6] But studio audiences were not yet welcome. Various performers requested them, but to no avail. Hegan produced two musical comedies for 3YA — 'To establish some atmosphere and audience reaction I wanted to do those musicals to a live audience and suggested we invite a small crowd to see the actual broadcast and supply applause and laughter. The suggestion was greeted with horrified disapproval.'[7]

The studio performances to which the RBC gave great attention and the *Radio Record* much publicity were those by the Wanganui Maori Concert Party. It was decided in 1927 to make a special feature of the next — the eighty-seventh — anniversary of the signing of the Treaty of Waitangi. A group of thirty performers from Wanganui took part and presented, from the main 2YA studio, a 'Radio Pageant of the Maori Race'. The two and a half hour programme was organised by Mete Kingi and the party was led by Hamiora Hakopa. All four stations carried the programme. The performers gave the programme again the following night at a later hour so that it could be heard on the Australian east coast and in the Pacific Islands. The programme was well received in New Zealand and also in Australia and the Pacific. The *Radio Record* saw the programme as attracting 'unquestionably the biggest audience ever concentrated on one single evening in this Dominion for any one event'.[8] The following year a further programme was presented by the same concert party to commemorate the second anniversary of 2YA. Again the programme was repeated on a second night, this time from 10.30pm to 1.30am, to permit reception in Australia, the Pacific Islands and, this time, also in Antarctica by members of the Byrd Expedition.[9] Again the public reception was enthusiastic. The following year the party presented a further pageant, again very well received.[10] For each of these broadcasts the party, already well rehearsed, would arrive at 2YA nearly a week before the event for final days of practice.[11]

These were the major but not the only broadcasts by Maori groups during the term of the RBC. The company also presented programmes concentrating on the pronunciation of Maori. They were started by J. F. Montague, the Auckland speech trainer who broadcast plays from 1YA. Montague, a Pakeha, was a Maori speaker and saw the radio as an appropriate instrument to improve pronunciation of Maori. 'Some of the common mispronunciations, many of them in fact, are atrocious, and it is time something was done about it' he argued.[12] In January 1929 these broadcasts were taken over by Henry H. Stowell (Hare Hongi), of Nga Puhi descent, whose broadcasts included discussions of correct pronunciation and talks on the meaning of Maori placenames and phrases.[13]

It would be a mistake to see the RBC as introducing any distinctive Maori broadcasting. Its programmes were not produced by and for Maori. They were designed to provide a New Zealand slant to its national programming. The few Maori programmes were exceptions rather than a continuing orientation. But by the standards of the time those few programmes were lavishly produced and regarded by the RBC as among its major broadcasts. Furthermore its attitude to the Maori language, which was not always followed by the company's successors, was essentially respectful. Its policy was that there was a correct pronunciation of Maori and broadcasters should assist in educating the general population in its usage. The company was no crusader for the Maori language nor for the Maori generally, but was in advance of contemporary opinion.

The years of the RBC were also years of improvement to the recording of music and to the broadcasting of recorded music. This later aspect notably included the 1928 introduction of the electric pick-up. In December 1928 the RBC considered the quality was such as to begin a series of operatic broadcasts from records.[14] In early 1929 the *Radio Record* noted, 'Electric recordings have now reached such a state of perfection that it can be safely said that all the world's greatest artists have been recorded, and in most cases can be reproduced by means of an electric pick-up with an efficiency which equals the actual performance.'[15] This statement is not without exaggeration, but does indicate the real advances made. However there were still considerable difficulties with sound reproduction. A major problem with the use of the electric pick-up (and it was to remain until the late 1940s) was associated with the steel needles used in contact with the records. The needles wore down rapidly and, for good quality sound reproduction, had to be changed for every side of a disc. At the start of a disc the needles were almost pinpoint sharp and, before they were ground down, they cut into the first few grooves of the record. Thus records could not be played many times before they became very noisy for the first few grooves. There was a reverse problem towards the end of each disc, for by then the needle had often worn so far down that distortion crept in. This problem was exacerbated in the early records which deliberately included a fine abrasive in the compound, which was intended to grind the needle to fit the groove of the disc.[16]

By early 1930 all four stations were fitted with two turntables, side by side, driven by electric motors, with the music broadcast via an electric pick-up. Records could follow one another without any perceptible break.[17] It was a large advance on the announcers' first experiences of hanging a microphone in front of a wind-up gramophone. These changes led to a greater acceptance of the broadcasting of records. A year later the *Radio Record* argued, 'Time was when the barbed shafts of critics were directed towards the gramophone items. The position now is different. The technical advances made in recording, and the great range of high class artists now available on the disc, have lead to gramophone items being among the — if not actually first — most popular items in the programme.'[18] A consequence of the change was an increase in the proportion of recorded music in programmes. Station directors saw this as an answer to the problems they experienced in finding sufficient quality live performances. The movement from local talent to imported records was also a change to a cheaper and more easily broadcast form of music. Stations were still governed by the 1925 regulations which restricted the use of mechanically operated musical instruments. This restriction was increasingly disregarded both by the stations and by the P&T Department, whose job it was to enforce the regulations. The B stations broadcast the greater proportion of recorded music, but the RBC's stations also increased their proportion.

The growing acceptability of records led to a further change in programming — the introduction of complete programmes on record, in the United States referred to as 'electrical transcriptions'. Harris made a trip to the United States in the winter of 1930 for various matters to do with his own company and the RBC. Prominent among the reasons for the journey was the quest for recorded programmes. The *Radio Record* stated the RBC's introduction of such programmes was a logical consequence of the improvements in recording techniques.[19] It was certainly relevant, but a further reason was to lessen the RBC's dependence on live performers and on New Zealand gramophone record distributors.[20] Even after the agreement reached between APRA and the RBC, the distributors were a continuing trial to broadcasters. Their view was that broadcasting of records adversely affected record sales and they regularly attempted to limit such broadcasts. Thus Harris was keen to lower the company's reliance on the local distributors.

Recorded programmes were initially developed in the United States, but the idea spread, with New Zealand later purchasing many programmes from Australia and England. The early United States recordings aroused considerable interest and criticism. The purchases included the usual broadcasting offerings, such as full musical programmes and comedy programmes. They also included specimen radio serials, a genre previously unknown in New Zealand. Harris, who had his early radio and business experience in the United States, saw the imported programmes as a natural orientation, but many regarded them as a foreign intrusion. Goodfellow was the first critic of the imports which Harris sent to him for appraisal: 'They are rather a mixed lot, some being exceedingly good, others such as *Cecil and Sally* are so ultra American that they would obviously never suit a 100 per cent British country like New Zealand.'[21] The RBC rejected *Cecil and Sally* and other programmes; but many were accepted. 'It is our purpose to include the *Ormandy Salon Orchestra*, the *International Singers*, and *Locharts* in our programmes, and these series have been ordered. With regard to *Cecil and Sally* . . . we have samples of several recordings which are worse, one of which features a murder including the victim's gurgles and gasps.'[22]

The American radio programmes were introduced in February 1931, starting from 2YA where they filled in the extra hours that came with the abolition there of the silent night. The programmes were the first 33⅓rpm records to be broadcast and required the installation of special turntables.[23] Soon programmes such as *The Japanese Houseboy*, *Eb and Zeb*, *Abroad with the Locharts*, *International Singers*, *Hy Wide and Handsome*, *Ambassadors of Melody Land*, and the *Eugene Ormandy Orchestra* became well known in New Zealand. They met a mixed reception. They were generally popular with listeners but there was little doubt that most established opinion was unimpressed. The decision to import and broadcast American programmes was discussed only privately, within the RBC, but their introduction signalled the start of a long debate on the wisdom of broadcasting foreign programmes. The first complaints centred on changes in the speech of New Zealand children. For instance, G. C. Munns, MP for Roskill, considered that American programmes were having a bad effect on New Zealand children's speech:

> I want to lodge a protest against the cheap American rubbish that we get over the air. I refer to the American slang terms such as 'Say bo,' 'gol darn,' 'O.K. baby,' and 'ma lil yello baby.' Is it not fine to teach our children such rubbish? Then we hear our children in the street using the expression 'Say, chief.' And where does it come from? Over the wireless. These terms are gradually getting into the young minds until they become a regular part of our children's daily talk. I think the influence on the child mind of what comes over the air is most important of all. I would like to hear good English broadcast.[24]

Munns had a point. It was obvious the slang had been introduced by the company's new programmes. In thousands of homes in the 1930s the

American expressions joined the established group of the unsavoury. Thus, for the then eight-year-old Janet Frame, 'bad language [was words] like "bugger," "O.K. , chief," "God Almighty," "bitch," "bastard"'.[25]

Furthermore the American programmes seemed at odds with the earlier stated policy of the company, which was conservative with regard to speech: in the opening issue of the *Radio Record* the RBC declared that it recognised 'the importance of correct English diction' and realised that 'broadcasting should be a powerful medium in speech training'.[26] Over the years the focus of concern grew to encompass considerably more than possible detrimental effects on speech. Harris, however, was satisfied with the general public's acceptance of the imported recorded programmes. Once again he showed himself more concerned with popular opinion than established standards.

RELAYS AND OUTSIDE BROADCASTS

Among the more notable and popular achievements of the RBC were its broadcasts from beyond its studios. Outside broadcasts did not begin with the company — they went back to Professor Jack; but the company developed the idea in earnest, and began a policy of broadcasting public events. Outside broadcasts soon became regular events, with commentaries on sporting events and the relaying of music as the major items. Goodfellow and Harris were keen to go beyond this type of material, and to broadcast public occasions such as political meetings and debates. Harris felt this would ensure that 'a greater number of people than hitherto is kept informed of the progress of events, and a greater interest in Dominion affairs is fostered and maintained.'[27] The ban on the broadcasting of controversial material inhibited this, and the company was in effect denied the right to broadcast most public meetings. Goodfellow and Harris saw public meetings as a ready and relatively cheap source of broadcasting material, and also as an appropriate use of the new medium to increase public regard for radio. They regarded such broadcasts as of benefit to New Zealand in that they would act to establish a national consciousness.[28]

The sites of the earlier outside broadcasts were chosen for their proximity to the studio; but the RBC engineers and their P&T Department counterparts gradually developed the ability and the confidence to venture further afield. The RBC relays presented technical and administrative difficulties. They were conducted over standard telephone lines, which meant that the sound quality diminished with distance from the studio. Over the years, as broadcasting equipment improved, the standard telephone line became the weak link.

In the RBC's first year, relays were generally over short distances, from theatres and concerts. Longer distances were achieved when sports broadcasting began: the 2YA broadcast in July 1927 of the Ranfurly Shield match from Solway Park in Masterton was the longest relay attempted to that date. Another long-distance relay took place in December 1927, when 2YA

Photographs taken at Richmond Aerodrome, near Sydney, during the departure of Moncrieff's and Hood's ill-fated flight. ALEXANDER TURNBULL LIBRARY

broadcast a Wanganui concert. In May 1928 a Wellington concert by the Port Nicholson Silver Band was relayed 450 miles by landline to Auckland and broadcast over 1YA. Also in May 1928, Sir Joseph Ward's speech in Invercargill at the opening of the Southland May Fair was relayed by landline to Dunedin and Christchurch and broadcast over both 4YA and 3YA. These occasions were celebrated in the *Radio Record*, but were soon regarded as standard.

A technically less difficult but still notable outside broadcast took place in February 1929 with the broadcast from 2YA, and rebroadcast from 3YA, of the Allan Wilkie Shakespearian Company's production of *The Merry Wives of Windsor*, to an estimated audience of 80,000. This was the first broadcast of Shakespeare direct from the stage; and it marked the introduction of condensor microphones, which quickly replaced the older carbon ones.[29]

In 1928 the country was united in mourning and in celebration by outside broadcasts of two major events — the first attempts to fly across the Tasman. On 10–11 January, Lieutenant J. R. Moncrieff and Captain George Hood made the first attempt, which cost them their lives. Their destination was the Trentham Racecourse, with an arrival time estimated to be between 7 and

Charles Kingsford-Smith and companions arrive at Wigram after the first flight over the Tasman. The seated figure on the hanger roof, fourth from right, is Alan Allardyce, giving a live broadcast of the event. ALEXANDER TURNBULL LIBRARY

8pm. 2YA covered the event, with John Ball broadcasting from the Trentham stand. Some 10–12,000 people were at Trentham, with cars lined up to light a runway. At 10pm, the usual closing time for 2YA, it was announced the station would remain on the air. Most of the crowd had left by midnight, but Ball remained, his broadcasting interspersed with gramophone records from the studio, until 2.30am. The desire to share in the triumph had been present in the large crowd and in the larger broadcasting audience. Instead it was a shared tragedy, made more personal and heartfelt for each listener by the long night's anxious listening.

On 10–11 September Charles Kingsford-Smith and his companions successfully flew across the Tasman, and again the event was shared through radio. Extensive arrangements had been made before the flight for radio coverage, and 3YA was on the air for 18 hours non-stop. After the plane reached New Zealand the company's four stations were linked into the simultaneous broadcast from 3YA. Kingsford-Smith's plane, the *Southern Cross*, first landed at Collingwood and then flew to and circled over Wellington before flying south to land at Wigram Airport. Alerted by the broadcasts of the plane's imminent arrival, a crowd of 30,000 were there to watch the landing. It was a triumph for New Zealand, for aviation, and for broadcasting.

The departure of the *Southern Cross* from Blenheim on the return flight was not such a grand occasion for broadcasting, though the audience did not realise it — and it demonstrated the readiness of broadcasters to deceive an audience. Clive Drummond broadcast from 2YA a commentary of the departure with vivid descriptions of the crowd, the aircraft and the crew. The engine sound was heard gaining in intensity as the plane took off and then fading into the distance. What was not stated was that the entire broadcast was done in the 2YA studio with the assistance of a special-effects record. The only contact with Blenheim was one phone call to ensure Kingsford-Smith actually was leaving.[30]

International broadcasts were possible after the advent of shortwave radio. The first was on 12 November 1927. A. L. Curry, the announcer at the time, later spoke of 'the wonderment of us all . . . when 3YA carried the first rebroadcast of the BBC empire service from G5SW Chelmsford, England.'[31] The broadcast of the Armistice Day service in Canterbury Cathedral included a ten-minute talk by Sir John Reith, director-general of the BBC. The following year an ambitious attempt was made to rebroadcast from 2XAD New York, General Electric's shortwave demonstration station, the commentary on the world heavyweight boxing championship. Gene Tunney was the holder and the challenger was a New Zealander, Tom Heeney. Naturally New Zealand interest in the event was high, and Harris went to considerable lengths to secure a commentary. The attempt was only a partial success and demonstrated the limits of radio communication at that time. A direct link was arranged so the ringside description could be broadcast; but Harris considered from the start (correctly as it turned out) that this would probably not be successful, given the time at which the fight was scheduled. He also made alternative arrangements, and a summary of the fight was broadcast from 2XAD at midnight (New York time). In Wellington the earlier shortwave receptions were picked up at various stations all linked with the 2YA studio. None was suitable for rebroadcasting, but sufficient detail could be deciphered to enable Clive Drummond to give a round-by-round description. The later shortwave summary was received successfully and was rebroadcast.[32]

By 1930, long-distance communication techniques had improved to the point where 4YA became one end of a chain of stations, in what the *Radio Record* called 'the most remarkable broadcast that has been enacted in the annals of radio'. The occasion was the return to Dunedin from Antarctica of the American explorer, Richard Byrd, and his party. Via landline from 4YA to 2YA, broadcast from 2YA to 2ME, Sydney, and by shortwave from 2ME to 2XAF, New York, and in return direct from 2XAF to 3YA, a two-way conversation was held between Rear-Admiral Byrd and various dignitaries in New Zealand and the United States. The conversations were broadcast over all the RBC stations and over many stations in Australia and the United States, as well as in Canada, England and Germany. It was an international occasion with a radio audience well into the millions, and a vivid demonstration of the growing capabilities of the medium.[33]

The following week Byrd gave two addresses to school children in the Dunedin Town Hall, each time to an audience of 3500. These talks were broadcast over the four company and other stations and gained the most extensive national coverage to that date. A total of 1274 miles of landline was used to connect from Invercargill to Auckland, with a branch connection to New Plymouth. Radio receiving sets were installed in schoolrooms, halls and parks for schoolchildren and others to hear, along with the radio-owning audience.[34] As Downes and Harcourt note, Byrd's 'was the first voice to echo simultaneously from Spirits Bay to Foveaux Strait'.[35]

DISASTER AND RADIO

Radio, both in its broadcasting and point-to-point forms, had grown considerably in the 1920s. Its position, and indeed indispensability, within New Zealand society were finally accepted during two disastrous earthquakes. On 17 June 1929, an earthquake centred near Murchison caused widespread damage there and along the West Coast down to Westport. Seventeen people were killed. On 3 February 1931 an even worse earthquake affected a region from Dannevirke to Gisborne and west to Taupo. More than 250 were killed, with the greatest toll being in Napier. In both earthquakes the traditional land-based means of communication were unusable and, amidst the chaos, radio was the medium which spread the news of the disasters and coordinated the response to the emergencies.

All forms of radio were valuable during this time. The first news from Napier came from sets on ships in the harbour and from local amateur radio transmitters. These continued to play a prominent role over the next few days. Broadcasting stations, both company and B stations, spread the news with a speed otherwise unobtainable and into areas that were closed to newspapers because the roads had been destroyed. Radio was also the major means of international transmission of the news. John Ball, via the P&T Department shortwave station, gave a fifteen-minute account of the Napier disaster to 2ME, Sydney; it was broadcast on five of the Australian A-class stations. His account also went via 2ME to W2XAF New York, and then to a coast-to-coast United States network.[36]

Broadcasting stations were pivotal in coordinating the rescue work. Harris placed the resources of the RBC at the disposal of the government. The services of 2YA, especially, were made available to the many organisations involved. For many days the station was on the air from 10am to midnight, hours previously unheard of. No relief was available for the staff, but none was sought. It was radio's first major emergency assignment, and the staff wanted

Wreckage in Napier after the earthquake of 1931. The flagpole on the left was used as a mast for the aerial wire of some of the first radio messages transmitted from the town. ALEXANDER TURNBULL LIBRARY

to be involved. The Red Cross, the Salvation Army, the St John Ambulance Association and other organisations, along with such government agencies as the Railway, Transport and P&T Departments, all needed the station to communicate with the general public and with their own staff. The station broadcast hundreds of personal messages as it made inquiries about missing people and helped reunite families and friends. It also broadcast long lists of the dead, and longer lists of the survivors.

Its advocates were quick to draw attention, sometimes not very tastefully, to the part played by radio. The first available *Radio Record* after the Napier disaster carried a full-page advertisement for one top-of-the-range receiving set which read, 'When Chaos reigned . . . radio came into its own . . . Quite apart from its use in times of emergency, a Gulbransen Champion could be the means of enabling you to secure endless hours of enjoyment.'[37] There was little doubt that, amidst the horrors of the earthquakes, radio had triumphed. Some newspapers, particularly the *Hawera Star*, argued that 'the press was the only reliable medium for the dissemination of news' and held that 'all sorts of rumours were floating about, most of them attributable to radio'.[38] This was a minority view: most people appreciated the services of radio. In the chaos of the Murchison and Napier earthquakes radio was the method of communication that could be relied on even when all else had failed. In New Zealand the disasters were the coming of age for radio.

B STATIONS

The Napier earthquake was the occasion for one of the more unusual broadcasting station inaugurations. The station concerned was 3ZR, Greymouth. The station had been tested and approved by the P&T Department, the call-sign allotted, and Mick Spiers, the founder and leading personality, was awaiting the official opening by the mayor when, on the morning of 3 February 1931, the region was hit by an earthquake. Thinking the centre of the earthquake was close to Greymouth and reasoning that the station could be of some use, Spiers started transmissions shortly before 11am, announcing that the station had come on the air early because of the earthquake, and would bring news of it as soon as possible. This was the start of two days and nights of continual transmissions. Spiers was soon contacted by amateur receiver and transmitter friends, who were in touch with other amateurs in Hawke's Bay, and thus he learnt of the Napier earthquake. Over the long hours of this first West Coast broadcast, Spiers and his colleague, Len Schaef, stayed in touch with the amateur receivers and transmitters. The broadcast, interspersed with record playing, conveyed the unfolding details of the disaster, and the pair were able to relay to listeners details of the fate of friends and relations.[39]

Spiers was a radio technician from Christchurch posted to Greymouth as the representative of his firm. He soon found the reception of New Zealand stations was of very low quality. The Australian stations were the ones listened to locally for usually they could be received clearly once darkness fell. Spiers founded the Greymouth Radio Society, later renamed the West Coast Radio Society, obtained his amateur receiver and transmitter's licence and then his broadcasting operator's certificate, and with the help of other enthusiasts, set to work to build a radio station.

Spiers resided in a house at 5 Cowper Street, where he rented the large front room. This became the station, with the equipment set up on a table. At first the equipment was little more than a wind-up gramophone with an electric pick-up, a double-button carbon microphone and a small amateur operator's transmitter of about 10 watts, which had been converted to cover the broadcast band. The components were obtained from a variety of sources — some parts, such as insulators, were salvaged from wireless sets of wrecked ships, the aerial was a long wire out the window to a cabbage tree and then to a pole at the back of the house, and the earthing system was a pipe driven into the ground.

Len Schaef, Spiers's friend and colleague, opened Greymouth's second station later in 1931. This was 3ZE, the first transmission of which was on 22 July, though its licensed broadcasts did not begin until 16 September. Schaef, a radio enthusiast who had built his first crystal set in 1921 so as to listen to the Morse from the ships approaching Greymouth, had £80 to spend on the complete apparatus for broadcasting. With these funds he set up a station with a power output of 45 watts. The station, licensed to Schaef's Garage Broadcasting Service, was in a corner of a dance hall above Schaef's business premises; and one of the features of 3ZE was its Saturday night broadcasts of the dance music.[40]

3ZE and 3ZR were both allotted the same 820 wavelength and cooperated to share the broadcasting hours so they were not on at the same time. Thus Greymouth had relatively extensive broadcasting hours including, on Mondays, one of the very early breakfast shows. Usually their broadcasting hours were: 3ZE Monday and Wednesday 8 to 10pm, Friday noon to 2pm, Saturday 9.15 to 11pm; 3ZR Monday 7 to 9am, Tuesday and Thursday 6 to 6.30pm and 7 to 10pm, Friday 6 to 10pm, Saturday 3 to 5pm, 6 to 6.30pm and 7 to 9pm, Sunday 3 to 9pm.

The Greymouth stations are indicative of an increase in the number of B stations during the final years of the RBC. These new stations were a result of broadcasting's growth, and of the RBC's decline in political favour. During 1930 a further thirteen stations were licensed, including an experimental shortwave station in Wellington owned by Western Electric (NZ). Thirteen stations in one year was the largest number yet, and indicated a dissatisfaction with the RBC and a mounting optimism that the government might finally favour B stations. Again most of the new stations were in provincial areas which did not have good reception from the YA stations. Three were in Auckland — which brought the total number of stations in that city to six by the end of 1930. The following month they were joined by a seventh — the first of a further nine stations licensed in 1931, including 5ZA in Apia, Western Samoa.

The station that Spiers and Schaef used as a model was 2ZW, Wellington — the most substantial of all the B stations. It was owned by the radio and music firm of Hamilton Nimmo & Sons Ltd. The studio was on the second floor of the Nimmo Building in Willis Street, with the transmitter in the tower and the antenna system surmounting the building. With a 1-kW power output, the station was the most powerful of the B stations, and its transmissions, which were of a higher quality than any other B station, were compared favourably with the local RBC station, 2YA. The 2ZW equipment was designed and manufactured by the local firm of Collier & Beale. The technician in charge at 2ZW during its two and a half year life was W. L. Harrison, who had been involved in radio engineering since 1922 when he was employed in the P&T Department. Harrison was an influential broadcasting engineer in New Zealand from this point and finished his career as director of engineering for the NZBC.[41]

Harrison was one of only two employees at 2ZW. The other — the dominant personality at the station — was L. E. Strachan, the station director, programme organiser and announcer. He was soon known throughout the Wellington region, as he had been known previously in Canterbury, as 'Mr Heigh-ho', after a phrase he regularly employed at the microphone. Strachan was an experienced broadcaster who came to Wellington after being an announcer at 3ZC in Christchurch, a station he had been instrumental in founding. He was relatively experienced in journalism, having worked for both the morning dailies in Christchurch. Importantly for 2ZW, he also had an extensive background in music. He had completed a course in music at Canterbury University College, held Trinity College teacher diplomas for pianoforte and organ, and had been organist and choirmaster at various Christchurch churches, a long-term music critic for Christchurch newspapers, and the first lecturer in music appointed for the Workers' Educational Association courses.

2ZW had comparatively long transmission hours: Sunday 10am to 12.15pm, 6.30 to 10pm; Monday and Tuesday 10am to 2pm, 7 to 11pm; Wednesday 10am to 2pm, 5 to 6pm, 7 to 11pm; Thursday 10am to 2pm, 3 to 4.15pm, 7 to 11pm; Friday 10am to 2pm; Saturday noon to 2pm, and 7pm to midnight. The morning broadcasts were an innovation that had long been requested of the RBC stations by radio dealers, who were unable to demonstrate radio sets until 3pm — the usual starting time for the YA stations. 2ZW's morning transmissions enabled dealers, especially Hamilton Nimmo & Sons, to demonstrate radio sets during normal working hours. The extended hours were also an attempt to outperform the YA stations. In response to the competition, 2YA (but not the other YA stations) extended its transmission hours to begin at 10am. 2ZW had also shown an interest in taking advantage of 2YA's silent day, but the RBC countered this by introducing the recorded American programmes on the silent day.

2ZW had regular outside broadcasts. Towards the end of 1931 it advertised them as follows.

> Permanent land lines link the studio and transmitting plant with the City Town Hall, the Town Hall Concert Chamber, St Paul's (Anglican) Pro-Cathedral, the De-Luxe Theatre, the Majestic Theatre and Lounge, and Messrs Kirkcaldie & Stains Ltd's Tearooms. From the Town Hall are relayed the wrestling contests, and Mr Bernard F. Page's organ recitals. From St Paul's (Anglican) Pro-Cathedral the evening service is relayed each Sunday at 7pm. From the De Luxe Theatre the Wurlitzer Organ is relayed, and occasional excerpts from 'talkie' productions. From the Majestic Lounge the dance orchestra, under Mr Frank Crowther, entertains the devotees of Terpsichore every Thursday from 10pm to 11pm, and on Saturdays from 10pm to Midnight. Every Thursday afternoon, from 3 till 4.15 a song recital is relayed from the tea-rooms of Messrs Kirkcaldie & Stains, Ltd.[42]

While 2ZW was the most technically sophisticated of the B stations, the most influential was 1ZR, Auckland, opened in 1930. In subsequent years, 1ZR introduced a style of broadcasting and a group of broadcasters who, there and later at 1ZB, were to change New Zealand broadcasting. 1ZR was licensed to the firm of Lewis Eady Ltd, which dealt in musical instruments, sheet music, gramophones and records as well as radio. It began selling crystal sets when the first Auckland radio station opened, and further developed its sales of radio parts and more sophisticated sets during the 1920s. The firm was also the box office in Auckland, selling tickets to the various concerts and shows.

Auckland in 1930 was well catered for with stations. The Eady brothers entered a still expanding field, and their station began with comparatively long transmission hours. On six days a week the station opened at 10.30am and continued until 2pm. Transmission resumed at 5pm, ending at around 7.30 to 8pm on Mondays and Wednesdays, 10pm on Tuesdays, Thursdays and Fridays and 11pm on Saturdays. Sunday broadcasts were 9 to 10.30am, 2.30 to 4pm and 6.30 to 9.30pm. As Ray Eady noted, these hours took 'a lot of filling in'. The day opened with a devotion session which generally ran for the first hour. In the late afternoons a children's session was run by the station's 'aunts' and 'uncles'. There was live music, and the firm, as the Auckland booking agents, had considerable success in persuading visiting artists to broadcast. But the station relied largely on the efforts of volunteers: 'We had to encourage anybody who was willing to come along and without charge.' 1ZR, like other stations before it, found this encouraged neither reliability nor quality of performance from its artists. As elsewhere, most of the sessions were talks and recorded music.[43]

BROADCASTING POLICY

The future of B stations and of the RBC increasingly became a topic of parliamentary debate in the final years of the company's contract. Even before the B stations' coordination and request for advertising rights they gained a champion in Clyde Carr, the ex-3YA announcer and Labour MP for Timaru who entered the House in November 1928. Carr rapidly became a prominent broadcasting spokesman for his party, a supporter of the B stations and opponent of his former employer. Carr painted the RBC as a Goliath, unfairly favoured by the department, and able to squash any innovation made by the B stations. In September 1929 Carr asked the government to take over the broadcasting service. He was supported in this by the Dunedin *Star* — one of the first newspapers to advocate this move — which suggested that 'such a splendid organisation as the Post and Telegraph Department could . . . give a very efficient service to listeners and would be in a position to install small relay stations . . . in the smaller towns or where reliable radio reception is not obtainable'. Thus the company's scheme for expansion was supported — but as a scheme to be implemented by the department. It was also at this time that the suggestion was first made in the House, by J. A. Nash, for control by a board: 'It is far better to have the broadcasting stations outside of Government control . . . a Board should be set up to control, in some measure, the programmes broadcast.'[44]

The minister stated, in October 1929, that he was not prepared to make a decision on the future of the RBC. He contended that, with the contract continuing to the end of 1931, he was in no position to indicate what broadcasting policy would be then.[45] As late as 1 October 1930 Donald continued to state that no decision had been reached on the matter.[46] Two days later, he announced during the P&T Department supply debate that the contract with the RBC would not be renewed.[47] On 6 October, perhaps stung into an early announcement by the continued criticism, such as that from Carr that 'it was general knowledge up and down the country that the Post and Telegraph Department was taking definite steps for controlling radio-broadcasting in the Dominion' and 'the House was entitled to know what was in the Minister's mind',[48] Donald gave some details of his future for broadcasting, offered as 'only a tentative suggestion'. The P&T Department 'which was well equipped technically for the control of broadcasting, would undertake the care of the equipment and the output of sound . . . With regard to programmes he had not yet got very far, but he would be prepared to receive suggestions from honourable members and from the outside public.' Donald proposed to increase the power of 1YA, 3YA and 4YA to 1500–2000 watts, and to start nine auxiliary stations, each with 500-watt power output, able to act as relay or independent stations. Although the power output of all stations was greater, the relay scheme was based on that proposed almost two years earlier by the company. Donald's proposal, however, had some major additions. The most notable innovations were the day-to-day control by the P&T Department of all technical matters to do with broadcasting, the separation of programme production from other broadcasting matters with this to be handled by organisations other than the department, and, finally, the acceptance of the B stations as a permanent part of New Zealand broadcasting: 'It was not . . . proposed to do away with Class B stations, for which there was still plenty of room in New Zealand . . . they would be asked to reach a certain standard. The Government wanted to help them to the fullest extent.'[49] Presumably this included the right of the B stations to have an income from broadcasting, though Donald did not state whether the B stations would receive a share of the licence fee or would be allowed to advertise. Much remained uncertain in the proposals, but it was the Australian model — advocated by McNamara, though without McNamara's unequivocal repudiation of the New Zealand B stations — which the minister favoured for future New Zealand broadcasting. It was eleven months before new legislation was brought before the House and in that time the proposal changed considerably.

Press reaction was largely negative. Most critical were those newspapers — led by the *New Zealand Herald*, the *Dominion* and the *Press* — which had long been opposed to government entry into areas they considered the preserve of private enterprise. The traditionally liberal papers, led by the *Auckland Star*, the *Christchurch Times* and the *Evening Star*, which generally supported the United Party, were also unhappy. They were particularly concerned with the costs of the proposal and with the thought of political control of broadcasting. Donald made further statements which were not always consistent, particularly on the matter of costs. The *New Zealand Herald* soon inferred that 'the original announcement was made before it had been sanctioned by the Government, and that consternation has been caused by his undertaking to involve the State in this hazardous experiment'.[50] The *Dominion* doubted whether the minister's scheme would be implemented: 'Much water will flow under the bridge and, incidentally, a general election will be held before January, 1932, when the present Government proposes to take over the radio broadcasting service. Listeners should therefore not worry overmuch at a hypothetical event still far removed in time.'[51] In their business arrangements and in their editorials the newspapers acted as if the postmaster-general's proposal was anything but a foregone conclusion. For example, the discussion and the agreements between the RBC and the Auckland newspapers on the question of payment for the printing of the daily radio programmes were held after the minister's announcement.

Goodfellow and Kent, the company solicitor, met with Donald on 10 November 1930. At this meeting the minister reiterated his scheme and Goodfellow could see no place for himself or Harris in the new plans. Donald 'made it quite clear that the Government intended to take over the stations and in future they would be operated by the P&T Dept . . . the Broadcasting Co. would have an equal opportunity with others of carrying on the broadcasting service, which, of course, would simply be an entertainment proposition, and, as far as I can see, would be of no interest to either of us.'[52] Goodfellow was not interested in involvement in the proposed new order, but he and Harris were yet to be convinced they should give up hope. Not only might there be political changes in 1931 but the RBC directors believed that the Postmaster-General had yet to gain the approval of his Cabinet colleagues for his scheme. By the end of 1930 they were stating as much and were arguing to journalists that the RBC's scheme for broadcasting was the one worthy of support.[53]

After Donald's October 1930 announcement that the RBC's contract would not be renewed, the B stations again requested a change in the ban on advertising. They argued that 'B class stations have always opposed anything in the nature of direct advertising over the air', and asked 'that sponsored programmes shall be permitted with one announcement at the beginning and one announcement at the end of the sponsored programme'.[54] Sponsorship had long been opposed by the anti-advertising-on-radio interests because of its similarity to advertising. The *Radio Record*, for example, characterised it as 'the thin end of the wedge in relation to the institution of advertising on the air in New Zealand'.[55] No rapid decision was given by the minister but most B-station proprietors were confident it would eventually be in their favour.

It was in this atmosphere that 2ZW in Wellington was brought to air. The founders were initially confident that the minister would deliver on his statements that B stations were to have a future. The station was a long time in the preparation, partly because the equipment took time to manufacture, but more because the directors were reluctant to hasten ahead without a definite statement from the minister as to the financial future for the B stations. By May, the 2ZW directors had prepared the most substantial and expensive B station yet, but were still without any acceptable source of income. It was too late to turn back, and the station was officially opened on 20 May 1931 by T. C. A. Hislop in his first official ceremony as mayor of Wellington.

Fortunately for 2ZW and the other B stations, the idea of sponsorship gradually won favour with the postmaster-general. On 10 June 1931, in a speech opening the Wellington Radio Exhibition, he announced the long-awaited assistance to the B stations. Donald distinguished sponsorship from advertising and made it clear that advertising as such was not permitted. What was allowed was sponsored programmes which could carry no more than the name of the sponsor, both before and after the programme. Stations were not to include additional information such as the sponsor's address or products. B station proprietors were less than satisfied — they wanted to be allowed to announce a sponsor's products as well as name. But they were encouraged by this first easing of the ban on advertising that had prevailed since the first broadcasting regulations were issued. They hoped it was an indication of things to come, when the new order for broadcasting was determined.

The B stations took immediate advantage of the new rule. In an article strongly critical of 'The Advertising Menace', the *Radio Record* reported that, within forty-eight hours of Donald's announcement, the advertising agency acting for 'a certain Wellington station', presumably 2ZW, had issued a circular seeking sponsored programmes and setting out a schedule of prices. On 15 June one of the Auckland stations was closed by the P&T Department for exceeding the bounds of permitted announcements.[56] The closure was temporary, but signalled that the department would monitor B stations closely.

Soon after the postmaster-general's October 1930 announcement that the P&T Department would take over the technical but not the programming side of broadcasting, it became apparent that this was a distinction difficult to uphold in practice. The *Evening Post*'s radio writer, 'Grid-Bias', probably the RBC's most consistent newspaper opponent, gave as illustrations of bad technical practices at 2YA — which P&T Department control would obviate — such 'glaring examples [as] voices that "blast" habitually, bad microphone placing, poor orchestral balance, false fluctuations in volume, uncompleted gramophone productions and objectionable "background" noises, none of which should be tolerated'.[57] As correspondents were quick to note, if the P&T Department were to have jurisdiction over such matters, it would be running the station and would be the broadcaster. Just what the minister and the department considered was the scope of the department's role was never clear. During 1931 this option appealed less and less to the minister.

A parliamentary committee to consider broadcasting met in June and July 1931. In a submission to the committee, the RBC defended its record and argued for a continuation of the existing system of private enterprise subject to government regulation, 'as experience has proved this is the best system for New Zealand'. It argued against the government's taking complete control because 'the Company has acquired a knowledge of the tastes and requirements of the public, which it is felt would take any Government Department, however zealous, many years to attain. On the revenue side, the Company is of the opinion that there would be inevitably a tendency on the part of artists and the proprietors of musical copyright and recordings to exploit the Government.' It argued against Donald's suggested system of dual control, with the government 'operating the technical side on its own account and assigning the programme side to private enterprise. Enquiries made where this system is in force show that it does not work satisfactorily. It is believed that such a system is impracticable, for the reason that the technical and programme sides are interdependent and inseparable.' The other possible method of control was that by a board. The RBC saw this as modelled on the BBC, and argued against it because a 'Board had no personal monetary responsibility'. This might be appropriate in England, where the BBC had no 'anxiety as to finance, the revenue available being virtually unlimited . . . [whereas] in New Zealand prudent business management is essential'.[58] This was scarcely a submission guaranteed to sway a government already inclined to change the existing order. The RBC's efforts in 1931 to secure its continuation did not inspire confidence in the company.

The company's major proposal for the future of broadcasting was issued in August 1931. It urged the formation of a public company with a capital of £150,000. This was to be divided into three parts, including £50,000 of A ordinary shares, some of which would represent the value of the RBC's assets, and the rest would be underwritten by the RBC directors. A further £50,000 of B ordinary shares would be offered to radio licence holders, with no licensee to hold more than 100 shares. A final £50,000 of C preference shares, which carried no voting rights, would be issued if and when required. The board would consist of eight directors: A and B shareholders were each to elect four, and one of the A shareholders' elected directors would become the chairman. The proposal included expenditure of approximately £100,000, divided equally between the purchase and upgrading of the four YA stations and the establishment of the relay station scheme.

Press reaction to the proposal was in the main favourable. Some saw the record of the RBC as good; and most saw the continuation of private control as desirable. The *Evening Star* of Dunedin gave the majority press view, 'There may be room for criticism, in detail, of the scheme which is now proposed, but it should be infinitely preferable to any which would make for complete Government control.'[59] However the postmaster-general was less impressed: he felt the proposal was designed to camouflage a continuation of control by the existing RBC directors. He described the listeners' shareholding as a gesture which would in no way lessen the overall control, via the chairman's casting vote, of the A shareholders — the existing RBC directors.[60]

The major press criticism of the proposal came from the *Evening Post*, which rebuked the other newspapers for seeing the issue as one between private enterprise and government control. It argued that corporation control was required. During the second half of August 1931 the *Evening Post* ran various articles and editorials attacking the RBC's record and proposals for the future, and advocating a new board control similar to that of the BBC. It also printed a long, anonymous article, which Harris suspected had been written by McNamara, the P&T Department secretary.

> 'Grid Bias', who prepares the *Evening Post* radio notes that are so hostile to the Company, is a sub-editor on the *Post*, by the name of Hardcastle, who lives next door to McNamara at Lower Hutt. There is little doubt, therefore, as to the origin of the recently contributed article which contains so much inside information respecting the Company's licence and suchlike.[61]

The *Evening Post*'s advocacy of corporation control was reflective of a change in the attitude of the postmaster-general that took place in 1931. In the face of the opposition to department control of broadcasting from both the newspapers and the Reform Party, with which the Forbes government was slowly moving to form a coalition, Donald abandoned the idea. Just what was to replace it was slow to emerge.

By mid 1931 R. H. Nimmo of 2ZW felt there was a strong chance that a contract for broadcasting would go to AWA. He was so opposed to this that he declared he would fight hard for the RBC to get a renewal, rather than see a contract go to AWA.[62] Nimmo and the other B-station proprietors would have preferred something similar to the American system, which offered full acceptance of private stations and their right to advertise. They acknowledged that this was politically unrealistic in New Zealand, and accepted the idea of corporation control as a compromise. In this they were in accord with the radio dealers. Many B-station owners were, of course, radio dealers. In July a deputation asked that broadcasting be placed under the control of a corporation of not more than five members, appointed by the government but entirely removed from political control. The members should include representatives of listeners, the trade and the government.[63] Other groups made similar statements. In Wellington, for example, a combined meeting of the local branch of the NZ Radio Listeners' League and the local section of the NZ Electrical Federation stated that, 'The best and only satisfactory system of control of broadcasting in New Zealand would be by means of an independent board of control to be set up by the Government on the same lines as those governing the British Broadcasting Corporation in England, which is entirely free of political control'.[64]

Corporation control was being advocated — but not everyone envisaged the same form of corporation. Some, such as the Listeners' League, wanted a New Zealand version of the BBC. Others, such as Nimmo, wanted a new organisation able to apportion the licence fees among YA and private stations: 'It should be possible for either the proposed board of control or whatever company was operating the service to subsidise selected B class stations in the main centres and in the more important provincial towns from the listeners' funds and so provide that choice of programmes which was definitely required by listeners.'[65]

The first public statement of the postmaster-general's change from his previous announcement came in July, when he told the secretary of the Chamber of Commerce, 'Possible methods of control which are receiving the consideration of Cabinet at the present time all definitely preclude any extension of the duties of the Post and Telegraph Department in the broadcasting field.'[66] From this point matters moved swiftly. The Cabinet endorsed the recommendations of the Special Committee on Radio Broadcasting that future control should be vested in a Radio Board modelled upon the BBC. There was an understanding that some adaptation to the needs in New Zealand would be required.[67] The Broadcasting Bill giving effect to the changes was introduced and given its first reading on 3 September.

Even at this point the end was not entirely certain for the RBC. In mid September a coalition was formed between the United and Reform parties and a new government was announced. Forbes remained as prime minister, but the Cabinet was made up of five members from each party, and the new postmaster-general was Adam Hamilton from the Reform Party. Donald was not included in the new Cabinet. He did not stand in the next election, which was held in early December, and retired from politics. With a private enterprise advocate as minister, the RBC again advanced its scheme for a new public company.[68] In October, Hamilton dashed these hopes. Unlike his predecessor, he at least notified the company before making the decision public. 'I took the opportunity yesterday of discussing the matter provisionally in Cabinet, and can see that the Government can come to only one decision. We all seem to be seized with the necessity for public ownership, and the matter is now left in my hands to finalise.'[69] Hamilton thus led the bill through its further readings and the RBC's contract ended on 31 December 1931.

ARBITRATION AND SETTLEMENT

The final act of the RBC was to come to a settlement with the government over its assets. Goodfellow was initially hopeful of an amicable settlement for 'it appears to me that when the matter was discussed by Cabinet the Reform Party agreed to the Government taking over the Plant on condition that a liberal settlement was made with the existing Company'.[70] Understandably the RBC directors were eager to receive as generous a settlement as possible, not least because they considered that this was how they would receive some profit from an ongoing venture from which they had been excluded. It soon became apparent that they would need to work hard to obtain a settlement to their liking. Once again they saw the P&T Department secretary as the problem: 'It is quite obvious from the information you have supplied that McNamara is out to cause trouble.'[71]

Harris assumed the department would carry out the minister's suggestion and allow RBC staff to accompany the department's valuers through the stations. This procedure was not followed. The department engaged Milne & Choice to value the furniture and the Bristol Piano Company to value the pianos; a team of its own officers attended to the plant and other technical matters. Goodfellow moved quickly to obtain the RBC's own favourable valuations and showed a prescient understanding of the final outcome. 'I think we should frankly state the position to the Valuers; let them know that the Government have obtained valuations without consulting us, and, in order to counter this, we desire our Valuers to place their highest possible value on the goods. No doubt the ultimate procedure will be to take both valuations and agree on a figure which will be probably about mid-way between.'[72]

Naturally the two valuations were far apart. The P&T Department's valuation of the assets of the RBC was £27,353. The company's own valuation was £85,812. The matter went to arbitration as had been provided for in the original agreement with the RBC. A Supreme Court judge, Mr Justice Blair, was appointed arbitrator. He heard evidence and personally inspected the four stations. His award gave the company £58,646 6s 2d. This sum was paid by the government. It was treated as an advance to the newly instituted Broadcasting Board and was to be repaid, with interest at 5 per cent, over ten years.

THE RBC IN RETROSPECT

Contemporary commentators always had difficulty determining the financial state of the RBC. As was their right, the directors let out little information — and it was not until August 1931, when they published their proposal for the future of broadcasting, that a general understanding of the finances of the company was made available. That proposal included a tabulated revenue account for the five years to 31 July 1931. The RBC was set up with a nominal capital of £20,000. Of this £17,000 was subscribed, mostly by the directors, and £13,896 paid up. The bank account was guaranteed by the directors to the extent of £15,000 and the government advanced £15,000 on mortgage at 6 per cent to build 2YA. The directors stated they had raised a total of £47,000.[73] All this was apart from the relationship with A. R. Harris Co Ltd, which provided the 3YA and head office accommodation and received rent, and which also built the stations and provided equipment for a 10 per cent commission, 5 per cent for the building of 2YA.

After providing for a 7.5 per cent dividend, the company sustained losses in the first four years. No directors' fees were paid in the first year and part fees only in the second and third year. Only in the fifth year did the company come into profit and even then it was a modest one of £188 3s 5d. During the final seventeen months of the RBC the business results were much better, for the company was marking time rather than expanding its operations and the licence fee income was rising with an average of close to 900 new licences being taken out each month. But the financial profit for the shareholders of the RBC was essentially in the settlement made when the company ended. As Goodfellow noted, 'The only way we came out of it well was we got good valuations.'[74]

It is difficult to assess the company's pattern of expenditure. The tabulated revenue account shows the expenditure rising from £6015 in the year ending 31 July 1926 to £56,214 in the year ending 31 July 1930. It is all in a general expenses category which includes, in the company's own listing, 'salaries and wages, talent, rent and rates, light, heat and power, printing and stationery, travelling, telephones, telegrams, and postages, advertising, copyright, insurance, royalties and taxes, repairs and maintenance, discount, interest and exchange, legal and audit, miscellaneous trade expenses'. As the *Evening Post* noted editorially, 'It is most unfortunate for the company's case that the most important item "general expenses", which keeps pace exactly with the growing revenue is not shown in any detail. It is precisely this figure which is puzzling both the friends and the opponents of the company — the former because they cannot give and the latter because they cannot get, the facts.'[75] The articles contributed to the *Evening Post*, which Harris believed were written with inside information from the P&T Department, contended that the payments to performers never exceeded one fourth of the total income and were possibly considerably less, that the total value of the equipment was one sixth of total revenue, and that copyright and other expenses connected with programmes brought these items up to about half the total revenue: 'The question then remains whether the balance is or is not an extravagant amount for the running expenses of the service and the overheads.'[76]

Of more significance than the profit made and the money received and spent by the RBC was the nature of the broadcasting service it founded, operated for six and a half years and handed on to its successor. In this the key person was Harris. Goodfellow, the chairman of directors, was always informed of affairs and was himself greatly influential. When he deemed necessary, as at the time of the settlement arrangements in December 1931, he was prepared to exert his authority and take charge. But the RBC was not Goodfellow's major business activity, and it is clear that he had considerable confidence in his fellow director. He willingly left the running of the RBC to Harris. Arnaud McKellar, the third director, appointed in 1927, hardly figures in the company records. He invested in the company — and was appointed a director on that account — but had little to do with the day-to-day activities of the RBC.

Harris was at times criticised as unsuited to run a broadcasting service. W. H. McIntyre, speaking in the Legislative Council, held that 'it would be difficult to find a man less qualified by nature and training to hold the important position of director of broadcasting in this Dominion. To be a successful director of broadcasting one should be a man of exceptional education and culture; a man not inclined to take the responsibility on his own shoulders.'[77] This was a criticism of Harris which misjudged both the man and the circumstances when he began the RBC. He began with a government contract to develop four stations under a central control. There was little understanding on anyone's part as to what broadcasting entailed. That a director should be 'a man of exceptional education and culture' was an understanding that appeared only after broadcasting had already become a significant part of New Zealand society. Goodfellow and Harris were selected by the Reform Government in 1925 because they were politically acceptable to that government, because they were capable of developing the stations as required and because there was little other choice. Goodfellow chose Harris as his partner because Harris understood the new and arcane topic of radio, convinced Goodfellow of its potential, and had the engineering and administrative ability to head its establishment. These abilities were regarded as more important in 1925 than cultural leadership.

As broadcasting developed from 1925 to 1931, Harris did display cultural leadership, though perhaps not of the sort desired by McIntyre. He recruited many people who were to continue careers in broadcasting long beyond the time of the RBC. He was very much the person in charge but, more than many of his successor heads of broadcasting, he was prepared to let his station directors and staff develop programmes as they saw fit. They and he laid the foundations of New Zealand broadcasting. As well as building the four major stations, they started the various programme categories and established a nationwide coordination of programming which long characterised New Zealand broadcasting. With this programme development and coordination, Harris encouraged popular acceptance of radio. His major concern was the spread of radio ownership which meant, of course, more revenue for the RBC. To this end he was concerned that radio should be a medium for popular entertainment. The development of sports broadcasting, the introduction of radio serials and the call for an emphasis on popular music were all aspects of Harris's shrewd understanding of what would be appreciated by the New Zealand public and of his acceptance of that taste. In no way, however, should this be seen as a resolute pushing of popular entertainment at the expense of the classics. For Harris was the man who established the musical and dramatic committees at each of the four stations, and allowed these representatives of local cultural groups to make genuine contributions to programme compilation. Harris accepted rather than challenged the social position of New Zealand's existing cultural associations. But he used his leadership to restrain the use of radio as a medium for cultural elitism.

The inadequacies of the RBC stemmed from the 1925 regulations, the nature of the company's contract and the uncertainties about its future, rather than from any flaw in the character of its principals. The regulations, particularly the ban on controversy, barred the RBC from any serious attempt at making radio a medium for news and public discussion. Both Goodfellow and Harris wanted this — they saw radio as a means of relieving the isolation of rural areas. Talks and news broadcasts were developed, and rural New Zealand began, during the years of the RBC, a change which drew it into contact with the cities, and which much advanced New Zealand's national cohesion. But the news and talks programmes were cautious and unchallenging. The radio regulations, as well as the pressure to have the contract renewed, forced the RBC directors into a timidity which weakened broadcasting.

The brevity of the five-year term of the contract made investors wary, and the venture was always undercapitalised. Goodfellow regretted that the company was unable to run at a greater loss for the first two years, in order to speed public acceptance of broadcasting.[78] Not until 1928 was the income from licence fees sufficient to allow significant development. It was at that time that the Reform Party lost power and the continuation of broadcasting as a private enterprise venture became doubtful. The last three years of the RBC were frustrating ones for the company and wasted ones for New Zealand broadcasting. While considerable programme development was made during these years, broadcasting was marking time with respect to station development and the expansion of coverage. The scheme for an extension of radio into the provinces via a series of relay stations which could also operate independently was a good plan, but the RBC had neither the time to build the stations, nor the cooperation of the P&T Department.

By the time the RBC's contract finished the organisation was employing some 114 staff, including the permanently retained musicians. From the initial total of 12 hours' broadcasting per week, 2YA was broadcasting 77 hours, and each of the other stations 41. The RBC regularly pointed out that these hours were considerably in excess of its contract requirements, but was slower to acknowledge that there was an understanding that the original number of broadcasting hours would be increased as revenue grew and circumstances warranted. The company was regularly exhorted by listeners, newspaper columnists, radio dealers and the P&T Department to increase its broadcasting time. While the total number of hours broadcast towards the end of the contract was well in excess of the original requirement, the increase was no more than circumstances warranted, and was generally made under pressure from the P&T Department or in response to competition from B stations.

The reluctance to increase programmes and extend transmission time came from the desire to safeguard profits by curtailing expenditure. This was a consequence of broadcasting's private enterprise organisation in an uncertain future. Goodfellow and Harris were concerned with developing broadcasting profitably within five years. Although initially confident of at least one five-year renewal of their contract, they never had the freedom to plan in the long term. They soon realised that one five-year term was all they would get. They were unable to implement the relay scheme — their plan to extend broadcasting — and they learnt that their worst fears — that, in Goodfellow's words, they would do 'the pioneering donkey work' and then have the concern taken over — were correct. In the circumstances, their reluctance to push ahead with the further development of broadcasting and their concentration on the immediate balance sheet were understandable. Neither Goodfellow and Harris nor private enterprise in general can be regarded as having had a free hand in the initial development and running of broadcasting — an innovation that needed to be planned for a much longer term than five years.

Broadcasting was handed over to private enterprise when it was a novel and seemingly trivial pursuit. The very growth and success of broadcasting and its emergence as a significant public institution led in New Zealand to its takeover by another, more public type of organisation. It remained to be seen what would be the approach of the new organisation, the Broadcasting Board.

6
New Zealand Broadcasting Board

THE BROADCASTING ACT 1931

The problem addressed by the New Zealand government in 1931 was the proper administrative structure for broadcasting. New Zealand had previously accepted the idea that all who could meet the regulations should be allowed to broadcast: the American system minus the advertising. New Zealand allowed private stations, but made it impossible for them to take the form adopted elsewhere. Another possibility was that all stations should be owned and operated by the government. This was the view of the Labour Party, and had been that of the United Party wing of the Coalition Government. It was opposed by United's coalition partner, the Reform Party, which favoured private enterprise. Reform, however, was not willing to allow the private stations to advertise and thus move into the financial preserve of the newspapers. The two other models available to the New Zealand government were provided by Australia and Britain. Australia had a dual system, with both government and private stations — the former dependent on licence fees, the latter on advertising. The British model was one of corporation control, with all stations owned and operated by one body that was neither of the government nor private, and with the revenue coming from licence fees.

The postmaster-general felt that the British example was the model for New Zealand: 'We propose to follow very similar lines to those adopted by Britain.'[1] The Broadcasting Act required a three-person board, the New Zealand Broadcasting Board (NZBB) to carry on the broadcasting service. The board acquired the RBC's assets on terms and conditions set by the minister, and received the portion of the licence fees previously paid to the RBC. The board was required to give the minister an annual report and statement of accounts, and could not borrow money or mortgage any of its property without his written consent. An advisory council with five North and three South Island members was to advise the board. Listener organisations had the opportunity to nominate council members. The board also had the power to appoint advisory committees as it saw fit. Board, council and committee members and employees of the board were not regarded as members of the public service.

It remained to be seen how the government and the board would act under the new legislation. As in Britain, the object was to have an organisation under public ownership and control but independent of political direction from the government of the day. In both countries there were always doubts as to how far the government wished to allow political independence. The critical relationships were those between the government and the board, and between the board and the executive officer, known in New Zealand as the general manager and in Britain as the director-general. In both countries the relationship between the government and the board was one of 'calculated impression'. In Britain the BBC experienced 'liberty on parole', the terms of which could be (and on various occasions were) altered immediately and without notice by the government.[2]

In the New Zealand legislation, as in Britain, the board was clearly supreme over the executive officer. But in Britain the director-general, Sir John Reith, had been able to stamp his authority on the BBC and on his board of governors beyond the apparent limit of his legislated power. Particularly since 1930 when John Whitley became chairman of governors, there had developed an understanding that the board had general and not particular responsibilities, while *de facto* authority belonged to the director-general.[3] Again in New Zealand it remained to be seen how the relationship would develop.

Unlike the BBC, the NZBB co-existed with B stations. Their continuing presence attested to the government's reluctance to move against the small but popular stations. However, their future was not made clear in the legislation or in parliamentary debate. Hamilton stated that 'the Government considers that broadcasting must of necessity be of a monopolistic nature, and in view of the fact that the State collects the revenue the service is one which does not lend itself to competition'.[4] This seemed to imply the demise of the B stations but, when asked if this were so, he stated, 'That will be a matter for the Board to determine.'[5] The ex-postmaster-general, J. B. Donald, speaking on the government's side in the debate, saw 'no reason why B stations should be interfered with very much. I have no doubt that the Board will make arrangements whereby a number of B class stations will continue their service.'[6] In the Legislative Council, the understanding advanced by Sir James Parr, introducing the bill for the government, was that the number of B stations should be reduced to twelve. Again, he felt the particular attitude to be adopted towards the B stations was a matter for the new board.[7] Savage was equally uncertain. He made it clear that his Labour Party, given the opportunity, would take over broadcasting but it was not obvious whether this included the B stations for he also stated that he did not think it would be right to legislate them out of existence.[8]

The B-station proprietors were not enthusiastic about the legislation or the views expressed in the debate. They had little reason for optimism: the minister had stated that broadcasting should be a monopoly, presumably an NZBB monopoly; and there was an apparent government understanding that there were too many B stations and that some two thirds should be closed. Furthermore, there was a lack of attention to the B stations in the legislation. Their future was to be decided by the new board which was set up as the B stations' competitor.

Board members were appointed by the government. Hamilton accepted that this was not ideal, but recognised the difficulty of forming any organisation that represented all members of the large broadcasting audience: 'There appears to be no effective method of electing a Board at present.'[9] Savage, leading the debate for the Labour Party, suggested the problem was the lack of government initiative: 'If the listeners had any encouragement at all they would soon bring an organisation into existence that would satisfy the Minister.'[10] Robert McKeen, member for Wellington South, argued, 'it may have been impossible for the listeners of the Dominion to form an organisation in the past, owing to the fact that most of the sixty-seven thousand now enrolled have taken out their licenses within the last two years. I think, however, that the time has now arrived when a listeners' organisation ought to be formed.'[11] McKeen and the other members could not foresee that the annual increase of the previous two years would be exceeded throughout the years of the NZBB.

Savage argued that listeners should have at least one representative on the board. However, more importantly, Savage and his party held that New Zealand broadcasting should be under the control of the government, and of the P&T Department in particular. As such the bill was 'a step in the right direction'. Savage would have preferred the government to take the extra step: 'If [the Minister] or his officers of the Post and Telegraph Department had taken control of the whole thing I do not think there would have been any dispute about advisory committees, Boards, or anything of the sort. The thing would have automatically worked without any difficulty.'[12]

THE BROADCASTING BOARD

The three board members continued in office to 1935, when board membership increased to seven. They were appointed on 18 December 1931 and took control on 1 January 1932. The chairman was H. D. Vickery, a public accountant from Wellington. The other members were G. R. Hutchinson, a company director from Auckland, and L. R. C. Macfarlane, a farmer from Culverden. Vickery was connected with a number of companies and was prominent in the New Zealand and Wellington Chambers of Commerce. He had acted for the government on the committee for the revision of the Companies Act and was a member of the Earthquake

From the left, G. R. Hutchinson, H. D. Vickery and L. R. C. Macfarlane, the members of the first Broadcasting Board. ALEXANDER TURNBULL LIBRARY

Adjustment Court. Hutchinson was managing director of Hutchinsons (Wholesale) Limited, chairman of the Auckland Harbour Board, and had a long record in Auckland local body politics. Macfarlane was president of the Royal Agricultural Society and of the Canterbury Agricultural and Pastoral Association, and had a long history of involvement in farmers' organisations. At the 1928 election he had been the Reform Party candidate in the Hurunui electorate against Forbes, the prime minister.

All three were capable and successful businessmen, but none had any other obvious qualifications for appointment to a broadcasting organisation. None had broadcasting experience; and Macfarlane had not owned a radio prior to his appointment. Rectifying this omission, he commented, 'I suppose I had better own one of these blessed things.'[13] Two had some contact with educational and cultural areas. Vickery had for many years acted as an examiner in accountancy subjects for the University of New Zealand. He was also an active member of a church choir and maintained an interest in Wellington musical life. Macfarlane was an avid reader and a writer of comic and serious verse, particularly on his great interests, horses and the turf. But clearly the board members had not been chosen for their educational or cultural expertise. The *Radio Record* argued that the government, realising that 'broadcasting was a business purely and simply', had appointed 'a Board of business men to control it'.[14] As Mackay notes, 'The appointments failed to draw any outward signs of enthusiasm from parliamentarians, the press, or listener organisations.'[15]

Vickery's first act as chairman was to arrange for the appointment of E. C.

E. C. Hands, general manager for NZBB and then business manager for the NBS. NBS

(Charles) Hands as acting general manager. Previously principal of the Telegraph Division at the General Post Office, Hands held the position of general manager throughout the years of the Broadcasting Board. His appointment indicated that the board was keen on maintaining a close relationship with the P&T Department.

The chairman was paid £650 per year, and the other members £350 per year. The payments to the board members were not designed to encourage them to give full attention to broadcasting. As Sir James Parr noted, 'It is not expected that it will be whole-time work for these gentlemen.'[16] Hands, on his appointment as general manager, was given a salary of £850. When the board was disbanded in 1936 his annual salary was £1500.[17]

The board was also keen to maintain staff continuity, and invited all employees of the RBC to apply for reappointment. The great majority did so. In this first major restructuring of broadcasting a pattern appeared: the comings and goings were at the very top level but otherwise there was little change in the personnel involved in broadcasting. The one immediate major change was that the head office moved to 2YA's premises in Wellington.

ADVISORY COUNCIL

All the members appointed to the Advisory Council served continuously until it was abolished at the time of the 1935 expansion of the board. The eight were A. B. Chappell, Auckland, J. S. Anchor, Hamilton, E. T. Davis, New Plymouth, W. A. Waters, Palmerston North, J. H. Owen, Wellington, C. R. Russell, Christchurch, H. Booth, Dunedin and A. W. Jones, Invercargill. The radio clubs and listeners' leagues forwarded nominations to the minister, and notice was taken of these. Waters was president of the Manawatu Radio Society, and Davis and Jones of the Taranaki and Southland counterparts respectively. Russell was the nominee of the Radio Society of Christchurch, Owen the president of the Wellington Listeners' League and Booth the founder and mainstay of the Listeners' League of New Zealand. Chappell was chairman of the 1YA Musical and Dramatic Committee. Anchor owned 1ZH, the Hamilton B station. The appointments were made with the required geographical ratio in mind, and also maintained a balance between main city and small centre members.

The Advisory Council was not a success as an independent source of advice to the board. Its resolutions were consistently in favour of the board's policies, but it was distrusted by the board, which considered some of its members too supportive of the B stations.[18] The council could meet only when requested to do so by the board who kept the meetings to three a year — a number which it thought would 'sound a little better' in the NZBB report to Parliament.[19] Nor was the council representative of listeners — one member later described the council as 'nothing but a sop to avoid the fulfilment of the Minister's promise of an elective Board'.[20] With the large increase in listeners in the 1930s, the difficulties of arranging any adequate election of listeners' representatives also grew. The 1935 disbanding of the council and increase of the board membership was in part an attempt to deal with the matter of listener representation.

THE RBC COMMITTEES

The Advisory Council legislated for in 1931 was not the board's only source of committee advice. It also inherited the various local committees set up by the RBC. The board's appreciation of these declined rapidly and, with the exception of the Church Service Committees, all were dissolved by the end of 1932. The problems were twofold: the board considered the RBC had handed over to the committees the responsibility and power for programme compilation which should reside in the broadcasting organisation itself; and the committees were regarded as having particular concerns rather than an understanding and acceptance of broadcasting's wider interests.

The board's problems were greatest with the music and dramatic committees, concerned with a majority of broadcasting time. Hands saw them as mainly interested in getting broadcasting time and payment for the organisations they represented. Hands had a point, but this was an unfortunate judgement. The music and dramatic committees were an administrative expression of the practice in previous years, when broadcasting was largely reliant on local performers who were expected to broadcast voluntarily. The need for the music and dramatic committees disappeared with the increasing emphasis on the broadcasting of records. In the new order, the difficulty the music and dramatic committees posed for the NZBB was their insistence that broadcasting should be a vehicle for highbrow entertainment and the cultural elevation of its audience. The differences in attitudes are exemplified in the following extracts from the transcript of the meeting of the 4YA Music and Dramatic Committee held on 3 June 1932:

A. F. O'Donoghue, Station Director, 4YA: 'There have been some complaints lately about the concerts being too "highbrow" . . .'

Chairman, J. Crossley Clitheroe, Dunedin Training College Choir: '. . . Would someone put this in the form of a motion — that we hope the Broadcasting Board will keep the programmes at a high level and not be swayed by letters that appear in the papers from people with an axe to grind, or representing but a few people.'

A. Cook, Dunedin Burns Club: '4YA should endeavour to keep up its high standard, particularly in gramophone records. It has a big difficulty to combat; take the listeners-in to the B stations. Those people listen in day after day to the records the B stations put out, and they are being educated to very inferior stuff. They naturally cannot appreciate 4YA.'

J. T. Leech, Society of Professional Musicians: 'Still, but for Wednesday night, 4YA puts on practically the same stuff.'

Cook: 'Oh, no. 4YA has a higher standard altogether than the other stations. Their library is of the better class of music. They never hear some of the records you hear from the B stations . . . I don't think as many people appreciate the B station type of music as we think. I hear comments from all sides that it is very low grade.'

O'Donoghue: 'Well, with regard to foxtrots and jazz records — we give them an hour and a half on Wednesdays and a full hour on Saturday nights. They are not deprived of it altogether!'

Chairman: 'I was going to mention about the Dance programme on Wednesday nights — last Wednesday we had a recital on Wagner, and directly it was finished, there was a foxtrot announced and played. It was the most rapid descent from the sublime to the ridiculous. I don't think the dance programme should follow the classical programme. It is too incongruous. The change took place in about thirty seconds.'

O'Donoghue: 'The reason for that is, of course, that we used to have two hours on Wednesday nights of what would be called heavy music — classical and operatic music; of course, there were complaints, so we shortened it to an hour and a half of the classical programme and gave them the dance programme to follow.'

Chairman: 'I think far too much prominence is given to those people who like jazz — the people who don't count.'

O'Donoghue: 'As far as we are concerned, every listener counts. Every man is entitled to some consideration and we have to try and please everybody. There is the man who wants his wireless entertainment as light as he can get it and the man who will appreciate good stuff. We have to consider the men away up in the country, perhaps shepherds and drovers away by themselves — we have to try and please them; perhaps they will appreciate good stuff, but it mustn't be above their heads. Our programmes are made up along lines that try to please everybody, or, at least, the majority . . .'

Chairman: 'Well, I hope the Board will not be swayed by those letters into altering its policy.'

O'Donoghue: 'I do not think it will. After all, besides those few letters in the Press, there are hundreds of letters going to Head Office expressing satisfaction with the service.'[21]

In Hands's view, the committees were denying the majority of listeners their choice of programmes. He responded by ignoring the committees' advice. After the Dunedin committee's June meeting, he wrote to the 4YA station director and informed him, 'We tend to drive many listeners to the B stations, and at the present juncture, for policy reasons, this is most undesirable. Unobtrusively you should lighten up the programmes as much as possible, without hinting of such an intention . . . Needless to say this letter is for your information only.' This instruction accompanied another letter from Hands to be read at the next meeting of the Musical and Dramatic Advisory Committee. The letter informed the committee that it could 'rest assured that no noteworthy alterations will be made without the fullest investigation and consideration'.[22] The conflict was with all the musical and dramatic committees. Hands's attitude to the 3YA committee's suggestions was that, 'If we should broadcast the classical programmes recommended by the committee we should not have a dozen listeners by the time we were half way through.'[23] Hutchinson, the Auckland board member, also had a low opinion of his local committee: 'I received yesterday', he wrote to Hands, 'a copy of the verbatim report of the meeting of the 1YA Musical and Dramatic Advisory Committee . . . I fail to grasp the angle from which these people direct their thoughts and energies. It is quite difficult to read through the pages of cheap sarcasm levelled at all and sundry.'[24]

Matters came to a head in October after the board rejected the 2YA Children's Advisory Committee's suggestion that the time devoted to the children's sessions be extended. In November the board resolved 'that the Advisory Council be informed that in the Board's view the Musical

Committees serve no useful purpose, also that there is no necessity for the Children's Committees.'[25] At its next meeting the Advisory Council followed suit, and the dissolution of the advisory committees, except for the church committees, was approved by the board in December.[26]

The committees had been appointed by the RBC with two aims: to offload some of the problems associated with programme compilation, and to engage the local community in the affairs of their station. The NZBB decided that it should take responsibility for its own programming; and it did not see the advisory committees as necessarily representative of the local communities. It regarded the cultural allegiances of the groups represented on the committees, particularly the musical and dramatic committees, as being in opposition to the preferences of the majority of listeners. The disbanding of the committees was further evidence of the difficulties in appointing listener representatives, a denial of any contribution from the listening audience, and a significant step in the centralisation of control of broadcasting.

RADIO COVERAGE

One of the Broadcasting Board's major tasks was to attend to radio coverage in New Zealand. Its first action in this matter was the appointment of a Radio Coverage Commission. Arthur Gibbs, by then chief telegraph engineer for the P&T Department, and Dr M. A. F. Barnett of the DSIR were chosen as commissioners. Both had extensive experience — Gibbs in the department since the first Wellington broadcasts, and Barnett from his graduate student days under Professor Jack, at the time of the first Otago broadcasts. The commission's report was received by the board in July 1932. Gibbs, who retired from the P&T Department at this point, was retained as an advisory engineer to the board.

Gibbs and Barnett's comprehensive report was produced after visiting most New Zealand cities and towns, and all the YA and B radio stations.[27] Their report covered three areas: the problem of radio interference, the standard of equipment at the existing YA stations and the provision of additional stations.

Radio interference is a topic that requires some explanation. The reception of some of the first stations was over so large an area that overly optimistic expectations had been held of the range of the YA stations. The reason for the long-distance reception of the first stations was the absence at that time of man-made interference. The later stations, in spite of their greater power, were often unable to match their earlier counterparts. The problem is well illustrated by an anecdote related by a P&T Department radio engineer, Tom Clarkson. In 1927, during the visit of the Duke and Duchess of York, the royal couple spent one night at Kowhai Flat, near Taupo. That night 1YA broadcast a special programme for the duke and duchess, and Clarkson was delegated to arrange the radio reception. In spite of his best efforts the static-ridden special programme was for his ears only, for the reception quality was such that he could not, in good conscience, relay it over the loudspeakers. The problem was the lighting plant installed for the comfort of the royal couple and large entourage. In the afternoon, with the lights off, reception of the usual afternoon gramophone programme was excellent, 'and the Duchess was good enough to say that she liked the pieces including "Bye Bye Blackbird"'.[28] Electric power was a two-edged sword. Its reticulation through New Zealand was the major event leading to the rapid growth of radio ownership, but the interference caused by a whole range of electrical apparatus, from high-tension lines to household appliances, took years to understand and counter.

Gibbs and Barnett saw radio interference as 'almost universal'. Increased strength for the YA stations was one answer. Their other suggestion was for collaboration between the NZBB, the P&T Department and the Public Works Department to deal with interference. At first there was little that could be done, unless the owner of the offending apparatus was willing to cooperate. The government took authority with the Post & Telegraph Amendment Act 1933, though even while introducing this bill the postmaster-general made it clear the new authority would not be used 'too drastically. Although radio reception may be an important social service, it could not be considered as important as the operation of an X-ray plant, which is perhaps one of the greatest causes of interference.'[29] However, it was after the passage of this bill that real reductions in the level of interference were made.

The old spark transmitters produced some of the worst radio interference. The P&T Department had changed all its transmitters to valve models soon after broadcasting began, but many spark transmitters remained and were owned particularly by medical practitioners, who used the high-frequency energy for the treatment known as diathermy. One spark transmitter, known in Christchurch in the first years of the 1930s as 'the roarer', ruined radio reception for thousands of listeners. It was traced to the surgery of Dr Henry Thacker, ex-mayor of Christchurch, and eliminated after the P&T Department purchased a new valve-type diathermy machine for Thacker. The spark transmitter was removed to the P&T Department Radio Section workshop in Wellington, where it gathered dust until it was used for a different type of interference — the 1935 jamming of Colin Scrimgeour's 1ZB.[30]

In 1932 the immediate problems were the standard of YA equipment and the provision of additional stations. The former was the most pressing. While the commission was deliberating, Bingham was giving Vickery and Hands 'disquieting statements . . . regarding his apprehensions of breakdowns in the [2YA] transmitting equipment'.[31] Bingham's report came after 2YA broke down during the broadcast of the ceremonies marking the opening of the Sydney Harbour Bridge and was off the air completely for three quarters of an hour.[32]

Gibbs and Barnett recommended considerable improvements to the YA stations. They called for new transmitting sites for 1YA, 3YA and 4YA, and for these 500-watt, 50 per cent modulated stations to be replaced by 100 per cent modulation transmitters equal to 2- to 2.5-kW rating. They proposed that 2YA should be rendered more stable in operation and doubled in power to 10 kW. They also suggested a 100-watt emergency transmitter was needed at 2YA.

Soon after receipt of the commission's report the board dismissed Bingham, and the position of chief engineer was accepted by J. R. Smith. The board intended to make large changes, and was keen to replace the engineer who had been responsible for the previous system. The appointment of Smith added a link between the NZBB and the P&T Department, where Smith had previously been in charge of the GPO laboratory. The position of South Island Engineer, stationed in Christchurch, was accepted by W. L. Harrison from 2ZW. This meant the transfer to the board of both engineer and station director/announcer from 2ZW, since the previous month L. E. Strachan had also been appointed by the board.

The board accepted Gibbs and Barnett's recommendations. 2YA was to be modernised and increased in power and an emergency plant supplied. New sites and new plant were to be found for the other YA transmitters. The board also decided to increase the hours of transmission of 1YA, 3YA and 4YA so that, like 2YA, they began in the morning. The afternoon sessions were extended and the board also abolished the silent days. By the end of 1932 all YA stations were transmitting from 10am to 10 or 11pm during the week, from 3pm to 11pm on Saturdays (except for 2YA which commenced at 10am) and from 2pm to 4. 30pm and 6pm to 10pm on Sundays.

In practice, Gibbs and Barnett's third area of investigation, the addition of further stations, meant addressing the board's relationship with the B stations. However, Gibbs and Barnett left this to the board and instead confined themselves to choosing geographically sensible sites for stations (see Appendix 2). They recommended four relay stations be established at Invercargill, Woodville, Tirau and an east coast North Island site. These stations were to use the YA stations' 500-watt transmitters once they had been replaced and modernised. Without suggesting the NZBB should support B stations or start its own stations, they further recommended assistance be given towards improving the service at Cromwell, Timaru, a West Coast South Island site, Nelson, Taranaki, Ohakune, Opotiki and Whangarei.

The board decided to provide assistance for certain provincial stations, both financially and by the provision of relays. After considering the Coverage Commission's report, it announced that it would not assist city B stations. The board had already discussed at earlier meetings which provincial stations it would assist. At its third meeting it resolved to continue the RBC's relationship with 2YB New Plymouth, and offered that station an annual £200 subsidy; and it decided 'to make any necessary arrangements to prevent station 4ZP Invercargill being closed'.[33] By July, when the Coverage Commission reported, subsidies were also being paid to 2ZF Palmerston North, 3ZR Greymouth, 1ZH Hamilton, and negotiations were under way regarding assisting 2ZD Masterton, 2ZH Napier and 2ZJ Gisborne. Thus the Coverage Commission's recommendation for additional stations, both relay and independent, to provide a nationwide coverage was not followed. Instead the board supported the above eight stations with the arrangements for this support being initiated and in many cases concluded before the Coverage Commission reported. The eight stations were in the eight centres designated by the Advisory Council in March (well before the commission reported) as requiring cooperation from the board.[34] Support for the eight stations, but for no others, continued during the tenure of the NZBB, except that, in 1935, the subsidy was stopped for 1ZH and a small one of £10 per month was started for 4ZC Cromwell. The arrangements varied with each station, but included general financial subsidies, grants to assist with increasing the power output and other capabilities of the stations, and payments of special expenses such as the purchase of records and the rental of landlines. Relay lines were also supplied, but for three stations only — 1ZH, 2ZF and 2ZD; the P&T Department felt it was not practicable to supply the other stations.[35] The subsidies for 1932 totalled £1098 and for the remainder of the board's term averaged just under £2600 a year.[36] These subsidies were not regarded as full financial support. This level of payment would have been only a minor contribution to the city B stations, but to the provincial stations which received the subsidies it was a substantial proportion of their costs. The secretary of the unsubsidised 2ZR, Nelson, for example, estimated that it required £400 per year to keep the station on the air.[37] The subsidised stations also profited from the relatively amicable relationship with APRA and the gramophone companies that the RBC, and then the NZBB, negotiated on behalf of their own stations.

Some of the subsidised B stations were in areas recommended for additional stations by Gibbs and Barnett, and others were reasonably close to the recommended areas. Some centres — in particular Cromwell, Timaru, Nelson and Whangarei — were mentioned by the commission but neglected by the board. There was to be some anger with the board's actions, especially from Nelson. Nor did it go unnoticed that the assisted stations included the four stations associated with the provincial members of the Advisory Council; one of them, 1ZH, was owned by a council member.

NZBB STATIONS CONSTRUCTION

The board accepted the Coverage Commission's recommendations regarding its main centre stations, and its programme of transmitter construction became a feature of its five and a half years in office. It took advantage of the significant advances in radio engineering since the YA stations were first built.

The motif of the New Zealand Broadcasting Board, a stylised rendition of the Mt Victoria transmitter building and masts, as on the new transmitter building at Titahi Bay.

By the time the board was dissolved, 1YA, 3YA and 4YA had been transformed, a new 2YA was under construction, and the four main centres each had auxiliary stations. To a lesser extent the board was also responsible for new studios. Auckland studios in Shortland Street were built and occupied during the board's term. In Christchurch a site in Gloucester Street was purchased, though the construction of studios there was left to the board's successor. In Wellington an option was obtained by the board in 1936 over the Hannah Estate property in Boulcott Street, but the final decision to purchase and erect studios there was left to the NBS.

The changes began in Christchurch. By September 1932 Gibbs and Smith had opted for a site on Gebbies Pass for a new transmitter. The site was purchased the following March and AWA's tender for a transmitting station accepted. The transmitter was of a type that could be added to, if required, to achieve a maximum power output of 10 kW. Two steel lattice-work towers, each 350 feet high (200 feet higher than 2YA's) were erected for the aerial system. The new 3-kW station was brought into service on 14 December 1933. The cost for the land, buildings, roading and plant was £23,272. The transmitter's power was raised to 10 kilowatts (twice that of 2YA) in August 1935. In May 1933 a property at 64–66 Gloucester Street was purchased for new studios.

On 23 January 1935 a new 1YA transmitter, also of 10 kilowatts and constructed by AWA, was brought into operation, as was the first of the new studios in Shortland Street. The Shortland Street property had been purchased in August 1933, and the following month a 16½ acre property at

The 1YA studios in Shortland Street brought into operation in 1935. RADIO NZ

Henderson was bought for the transmitter site. The distinctive feature of the Henderson site was the umbrella-type aerial system — the first in New Zealand. It used a single 500-foot-high steel lattice-work mast with three radiating conductors hung from the top of the mast and anchored at the ground some distance from the base. The object was to radiate the maximum energy in a horizontal direction.[38] The Shortland Street building was four storeys high, and contained a variety of studios. The largest, to be used for concert presentations, was 60 by 40 feet and 25 feet high, and occupied two floors in the building. Amongst other innovations, the building had an 'echo' room, used for the generation of special sound-effects. The total cost for the Shortland Street and Henderson sites and associated buildings and plant was £87,000.[39]

In September 1935 another AWA-constructed 10-kW transmitter was brought into operation for 4YA, though it was not officially opened until 2 December. The transmitter was installed at Highcliff on the Otago peninsula, 1200 feet above sea level, on a 20-acre site which the board had purchased in June 1934. The aerial system was a replica of that used for 1YA.

The 2YA transmitter was also renewed by the board, although this did not come into operation until January 1937. It was a 60-kW transmitter, easily the

The Titahi Bay 60-kW transmitter, pictured in 1937 with technician E. Hodgekinson on duty.

most powerful in the country, constructed by AWA and installed on a 100-acre site at Titahi Bay, north of Wellington. The umbrella aerial was suspended from a single mast 700 feet high. The total cost for the Titahi Bay site, buildings and equipment was £101,149.[40]

As well as increasing the power of the YA stations, the board founded four auxiliary stations. The first was 2YC, Wellington, built by AWA with an aerial power input of 250 watts. It commenced transmissions in March 1933. The plan was for it to remain an emergency transmitter for 2YA. To keep it in good running order, it was to be used for one hour each evening.[41] The hour chosen was 7pm to 8pm, a time designed to offer listeners a non-B station alternative to the News and Talks hour. The second auxiliary station was the 100-watt 1YX, which commenced transmissions in July 1933. While the auxiliary stations remained as emergency substitutes for the YA stations, they were brought more completely into the programming schedule in December 1933. 3YL Christchurch, using the ex-3YA 500-watt transmitter, commenced in January 1934; 4YO Dunedin, with a power of 150 watts, opened in March 1934.

RADIO REGULATIONS 1932

Along with the Broadcasting Act, the Radio Regulations determined what broadcasters could and could not do. These were rewritten as part of a reordering of the radio regulations generally and gazetted on 21 July. There were numerous changes regarding the technical capabilities of stations that took notice of the advances in design that had occurred since the 1925 regulations. For the first time television was recognised, and provision was made for its transmission.

For the broadcaster there were significant changes and repetitions. The previous regulation that records occupy only a minority of programming time was amended to permit their use for 'the whole or a stipulated part of the time'. This had considerable consequences for the broadcasting of local talent. The 1925 introduction of the regulation had not been intended as a promotion of local talent, but had that effect. Its removal legitimated the growing reliance on records, and meant that New Zealand broadcasting, from this point, transmitted predominantly imported records rather than live local entertainment.

The regulation concerning Sunday programming gazetted in 1922 but omitted in 1925 was returned: Sunday programming from 11am to 12.30pm and 6.30pm to 8pm had to give preference to 'religious services and kindred matter'.

The ban on the broadcasting of controversial material was continued. This cast doubt on the intended modelling of the NZBB on the BBC, which had begun with a similar ban, but had it lifted in 1928. The BBC was able to broadcast controversial material, though it was not allowed to express an editorial opinion. The New Zealand government was not prepared to give a similar latitude to its broadcasters.

The number of hours a station could transmit per day remained under the minister's control. In 1925 the regulation had been that a weekly minimum of twelve hours had to be maintained and the minister would determine the daily duration of a station's programme. In 1932, although the reference to twelve hours was omitted, a station had to be operated for 'such minimum period as may be determined by the Minister'. In the 1932 regulations the minimum period was also 'regarded as the normal service, and no variation of such service shall be made without the authority of the Minister'. B station proprietors wishing to extend their hours found this authority difficult to obtain. Ominously for the B stations, sponsoring received no direct mention. Instead the relevant regulation stated that 'no form of advertising shall be undertaken . . . unless specially authorised by the Minister'.

Finally, ministerial control over stations was extended to the announcing staff. The minister was given the power to debar from announcing duties anyone who, in his opinion, was unsuitable for such duties.

1932 LISTENERS' QUESTIONNAIRE

At its second meeting the board resolved 'to arrange for the taking of a plebiscite on the question of programme requirements and other matters in connection with the broadcasting service'.[42] The board prepared and printed 50,000 copies of a questionnaire which was given to listeners between mid March and the end of April, as they renewed their licences. The period was extended to the end of May and 24,000 questionnaires were returned. This represented some 40 per cent of listeners. The results, published in September, indicated distinct listening patterns and preferences.[43] They showed that the days of crystal sets were past: over 93 per cent of respondents were using valve sets. Of major interest to the board from the returns was the New Zealand audience's distinct preference for popular music and light entertainment. The NZBB was given a clear message that its majority audience did not favour the use of radio as an educative medium in highbrow culture. The audience also indicated a desire for recorded music as opposed to local artists. This may have been a vote against the type of music performed by local artists rather than against the local artists themselves, but it remained a vote against the preponderance of local talent that had been required in the 1925 regulations.

The board's questionnaire had flaws in its design and execution, but it represents the first systematic listener research in New Zealand broadcasting. Disappointingly, there were no further attempts by the board to study listeners' preferences. Not until the advent of commercial broadcasting would listener research become a normal part of broadcasting.

PROGRAMMING POLICY

The NZBB took notice of its plebiscite. Like Harris before him, Hands worked against what he saw as an over-emphasis on classical music. The board had a dual policy. It included both a need to 'cater for every legitimate taste' and that of 'increasing appreciation of good music'. Here was a clash between the desire to give an audience what it wanted and to educate an audience to want superior fare — and this at a time when many listeners were able to tune in to the YA stations' competitors, the B stations. The board's approach to the problem was, in regard to 'good music', to emphasise quality rather than quantity. In particular this meant limiting the amount of local content.[44]

The removal of the musical and dramatic committees was related to this reduction in local content. Not only did the board and Hands wish to bring programme compilation firmly under their own control, they were particularly concerned to end the dominance of the various local cultural groups, most of which were represented on the committees. The result was an increase, as permitted under the new regulations, in recorded music broadcast. There was also a lesser emphasis on classical music, though the NZBB was never to rival the B stations in the amount of time given to popular music. The YA stations continued to be considered of a higher cultural level than their private station counterparts.[45]

The board accepted that there should be a general continuation of the programming established by the RBC; however, one of its first actions was to remove the programmes and practices it considered objectionable. At the first board meeting it resolved to replace the recorded American programmes with British ones. A series of recordings of talks by British authors were broadcast in 1932 and, on 4 June 1933, 1YA and 2YA broadcast their first BBC transcription — C. Whitaker-Wilson's play on the life of Sir Christopher Wren, produced to mark the tercentenary of Wren's birth. The play marked

The staff of 2YA in 1932. ALEXANDER TURNBULL LIBRARY

the beginning of a long reliance on Britain for radio programming. The price paid increased steeply, but with the increase in transmission hours, Hands considered the transcriptions to be 'something of a godsend, no matter what the price'.[46]

The inclusion of British recordings did not mean the permanent removal of American ones. The need for programming and a realisation that their removal had been somewhat headstrong led to their return in April 1934. They were joined in the same year by Australian radio serials.

The board also attempted early on to curb the individuality of announcers. Announcers were never spoken of by name, but were addressed as 'Mr Announcer'. In a 1951 programme celebrating the first twenty-five years of 2YA, A. L. Curry introduced himself, and said it was the first time in his long career he had announced his name.[47] Not broadcasting their names did not, however, give anonymity to announcers. Clive Drummond, the 2YA chief announcer, was the best known. His colleague, Ken Collins, wrote, 'To walk down Lambton Quay with him was like walking with royalty. People turned and pointed and you could hear loud asides, "There's Clive Drummond!"'[48] All announcers were well-known public personalities, the first to experience the mass recognition that broadcasting conferred. This gave them unusual advantages: it was evident that Clyde Carr's announcing position at 3YA had helped him win the Timaru seat in 1928. Whereas Harris had viewed the announcers' popularity positively, as adding to the public regard for the RBC, and had accepted their individual styles, the board attempted to standardise announcements. This did not please listeners. The change that caused the most protest was Hands's banning of Drummond's distinctive, elongated 'Goodnight everybody, goodnight' at the close of transmissions.[49] Hands was unmoved by the protest, which he regarded as emanating 'from juvenile minds or from hysterical females'.[50] Hands was later to retreat somewhat from the hopeless attempt to make all announcers sound alike, but it was the first sign of concern over the development of radio personality, later most evident in the board's attitude to Colin Scrimgeour.

Head office control was increased over each station's programming. Stations had never had a completely free hand to determine their programming, and particularly since the introduction of complementary programming, there had been central coordination. In 1932 this was extended considerably. Sometimes this was for financial reasons — 'In all cases where orchestral accompaniment is desired to vocal items and where special scoring will be necessitated as a result, such items should be submitted to this office for consideration prior to being scored or requisitioned for, and we will then advise whether the work should be proceeded with.'[51] Sometimes the problem was the rule against controversial broadcasting. 'On no account are any plays whatever to be broadcast without full details having been submitted to this office beforehand and prior approval obtained. This applies not only to published versions of well known overseas authors, but also to locally written plays.'[52] There was a continuing assertion of central authority over the stations: 'As from Monday, April 4th, you will be relieved of the preparation of Dinner and Dance programmes, this work being undertaken at Head Office. Details will be forwarded to you with your approved programmes each week.'[53] The major exception was a 15-minute period from 9pm to 9.15pm, the content of which could be determined at each station. This was allowed after a letter from John Ball, the 2YA director, asking for 'a certain element of freedom in our evening concert programmes . . . under our present system every minute of our programme time is allocated'.[54]

The central control of programming was accompanied by an accentuation of the YA stations as national stations rather than the voice of the local community. The centralisation policy had various effects. It led to the introduction of tours of the board's stations by artists. Under the RBC, performers had only occasionally visited other centres. Normally they were identified with one station. Starting with Annette Chapman, the New Zealand mezzo-soprano, who sang from all the YA stations in October and November 1932, the better artists began to be broadcast nationally.[55] Centralisation also meant the end of the YA stations as local noticeboards. As an example of what had been common but was no longer to be permitted, Hands referred to a request for a broadcast appeal for a pram from a woman whose husband was on relief work and who had five children, including three-week-old twins. The growth in radio listening and the increase in the numbers of people affected by the depression meant that the number of appeals had escalated. Station directors were instructed to confine appeals to one a week and to give preference to those from organisations such as the Red Cross.[56] This policy meant that involvement with the local community became more and more the preserve of the B stations.

As part of the centralisation policy, Hands extended the programme coordination first introduced by the RBC. His first 'schematic programme plan' — coordinating the four YA stations — operated from the start of 1934. It was replaced, from the start of March 1935, by another plan coordinating the YA and the four auxiliary stations. The blueprint, as it came to be known by NZBB (and later NBS and NZBS) employees, allowed listeners, according to their location and the power of their receivers, to select particular types of programming. By offering a choice it muted the complaints from listeners about the programme offerings.

The *Radio Record* noted that, in the programme plan, 'the proportion of highbrow music and lowbrow music is evenly balanced'.[57] The terms 'highbrow', 'middlebrow' and 'lowbrow' were widely used, and indicated the supposed levels of intellectual ability and cultural appreciation of sections of the audience. The sorting of programmes into these levels, as well as into dance music, chamber music, band music, spoken entertainment and so on, was an important aspect of the programme coordination — it allowed listeners to concentrate their attention on highbrow material; or, of course, to evade it. This major reason for coordinating programmes was not publicised, since it ran contrary to the educational policy of the NZBB, which was to gradually raise listeners' tastes. Like Harris before him, Hands regarded it as more important to keep listeners tuned to the YA stations than to concentrate on highbrow material and lose them to the B stations.

SPORTS BROADCASTS

The change to board control meant the end of Alan Allardyce's commentary career. The board decided it did not need a permanent head of sports broadcasting. Allardyce also fell out of favour with the board, particularly its Canterbury member, L. R. C. Macfarlane, who regarded him as sympathetic to the B stations and an 'agitator'. Although the board recognised his abilities as an announcer, Canterbury commentaries were increasingly given by others.[58]

Sports broadcasting, although it had no permanent head, became an even more prominent programme mainstay under the board. The board accepted the need to pay sports bodies for the right to broadcast events, and after this change in policy, rugby and racing commentaries returned to radio.

In February 1932 Hands urged board members to seriously consider

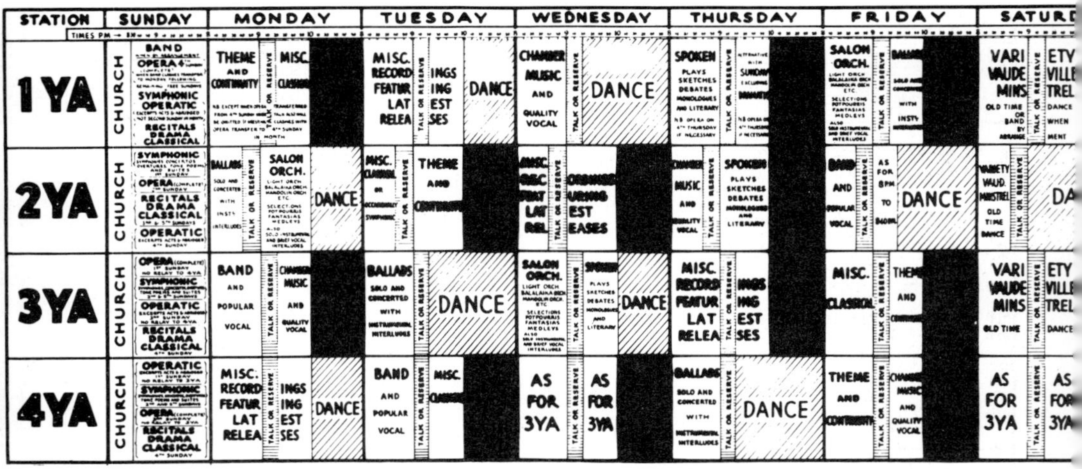

The 1934 coordination of programmes schedule as published in the *Radio Record* for the guidance of listeners. RADIO RECORD

paying for the rights to broadcast horse-racing commentaries, arguing that any reasonable costs would be recouped in licence fees.[59] Agreement was reached in time for broadcasts to resume at the meetings during National Week in Christchurch in August. The return of the commentaries was expensive for the board. The racing and trotting authorities accepted the board's offer of '£1000 per annum for the privilege of broadcasting running descriptions of trotting and galloping events in the four metropolitan areas'.[60] At the end of the board's years the price had risen to £1500 per year.[61] Allardyce argued that permission for the commentaries was regained because of the NZBB's semi-government status and the racing and trotting organisations' desire to be on good terms with the government.[62] In addition, unlike the RBC, the board accepted the need to pay for permission for the commentaries. Also, the four years without commentaries had been long enough for many racing and trotting officials to change their views.

Rugby broadcasts were also made in the winter of 1932, though not from 3YA. The Auckland, Wellington and Otago unions were paid £50 each for permission to broadcast national and provincial matches. The unions remained opposed to the broadcasting of club matches. The Canterbury Union still withheld permission entirely, and the NZBB resolved instead to broadcast soccer and hockey matches in Christchurch.[63] 3YA also broadcast the rugby league test between New Zealand and England. Canterbury opposition began to fade at the time of the Canterbury v. Southland match in August 1933. The prime minister and the postmaster-general received many telegrams from irate Southland rugby followers. Vickery related to the postmaster-general the board's lack of success with the Canterbury Union and Hamilton himself took over the negotiations. The following week's

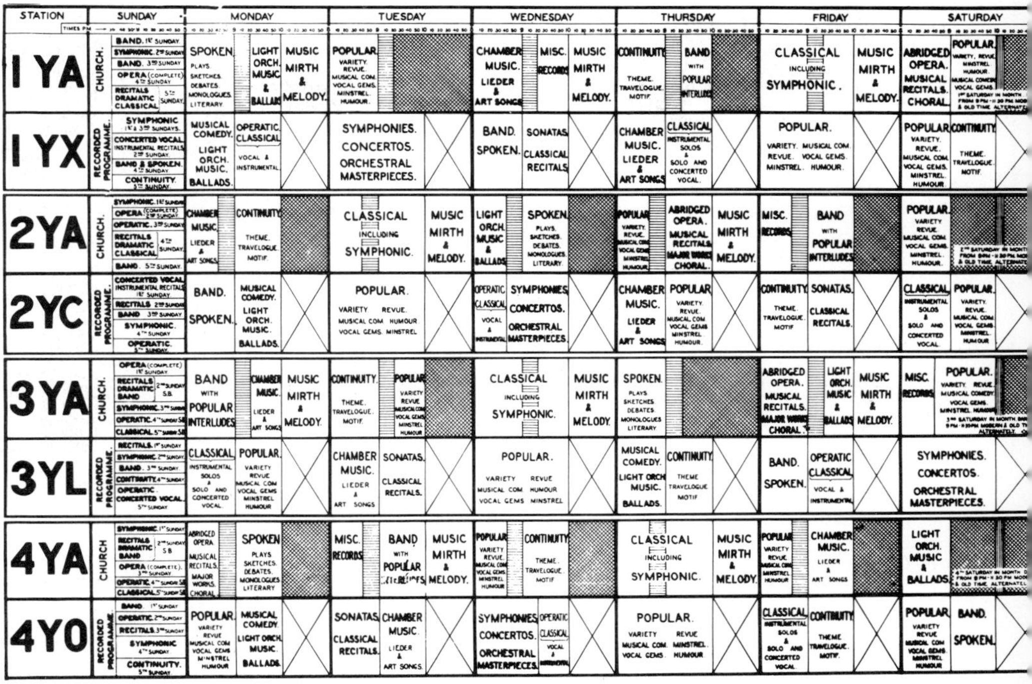

The 1935 programmes schedule in which the Broadcasting Board extended its coordination to its eight stations. *RADIO RECORD*

Canterbury v. Otago match was broadcast, and the following year the Canterbury Union accepted broadcasting of match commentaries. It and the other main center unions received £50 in payment. In that year the New Zealand Rugby Union was paid £500 for the right to broadcast matches under its control. With this arrangement, the NZBB turned its back on rugby league and resolved 'that as the Board is committed to the broadcasting of the national game it cannot agree to broadcast descriptions of matches played under the League code'.[64]

International broadcasts of rugby commentaries were started by the NZBB in 1932 when the All Blacks toured Australia. AWA, with its Sydney shortwave station 2ME, had the ability to transmit commentaries to New Zealand, where they were broadcast from 2YA, via the P&T Department Mount Crawford receiving station. In 1932 the first and third tests and the New South Wales match — all played in Sydney — were broadcast.[65]

Regular broadcast commentaries were arranged for wrestling also. Wrestling was a sport that attracted great public interest in the 1930s, and bouts were usually well attended. Wrestling commentaries had been started by the RBC and proved very popular, and the board was keen for them to continue. It made its first arrangements for 1YA. The board began by paying the Auckland Wrestling Association £1 per match, but in August it paid £75 for broadcasting rights for all matches for the season.[66] This was extended, and by 1934 the four main center wrestling associations were each paid £75 for the season's broadcasting rights. Wrestling commentaries remained among the most popular programmes and the sports commentators were among the most popular broadcasters. Gordon Hutter, the 1YA wrestling commentator, was at one stage reported as receiving 500 letters a week from fans.[67]

SCHOOL BROADCASTS

School broadcasts were regarded as important by the NZBB. Their continuation and expansion was a considerable achievement, given the economic depression and a general policy of retrenchment in the Department of Education. The programme was championed particularly by Macfarlane, the rural member of the board, who saw it as well suited to children in isolated areas: 'We might with profit take up the matter of more educational broadcasts in the near future. This country correspondence school course is gaining ground all over New Zealand and my thought is that a well organized course of talks to these lonely scattered children would be very beneficial to them.'[68] The broadcasts soon became part of education for all school pupils.

In 1932, its first year, the NZBB resumed the broadcasts from 2YA, again for one hour a week. The programmes were prepared by F. C. Brew and Ernest Jenner, vice-principal and lecturer in music respectively at the Wellington Teachers' Training College. Following examples set by the BBC and the Australian Broadcasting Commission (ABC), the NZBB issued a series of booklets to be distributed among schools to accompany the broadcasts.[69] This was the start of a long-lived series of publications which gained an enduring reputation for the quality of their content and the standard of their production. The booklets were designed to involve the pupils in their radio learning. As the *Radio Record* noted, 'The busy cooperation of the pupils is sought and encouraged rather than passive listening.'[70] The Department of Education published programme details in its *Education Gazette*, but was not involved in, and declined to share the cost of, publication of the booklets.[71] Publication of the booklets became a substantial undertaking — 64,000 were printed during the board's final year.[72]

In 1933, 1YA and 3YA began weekly one-hour school broadcasts. The programme would have collapsed otherwise, as the cost-saving closure of Wellington Teachers' Training College at the end of 1932 brought a halt to school broadcasts from 2YA. Wellington Teachers' Training College, along with its counterpart in Dunedin, remained closed for all of 1933, 1934 and 1935. In 1934 the Auckland and Christchurch Teachers' Training Colleges were also closed.[73] From October 1933 the 3YA broadcasts were rebroadcast from 2YA. In 1934 with all the training colleges closed, the broadcasts to schools programme was continued largely through the efforts of James Shelley, professor of education at Canterbury University College. Shelley had already been involved in broadcasting, mainly through talks to the WEA, and was to be even more closely involved in the future.

Otago was regarded as lagging behind in the matter of school broadcasts and, from 1934, Hands arranged for a rebroadcasting of other stations' sessions from 4YA in the hope of inaugurating a similar session there. Hands's comments to his board members while arranging these rebroadcasts indicate that the NZBB was not concerned solely with education in its support of school broadcasts. He argued to the board, 'The more listeners the better for it will tend to make "Mum and Dad" radio minded.'[74] School broadcasts would thus encourage the spread of radio. School broadcasts were encouraged by the board, with this dual agenda, and by the end of its term 576 schools were participating in the programmes.[75] Under the NZBB, school broadcasting was taken from a stage of tentative experimentation to one of long-term acceptance in New Zealand education.

BROADCASTING, NEWSPAPERS AND NEWS

After it took office, the NZBB was notified of the agreement the RBC had reached with the Auckland newspapers to pay for publication of its programmes. The *Auckland Star* pointedly suggested that 'This newspaper affords the local broadcasting station facilities that would be cheaply paid for at £1000 per annum.'[76] The board, which had not been party to the RBC's agreement, declined to proceed with it. As a consequence the Auckland newspapers stopped printing the YA station programmes. The board's position was that 'Programme matter has a very definite news value to newspapers.' The newspaper proprietors in return argued that 'All classes of amusement promoters, and notably the picture companies, could put forward with equal force the same claim.'[77] While there was logic on both sides, the NZBB did not change its position. Not only was it convinced that newspapers would benefit from publishing radio programmes, it considered that the board could never afford to pay for publication of its programmes every day in all New Zealand newspapers. Eventually the sheer number of radio licence-holders convinced the Auckland newspaper proprietors that the publication of radio programmes was a general news service rather than an advertisement.

The negotiations between the newspaper proprietors and the NZBB were protracted. The proprietors proposed to link news and advertising: if the board would pay for advertising, the UPA (the country's press association) would provide cable news to the board. In late 1934 the board resolved against this; it stated that it would prefer to deal on a cash basis for press cables.[78] In early 1935 the two parties finally agreed and the NZBB began paying the UPA £600 per year for a cable service of 200 words per day which the board, like its predecessor, consented not to broadcast until after 7pm.[79]

The news service of the NZBB was little better than that of the RBC. The board remained dependent on the newspapers' news service and did not broadcast its own news programme until the evening, when the afternoon newspapers had been purchased and read. It was a problem the board recognised early. In 1932 Hutchinson wrote, 'Everybody knows what is coming for the next half hour almost — the headings out of the *Star* which they have read . . . We should make every endeavour to overcome this disability.'[80] However, the board accepted the limitations of its news service and argued that it was primarily a service for rural dwellers — in other words, those beyond the reach of the daily newspapers.[81] As with the RBC, the regulation against controversial material would have made it difficult for the board to develop its news and commentary programmes. There was no serious pursuit of the matter — the possibility of instituting a full news service was a threat which could be held against the press,[82] but it was only a threat. The board did not intrude on what it accepted as the newspapers' territory.

THE *RADIO RECORD*

The *Radio Record*, begun as the voice of the RBC, did not have so close a relationship with the NZBB. By the time the board took office there had been a change in the ownership structure of the journal. Charles Earle, managing director of the *Dominion* (the Wellington morning daily) became a further director of the Radio Publishing Company. Arthur Heighway continued as editor. The board's general policy of continuing existing arrangements did not apply to the *Radio Record*: the annual £2000 subsidy was stopped at the board's first meeting.[83] Heighway and Earle, at least in the meantime, were forced to continue the *Radio Record* without support. In 1932 the journal again received a subsidy, but at the much reduced rate of £10 per week. All payments from the board ceased in December 1932. As conditions of the subsidy the *Radio Record* was required to continue publication in full of YA station programmes, to publish news and statements as supplied by the board and to print editorials on policy matters in a form approved by the board.[84]

In November 1933 the ownership of the *Radio Record* was taken over by the newly formed National Magazines Limited. Heighway sold his interest in the New Zealand Dairy Produce Exporter Newspaper Company and used his funds to form the new company. Earle was also a director, representing on the board the interests of the Wellington Publishing Company. This remained the ownership structure until October 1938 when National Magazines and the Wellington Publishing Company severed their connection.

From the end of 1932 Heighway and Earle were in an ambivalent relationship with the board. They were not subsidised and the *Radio Record* was no longer the official publication of the board, as it had been for the

RBC. But they were beholden to the board in that they remained favoured with copyright of broadcasting programmes which only they could print for the week ahead.

The tenure of the *Radio Record* was regularly under threat. The NZBB wanted to have its own weekly journal, and it investigated the matter during 1932, and again in 1934 and 1935. In 1935 the board looked at printing programmes each week and distributing copies free to all listeners. After costing the proposal, it dropped the idea.

Apart from the NZBB's investigations into programme publication, various newspaper proprietors attempted to move into the field. The proprietors of the *New Zealand Herald* and the *Auckland Star* had discussions with the board regarding replacing, or competing with, the *Radio Record*. There were also requests to abandon the copyright of the *Radio Record*. The *Truth* editor argued, 'No private concern should have the right to solely announce the programmes of publicly maintained organizations like the A broadcasting stations.'[85] The *Radio Record* retained the copyright throughout the years of the NZBB, but they were years of continuing uncertainty for the proprietors.

Heighway and Earle's position was insecure, and the result was that, while it no longer the board's mouthpiece, the *Radio Record* had at best a muted independence and remained an editorial defender of the NZBB. The main change in its policy was a favourable recognition of the B stations, which previously had seldom been mentioned. The *Radio Record* became a journal dealing with the full range of New Zealand broadcasting. It also attempted to develop a wider appeal in tune with what it saw as the home and family orientation of radio:

> The paper's chief purpose is to present radio programmes and full details of coming events and interesting happenings in the broadcasting world. But it is essentially a home paper — for wireless is a part of the family fireside — and for this reason it retails other news of the entertainment world; the professional stage, the talkies, repertory, gramophone recordings: there is news, too, for women and comments on modern literature.[86]

APRA AND THE RECORD COMPANIES

The NZBB inherited and, with minor alterations, continued the agreement which the RBC had made with APRA. It also entered into protracted negotiations with the various record companies. The companies considered that the broadcasting of records adversely affected sales, while the board argued that such broadcasting amounted to free advertising for the record companies. The eventual agreement included many provisions 'to prevent excessive broadcasting'. In particular, a new record could be played only once in the first week of release, and for the rest of the first six months, no more than twice a week.[87] From August 1933 this was changed so that new releases could be played once daily for the first week and then twice weekly for six months. The board also agreed to announce the name of the manufacturer of the record being broadcast, and in addition paid the combined companies £2000 for the following twelve months.[88] This agreement was renewed annually during the board's term, although the annual price paid by the NZBB increased. The final payment, for the year from 1 August 1935, was £3500. The agreement covered the board's stations plus the B stations subsidised by the board.

The agreements with the record companies and with APRA did not include the non-subsidised B stations — their difficulties with these organisations continued during the years of the NZBB. The disputes flared intermittently, and some stations closed temporarily or restricted their record-playing.[89] There was no solution — the B stations were too impoverished to come to any agreement acceptable to APRA and the companies; and those organisations were unwilling to agree on anything less while there was a possibility that the B stations might in the future be permitted to earn an income.

B STATIONS

There were thirty-six B stations in operation when the NZBB began. In spite of the government's statement during the Broadcasting Bill debate that there were too many B stations, three more were started before the issuing of licences was stopped in 1933: 2ZR Nelson, 4ZC Lowburn Ferry (Cromwell) and 3ZM Christchurch. All three were set up after the report by the Coverage Commission, and the Nelson and Cromwell stations were encouraged by the statements in that report that stations were needed in the two centres.

2ZR NELSON

The experience of the Nelson B station illustrates the difficulties under which the smaller private stations operated in the 1930s, and the methods they adopted to finance their broadcasting.[90] The station, 'the voice of the city of sunshine', broadcast for the first time on 4 July 1932. The licence holder and station owner was N. D. Braithwaite, a recent arrival in the city, who stated he had moved to Nelson for health reasons and that broadcasting was his hobby. Nelson, because of its topography, did not enjoy good radio reception and the new local station was a welcome addition. 2ZR was given rent-free accommodation in the premises of Charles Begg & Company in Trafalgar Street. The generosity of Charles Begg & Company was not entirely altruistic. The firm sold sheet music, records, radios and gramophones, and hoped a local radio station would promote its products. But the provision of premises, particularly of the standard now required by the regulations, was a major gift to 2ZR.

According to Mackay, 'it was soon realised that if the station was to remain on the air, the listeners would have to assist in some way'.[91] Nine days after 2ZR began transmissions, a public meeting was held at which the 2ZR Radio Club was formed. The club appointed a committee of management to assist Braithwaite, and fixed club subscriptions at 5s per member per year. In April 1933 the club membership was reported to be 600, which would have given the station around £150 a year.[92]

2ZR had the support of the city. The local MP, Harry Atmore, was the club's patron. Both he and the Motueka MP, Keith Holyoake, were ready to assist and the various public meetings concerning the station were well attended affairs usually presided over by the mayor. Both the Nelson City Council and the Nelson Harbour Board financially assisted the club — the council provided an annual grant of £60.[93]

Mackay estimated it cost around £400 per year to keep 2ZR on the air.[94] With subscriptions and grants totalling little more than half this amount, the quest for funds was never-ending: 'In addition to club membership fees, we organised raffles, hampers, Hokitika swindles, and art unions. Many were the methods used to separate coins of the realm from their rightful owners. On one occasion we ran a sweepstake nominating the day when a leading citizen's offspring would enter the world.'[95]

Sponsored programmes were permitted, but with the limited information that was allowed to be broadcast, they were never a large revenue-earner for the B stations. The financial pressures drove B-station operators beyond legal requirements and community sensibilities in their search for money. Mackay records one of the more notorious incidents at 2ZR:

> An evangelist with all the hallmarks of a high pressure salesman . . . made us a proposition to broadcast his meetings each night for a week. Legally we could not sell time, but a mere act of Parliament was not going to prevent us from making some much needed money to purchase new equipment . . . Our evangelist received value for his money and we nearly went out of business. In the first broadcast he lambasted the Roman Catholic Church from the Pope down. Needless to say, the next morning I had a visit from a local Catholic priest who was a very good friend to the station. We were in a quandary. We had already spent the man's money, but it was obvious that we could not permit this sort of thing. We promised our priest that whatever happened there would be no repetition of these statements. I interviewed the evangelist and he promised that he would not mention the Catholic Church again. We were quite happy but soon realised that we must be simple folk at heart. Our friend did not mention the Roman Catholic Church in his next broadcast — he attacked Presbyterians and Methodists . . . We thought the wrath of God had descended upon us in the guise of numerous men of the cloth who protested in no uncertain terms. We spent a series of very uncomfortable interviews. In the end it was decided that we would cut the broadcast if any offensive references were made by the speaker. This resulted in his next broadcast being cut on three or four occasions, which in turn brought him capacity houses for the remainder of the week and many curious people found their way to his theatre.[96]

2ZR attempted, without success, to ensure its survival by becoming one of the B stations subsidised by the NZBB. A first application from Braithwaite was turned down by the board in August 1932.[97] In October Braithwaite retired and the club executive assumed control and responsibility. Another subsidy application following this change was also turned down by the board.[98]

Subsidy applications continued during 1933 and included a club deputation travelling to Wellington to interview the postmaster-general, the P&T Department secretary and Hands. After this application was turned down at the board's April 1933 meeting, a well-attended public meeting of protest was organised in Nelson by the Nelson Listeners' League. At this meeting it was noted that the four provincial members of the Advisory Council had obtained subsidies for their stations.[99] Such remarks were not received happily by the Advisory Council and, from this point, relationships between the board and 2ZR were polite but cold. Deputations despatched from Nelson were not made to feel welcome. Before one deputation was received, the board discussed the matter and 'proposed to listen to them and avoid discussion as much as possible'.[100]

The Nelson station was one of the more persistent in its subsidy applications, but it was joined by many others. Applications came both from stations in centres without a YA or a subsidised station, and from those in centres with a subsidised station asking for that subsidy to be shared. The board was unmoved.

B STATIONS' REVENUE

With the main-centre B stations specifically excluded from subsidies and few of their provincial counterparts receiving recognition from the NZBB, the B stations were forced to try other means of increasing their revenue. Their choices were few: relying on the generosity of their listeners and raising funds through the sale of sponsored programmes.

As at 2ZR, many B stations formed radio clubs, with an annual membership subscription paid to the station. Membership subscriptions were, of course, in addition to the annual licence fee all listeners were required to pay, and they thus indicate a commitment to the particular B station. As in Nelson, most of the supporters' clubs were formed in 1932 after it became obvious that the NZBB was unlikely to add to the number of sponsored stations. The 2ZR club subscription of 5s was a large one — most clubs asked for 2/6d. Membership numbers of the various clubs fluctuated considerably over time. Numbers were highest in the early days of each club or when membership campaigns were held. Membership lists tended to be exaggerated by the clubs, which could then point to a high level of support within their communities. In spite of this tendency it is clear that the B stations did enjoy real support in their local communities. However, the memberships of the clubs were never of a size approximating the actual listening audience. Just as the postmaster-general was unable to organise the radio audience to elect representatives, so the B stations were unable to persuade more than a fraction of their listeners to become club members. The huge and still growing radio audience was not to be organised into traditional voluntary associations.

The city stations, especially 2ZW and 1ZR, had the largest clubs and the highest running costs. Walter Nash stated it cost £980 a year to run 2ZW; other estimates ranged from £1200 to £1500.[101] The estimates of club membership varied widely — the 1ZR membership was estimated at between several hundred and four thousand; in August 1934 it was reported as having a paid-up membership of 2000.[102] Two months later 4ZM, the station with the largest Dunedin club, had a membership of 1370.[103] 3ZM began its supporters' club in June 1933 and enrolled 400 members in a week.[104] Membership grew to 2400 by the following January.[105] By June, after a membership drive, it had grown to 4000, but the station owner estimated that would need to be doubled for the station to survive.[106] Club subscriptions provided a continuing support for the stations, but only occasionally would they have been enough to pay all costs. In no sense were the supporters' clubs sufficient to relieve any B station from constant financial worry.

COLIN SCRIMGEOUR AND 1ZR RELIGIOUS BROADCASTING

In the government and board plans for the development of broadcasting, the B stations had no long-term future. They were gradually to be replaced by board stations as the board enlarged its operation. This intention, however, was seldom voiced openly by government spokesmen for fear of arousing the considerable public opinion which held the B stations in high regard. This supportive public opinion was not the only weapon in the armoury of the B stations. In the first years of the 1930s there developed a form of religious broadcasting at odds with the New Zealand regulatory understanding of appropriate broadcasting and, in Colin Scrimgeour, religious broadcasting's most accomplished practitioner, the B stations found a champion who challenged the power of the NZBB and of the government itself.

Religious programming was the area of broadcasting least amenable to the general radio ban on controversial material. Non-religious programming could be banned outright if it seemed potentially controversial. But religious broadcasts were expressive of the nation's supposed basic Christian values and were an accepted programming division. Religious broadcasting was given a relative freedom of expression that was not available in other programmes. The RBC and NZBB religious broadcasts were under the control of the various denominations. In the B station 1ZR, a new type of religious broadcasting was developed. The 1ZR religious programming was taken particularly seriously. The owners of the station wanted a religious programming different to that on the NZBB stations — essentially separate services under the control of one or other of the religious denominations. They found in the Reverend Thomas Garland the man they needed. Garland had been associated with 1YA religious broadcasts and children's programmes almost since that station began. Independently he had been advocating the inauguration of non-denominational religious programmes of a more informal nature, for both children and adults. When he was offered a free hand to organise Sunday sessions at 1ZR, he seized the opportunity and resigned from 1YA.[107] The non-denominational character of his work was summed up in his motto: 'You go to your church. I'll go to mine, but let us walk together.' Garland carried out not just the religious programming but also the children's session work with children's choirs from 1ZR.

The station had been on the air only a matter of weeks when Garland began to be assisted by the man who was to take religious broadcasting and, in fact broadcasting generally, beyond anything yet envisaged. He was the Reverend Colin Graham Scrimgeour, at the time minister of the Airedale Street Methodist Mission. Scrimgeour and Garland met when Scrimgeour, who had his pilot's licence, appeared on the 1YA children's session at Garland's request to talk about the experience of flying through a rainbow.[108] The two became friends through their work for the Methodist Mission — Garland conducted the singing for the mission's Sunday evening services which were held in the Strand Theatre in Queen Street. In return, Reverend Scrimgeour began assisting at 1ZR with the children's session, where he was named 'Uncle Scrim'.

Colin Scrimgeour was born at Wairoa in 1903. One of six children, Scrimgeour left school at the age of eleven and began work at the Gisborne freezing-works. Both his parents died from cancer after considerable suffering. After his mother's death he resolved, at twenty, to become a minister. Scrimgeour's was one of the more unusual callings to the ministry, for he had little previous church experience (he said he had never been in one) and no preference for any particular denomination. He acknowledged that both his Christianity and his social thought had been influenced mainly by his reading of Tolstoy. Reasoning that neither the Catholics nor the Anglicans would accept him, he approached the Methodists. He went to their theological college but, after four months, left by mutual agreement and instead became a home missioner probationer — the junior grade of the Methodist ministry. Scrimgeour was never ordained, but he qualified as a home missioner and at the end of 1926 was appointed to establish the Methodist Central Mission in Auckland and organise its social work. He began this ministry with a brief to care for the poor of Auckland just as their numbers and their distress started to swell with the onset of the depression. He 'preached the Jesus-type of socialism' and his religious beliefs were distinguished from those of most of his clerical colleagues by his emphasis on offering practical, worldly help.[109] John A. Lee, the Labour MP, recalled being first introduced to Scrimgeour by Pat Hickey, the union leader and radical writer, who described him as 'a parson for whom I have considerable respect, a fellow who is more concerned about feeding the people than about praying for them'.[110]

Scrimgeour was a well-known figure in Auckland before 1ZR began. By that time he had been the Methodist city missioner for four years. His exposure of the nature and extent of poverty in Auckland had been well publicised, particularly in the *Sun*. In an attempt to take his church to a new congregation Scrimgeour had also begun holding his Sunday services in the Strand Theatre, where films were included, along with singing and sermons, as part of the service. The 1500-seat theatre regularly had a capacity congregation.

It was on the radio that Scrimgeour found his largest audience and his own métier. 1ZR had a devotional session every morning. Scrimgeour soon began taking one of these sessions. Partly because he lived close to the station, he found himself taking more and more of the morning sessions. Lewis Eady's would

> ring me and ask me to come down straight away because the Reverend so and so was supposed to take the devotional service and he's ill. So it was a few hundred yards' sprint and I'd get down there. I'd no idea what I was going to say. I looked along a shelf of books and I would see there 'The Song of the Earth'. That's all I had, just 'The Song of the Earth', Nature's perpetual renewance — and just go on from there and talk for perhaps twelve minutes — and the next day go along and do the same thing because the other minister hadn't turned up, until finally I found out I was going five days a week.[111]

Scrimgeour, who was less than kind about many of his colleagues in religious broadcasting, felt that his frequent broadcasts were needed not because of others' sickness but because of their inability: 'They couldn't understand this microphone business, they were hopeless failures. So then they started to fail, they started not to turn up, there were excuses of sickness.'[112] Scrimgeour, on the other hand, realised from the beginning that broadcasting required a new type of presentation and, during 1931 and 1932, developed in the 1ZR morning devotion sessions a broadcasting style that was to make him the most influential of all who had access to the microphone.

Two references Scrimgeour received in 1932 attest to the high regard in which he was held and to his social and political sympathies. The first, from the 1ZR station director, introduces Scrimgeour to his 2ZW Wellington counterpart.

> Our Uncle Scrim (Rev. C. G. Scrimgeour) is an outstanding personality of Auckland broadcasting, conducting the 1ZR Devotional Services, and doing more

good to the square inch than all the politicians put into one bag, to say nothing of Radio Boards and the like . . . Open the gate wide — give him the freedom of the city and of 2ZW, and the keys of the treasury or family vault — anything is safe with him.[113]

The other is from Jim Edwards, the labour leader at one stage sought by authorities for his part in the 1932 Queen Street riots.

Dear Comrades, The Rev Scrimgeour is a personal friend of mine. Whilst I was under cover he was trusted and helped me in every way. Please trust him with any literature he requires and give heed to any suggestion he may bring forward. Yours in revolt, Jim Edwards.[114]

B-STATION SPONSORING

Scrimgeour's first political impact on broadcasting came late in 1932 when he organised and presented to Parliament a petition to extend the B stations' allowable source of income. Sponsoring of programmes was the main method by which the B stations could legitimately earn money, but the nature of the sponsoring that was permitted meant that, within the confines allowed, little was raised from this source. It did not turn out to be the financial salvation the B stations had hoped. In practice the confines of permitted sponsoring were regularly exceeded. Hands referred to many B-station programmes as 'advertising talks' and drew attention to such matters as the naming of the 2ZW orchestra as 'The Plume Melody Makers' (they were sponsored by the Plume Motor Oil Company). The station owners were eager for permission to announce more information about their sponsors. In particular they wished to state the nature of the sponsors' products along with the sponsors' names. The petition Scrimgeour organised requested that either B stations receive a share of the licence-fee revenue, the suggested share being 2/6d per licence fee, or — and this was the alternative recommended by the parliamentary petitions committee — 'that approved B stations be allowed to indicate in their announcements of an unlimited number of programmes the names, addresses, and products of the sponsor's programmes'.

Just what was being asked for is indicated in the following exchange between the leader of the Opposition and the postmaster-general:

Mr Savage: 'I am asking the Postmaster-General . . . that an unlimited number of sponsored programmes be allowed.'
The Hon Mr Hamilton: 'That is allowed now.'
Mr Savage: 'No, it is not allowed now. There is at present allowed an unlimited number of limited sponsored programmes, but not an unlimited number of sponsored programmes that mean anything . . . The announcement that "John Jones is providing this programme" does not mean anything to the average listener; but the announcement that "John Jones, Boot-manufacturer, of Auckland, is providing this programme" means everything to the advertiser . . . The announce-

> ment that a programme is sponsored by Quigleys Limited, Auckland, is practically of no use to the sponsor; and an announcement that the programme is sponsored by Quigleys Limited, Furniture-Manufacturers, Anzac Avenue, Auckland, makes all the difference in the world.[115]

The petition was referred to the government for favourable consideration, though the postmaster-general made it clear he did not favour any change. He argued that the current scope of sponsoring was sufficient, and if the petition were allowed, the B stations would be encouraged to go further. The government's reluctance to allow the B stations any extension to their limited permission to advertise was widely regarded as bowing to pressure from the newspaper proprietors. Savage, for example, saw these 'vested interests' as attempting to 'muzzle broadcasting in every phase as far as they can do it'.[116] While the government may not have wished to tamper with the close support it received from the press, there was another reason for its reluctance to grant the petition. The petition ran counter to the government's own plan to model New Zealand broadcasting on the BBC, a model which excluded private stations. While the petition was being prepared, the NZBB reminded the postmaster-general of the fact and argued that the furtherance of the board's coverage plans would gradually render the B stations superfluous. 'They will, in fact, tend to hinder rather than help the national service.' Rather than assisting B stations, the board advised the government to issue no further licences and gradually to withdraw the current licences from the existing B stations.[117]

However, along with their championing by the Labour Opposition, the B stations enjoyed the regard and affection of the New Zealand public. Along with the petition, Hamilton received various deputations urging assistance for the B stations. The government felt unable to follow the board's advice and its own policy. At the end of the year Hamilton moved a little towards meeting the petitioners' request. The changes made were such that an unlimited number of sponsored programmes of music or elocutionary items, but not of talks or lectures, were permitted. An announcement could be made the day before the broadcast of the sponsored programme, and immediately before and after the programme. These were not to differ in any way from a model announcement supplied by the P&T Department, which stated that the programme was provided 'by Messrs Blank and Company, Limited, bakers, 61 Queen Street, Masterton.'[118] Thus the B stations ended 1932, the first year of the NZBB, with an apparent concession from the government. The changes to the sponsoring permission came too late for some B stations: during 1932 six stations ceased transmission. Those remaining considered they had won a victory and looked forward optimistically to a future when B stations, supported by a dependable income, would make a greatly increased contribution to New Zealand broadcasting.

7
Victory for the B Stations and Labour

THE FELLOWSHIP OF THE FRIENDLY ROAD

At the end of 1932 Tom Garland and Colin Scrimgeour resigned from their previous employment to work full-time with 1ZR. Thus began a type of broadcasting increasingly at odds with officialdom, which, after the Labour Party's 1935 electoral victory, reached a new kind of accommodation with the state. With the support of Lewis Eady's, the station's owners, Garland and Scrimgeour extended Garland's original proposal for non-denominational services and began the Fellowship of the Friendly Road, a radio church not tied to any particular denomination. The *Radio Record* reported it as the first of its kind in the world.[1] Previous religious broadcasting had always been under the control of one of the religious denominations.

Scrimgeour chose the name for the radio church: 'I didn't dream it up in the middle of the night. I read somewhere a little poem . . . "After all, Life is a road we must all travel. We begin the journey with our birth" . . . I took that title as a road that we must all travel. Full stop. Let's make it a Friendly Road.'[2]

Among the new programmes started on 1ZR early in 1933 was Scrimgeour's Sunday evening 'Man in the Street' session. It was a weekly talk which rapidly became Auckland's most popular radio session. By now an experienced broadcaster confident of his ability, Scrimgeour set out to make his weekly programme appropriate to the lives and times of those caught in the suffering of the depression. The session was at times criticised as mild in tone and saccharine in philosophy, but more important was the comfort and encouragement given to the many who were facing great hardship. Although heard principally in Auckland, the 'Man in the Street' session during 1933 became the voice of the people, and Scrimgeour the country's major broadcasting voice.

B STATIONS 1933

There was uncertainty among B-station proprietors as to what was allowed under the changed sponsoring regulations but most went ahead on the understanding that products could now be named. In effect, a sponsored programme was usually one or two items taken from records or performed live in the studio. At one stage 1ZR had thirty sponsored programmes in its breakfast session alone.[3]

During 1933 the official view hardened, and in November it was announced that sponsoring was soon to end. In February the B stations were informed that the changes to the regulations did not allow the naming of products, and the number of sponsored programmes dropped accordingly. Hands was told by his P&T Department informants that the changes limited the B stations. With the nature of permissible sponsorship so exactly delineated it was thought B stations could no longer exceed the permitted limits.[4] B stations were closely monitored and 1ZQ was closed at the end of March. The owner, Roy Keith, had his licence revoked for continual disregard of the limits to sponsoring.

On 7 June 1933 the 1ZR proprietors were similarly ordered to close their station. The main charge was that 'the licensees of 1ZR have been making a practice of deliberately altering the form of announcement of sponsored programmes'.[5] The station also was accused of advertising during its programmes. A schedule of examples of irregularities was attached to the official letter. They included this sentence taken from a gardening talk: 'I import my particular goods from England and supply them all over New Zealand.' Another example was, 'We will now have Mr Rimmer, who is entertaining us with "The Romance of the Coffee Cup"'. The P&T Department argued that the announcement of Rimmer's name (he was a coffee processor) was in breach of the regulations.[6]

1ZR transmissions were restored after a week of negotiations and an assurance of future compliance with the regulations. The station was supported by such bodies as the Auckland Chamber of Commerce which protested to the postmaster-general over the closure.[7] To his P&T Department contacts, Hands wrote that, 'The Minister is merely giving them [1ZR] a lesson and letting them see that he has the necessary power to prohibit broadcasting.'[8] Permission to restart broadcasts came on the eve of a protest meeting in the Auckland Town Hall on Thursday 15 June. Scrimgeour's biographer argues that the reprieve for 1ZR came after intervention from Gordon Coates, who knew of the impending meeting and of the following Scrimgeour, in particular, was attracting.[9] The meeting went ahead and the *New Zealand Herald* reported that 1500 people attended it.[10] The meeting gave further notice of the public support for the B stations and the ability of 1ZR to call out supporting crowds in protest at official action. Garland and Scrimgeour had considerable experience at attracting large crowds. This was the first time they did so in opposition to a government action.

After the resumption of 1ZR, the NZBB developed a tougher stance against the station and against Colin Scrimgeour personally. Hitherto the board, and particularly Hutchinson, the Auckland member, had a grudging admiration for Scrimgeour. On the closing of 1ZR he had told Hands, 'I can tell you the public will miss on Sunday morning and evening the religious services which are a little out of the orthodox by the Rev. Scrimgeour . . . He is certainly a man with a fund of sympathy for the despondent and down-and-out.'[11] When 1ZR resumed Scrimgeour was regarded as dangerous. Hutchinson's changed view is shown in his complaint to Hands about a Scrimgeour sermon entitled 'The Carpenter at the Conference': 'It seems to me that we have enough unrest without people's imaginations being inflamed in this manner over the air.'[12]

During the winter of 1933 B station proprietors sought public support for their cause. New clubs were formed, membership drives were held for existing clubs and the activities of stations and clubs were coordinated. The Dominion Alliance of B Class Stations was formed at a May conference held in Nimmo's studios. The Alliance was dissolved in February 1935 after foundering on disagreements and the difficulties of maintaining communication among widely dispersed stations.[13]

The government response to the B stations' activities was twofold. In September the postmaster-general decided to end any form of sponsored programme. Because 'the regulations were being exceeded constantly' sponsoring would cease from 31 March 1934 when the current licences expired.[14] Although the decision was not officially announced until November, the government made it known earlier to the B-station proprietors. The second part of the government's action was a selective purchase of B stations.

Lewis Eady's 1ZR and Hamilton Nimmo's 2ZW were the two major B stations: they were in the two main cities, were owned by established and relatively large businesses, employed station staff, broadcast for the most hours and had large and active supporters' clubs. If they could be removed, the power of the B-station movement would be countered. In September Eady and Nimmo were told that all sponsoring was to cease. Realising that the B stations were facing further claims from AWA and from the record companies, they sold their stations to the government — 1ZR for £2200 and 2ZW for £2300. The decisions to purchase were made and the negotiations conducted within the P&T Department, not the NZBB. The first reference to the purchases in the board files is when McNamara informed Hands that the Lewis Eady directors had decided to sell and that he, McNamara, was leaving that day for Auckland to settle the details.[15] A third station, 3ZC, was sold to the government for £600. The decision announced in the House in November was that the government would keep 1ZR and 2ZW on the air for some time with their existing staff and programmes.[16] This decision was not implemented and the three stations ceased transmissions on 4 December

1933 — the date they were taken over by the NZBB.

In a sense two of the stations continued, however, for the plant of both 1ZR and 2ZW was taken over by the NZBB. The 1ZR plant, with its 150 watt transmitter, was substituted for the existing weaker Auckland auxiliary station, and thus became 1YX. The 2ZW plant was taken to Dunedin where it became the auxiliary station 4Y0, and opened transmissions in March 1934.

With the closure of the three B stations, the NZBB filled the gap by increasing its own transmissions. The NZBB auxiliary stations' transmissions were extended from one to four hours per day. In particular the board started its own breakfast sessions, an innovation introduced by the B stations. The board had previously opposed breakfast sessions, following the lead of the BBC which refused to open at that hour. Macfarlane was the first board member to break ranks and argue both for a breakfast session and, more generally, that, 'We will be wise to develop along our own lines in this country.'[17] His view did not prevail until the end of 1933.

By late 1933 it looked as though the B stations were collapsing. Three B stations had been purchased by the government and closed. Notice had been given that sponsored programmes were to cease. It seemed only a matter of months before the remaining B stations, other than those subsidised by the NZBB, would also end their transmissions. The policy of a central board controlling a nationwide group of publicly owned stations seemed close to realisation.

1ZR TO 1ZB

Both Garland and Scrimgeour sought to continue the Friendly Road programmes. It had been announced that no more B-station licences would be issued, so the only hope for the Friendly Road was to transfer to an existing station. There were two offers. E. R. Boucher, the 1ZS owner, offered the Friendly Road the free use of his station.[18] The option which was eventually taken was that of Charles Pearson, owner of 1ZB, to sell his station. Pearson had started the station as 1YB in 1923. It was one of the oldest in the country, after the YA stations and 4ZB.

Garland has been credited with purchasing 1ZB for £50, sealing the deal with a deposit of £5.[19] Scrimgeour stated he arranged the sale, buying the station for £100 and raising the money by selling his wife's piano.[20] His biographer records him as buying 1ZB for £40.[21] Whoever arranged the sale and whatever the price paid, the purchase was the lesser part of returning the Friendly Road to the air. The more difficult task was persuading the government to allow the transfer of the 1ZB licence to the Friendly Road.

The closure of three city B stations was very much in line with the NZBB policy and was greeted happily by the board. Thus it was not in favour of allowing the Friendly Road to restart broadcasts and counselled strongly against transferring the licence. While the presence of city B stations was against the board's policy, it made it clear that its strongest objection concerned Colin Scrimgeour personally.

> No individual or religious body should be given a monopoly or quasi-monopoly in broadcasting from any station. The potential power given to any person granted such a facility can be realised, we feel, only by those experienced in broadcasting. Already the Board has had several instances of the danger of permitting any individual to develop what — for want of a better term — is usually called radio personality . . . Once a sufficient following is obtained, it becomes a difficult, if not an impossible task, to exercise reasonable control over the individual. We feel strongly that it is impossible to gauge the reaction on both the Government and the Board should the gentleman in question be given the desired facilities. My Board . . . considers it its duty to stress the extreme danger of establishing a precedent inimical to broadcasting and full of potential trouble to established authority.[22]

No doubt the government agreed with the sentiments expressed by Vickery, but Garland and Scrimgeour demonstrated they already had 'a sufficient following'. The Friendly Road's own record of the demonstration includes the following description:

> A public meeting was called for a Sunday afternoon in December 1933, where an estimated gathering of over 14,000 people met in Carlaw Park. The combined brass bands of Auckland turned out in full force. The day was fine. Uncle Tom led the band and the people in community hymn singing. Uncle Scrim addressed the multitude, receiving a most enthusiastic hearing . . . We faced a packed Town Hall that evening, while outside in Grey's Avenue, according to the Auckland press, there was an estimated crowd of over six thousand sharing the service through a public address system. It was a day of thrills.[23]

The Carlaw Park and Town Hall meetings were followed by a letter-writing campaign to the prime minister and MPs generally. By 13 December MPs were described as 'inundated with letters from correspondents urging the reinstatement of the "Friendly Road" broadcasting services'. The prime minister acknowledged receiving 2620 letters on the subject to that date.[24] Not only was the Friendly Road supported by the general public, leading Auckland businessmen also wanted it to continue. George Forbes received 'a petition signed by eighteen Auckland citizens of renown, a knight among them and four future knights. All eighteen were men of substance, the kind on whom conservative parties rely for finance'.[25]

The Friendly Road was offered, but refused, a position on the Church Services Committee so that it could join the roster for a share of the 1YA religious broadcasts. This was very much a lesser alternative to its own station. Forbes described the Friendly Road application for 'a separate broadcasting station . . . [as] a very serious step in broadcasting, and one that later on

"KAMERAD!"

A cartoon from the *New Zealand Observer* of 7 December 1933: prime minister Forbes changes his policy to the Friendly Road.

might have far-reaching effects, especially when applications come in from other similar organizations.'[26] However the pressure on the government was such that the application could not be resisted. The transfer of the 1ZB licence to the Friendly Road was approved, permission was given for the power output to be raised to 200 watts and the station was officially opened on 10 March 1934 by Gordon Coates, the deputy prime minister.

When it reopened 1ZB applied for longer hours and particularly for the right to broadcast every evening.[27] Again public support was considerable. The 1ZB cause was taken up by Gordon Coates who was 'inclined to do something'. Coates consulted with Hands who 'went fully into the question of the likelihood of Scrimgeour becoming a political menace as Father Coughlin is in America'.[28] Longer hours were granted and from 15 May 1934 1ZB was allowed to transmit the following hours: Monday to Saturday, 9am to 9.30am, 10.15am to 11am. Tuesday to Friday, 6pm to 7pm, 8pm to 10pm. Sunday, 9am to noon, 6.30pm to 9.30pm.[29] It was a grudging permission, with breaks in the morning and evening transmissions which would not help the station retain an audience.

The Fellowship of the Friendly Road functioned as more than a broadcasting station. Funerals, marriages and baptisms were performed by Garland and Scrimgeour. The first Sunday of each month was devoted to baptisms. In April 1935, in a typical dedication service, Scrimgeour was reported as baptising sixty-five infants: 'Some infants laugh and others cry and everything that occurs is broadcast.'[30] At first the Friendly Road marriages were also broadcast, but there were soon too many for all to be on the air. As well as performing these church services, the Friendly Road also operated as a social service centre, collecting and distributing food, clothing and money to the needy. Daisy Basham records Scrimgeour as very much in charge of this aspect of the station's work, vetting everybody and well able to pick the genuine from the undeserving.[31]

Existing precariously on the generosity of listeners, 1ZB employed a small staff and also paid part-time announcers. As well as Scrimgeour and Garland and two technical operators, C. McCarthy and S. Catton, the staff included W. Hindman as sports announcer, Rod Talbot who ran a weekly Diggers' Session, Daisy Basham and Dudley Wrathall. It was an experienced staff — Garland, Basham, Talbot and Wrathall all had broadcasting careers extending back to before the RBC gained its contract. All the 1ZB announcers became public identities but Basham, in particular, began on 1ZB a broadcasting style that rivalled Scrimgeour in prominence and considerably exceeded him in longevity. She was the announcer for the 1ZB morning sessions. Her regular opening greeting, a hearty 'Good *morning*, everybody', was in deliberate contrast to the anonymous style of the YA stations and was hers, and New Zealand's, for the rest of her career. The other sessions were also operated on a cheerful and informal format with a policy of endeavouring to keep up the spirits of a people deeply troubled by the depression.

Tom Garland expanded his choir work, and by the time 1ZB became the founding station of the NCBS, he had four choirs broadcasting. The children's choir went on the air at 9am on Sundays, the Service of Song adult choir broadcast at 11am on Sundays, the Sankey Singers were on from 6.30pm to 7pm on Sunday evenings and the Merry Makers choir, formed in 1935, broadcast on Friday evenings from 6pm to 6.30pm. Garland was an extraordinarily innovative broadcaster, not only with his choirs. Among his most popular broadcasts were his talks with Ian Shaw. The five-year-old schoolboy visited 1ZB and was taken to meet Garland, whose session was in progress. They conversed about Shaw's week, at school and at home, and Garland invited him to return for another chat. Shaw never realised at the time that the chats, which became a weekly feature, were being broadcast — but he was soon as much a household name as the other luminaries of the station.[32]

B STATIONS 1934-35

The close attention 1ZB received from the P&T Department was not peculiar to that station. All B stations were under scrutiny with continuing attention paid to any attempts to evade the prohibition on advertising. Transmission hours were also monitored and any broadcasting beyond them was punished by the department. The usual procedure was to require the station transmissions to cease for an equivalent period of time. This was particularly troublesome for sports broadcasting. 1ZB, for instance, often broadcast

The audience in, as usual, a full Auckland Town Hall stands for the playing of the national anthem at a concert given by the Friendly Road combined choirs. Tom Garland is at the microphone.

rugby league commentaries — a sport the NZBB would not allow on its stations — but such games were seldom played during the station's regular hours. The most publicised incident was in Christchurch and concerned 3ZM's 1934 rebroadcast, from 2UE Sydney, of the commentary of the first day of the Australia versus England cricket test match played at Lords, London. It was the first time such a commentary had been accomplished. The station remained on the air until 5am but its innovation was not appreciated by the department. The next day 3ZM was ordered not to repeat the commentaries and to forfeit five hours of broadcasting time. Among the considerable public comment was a verse which appeared under the title 'We Can't Let You'. It encapsulated the attitudes of many to the board and B stations and the uneven contest between them. As ever in public opinion, it failed to distinguish between the NZBB and the P&T Department.

What let you go upon the air
Delighting all the listeners there
With cricket tests, which are a sort of
Thing that we had never thought of.

Permit your prestige thus to grow
While others languish? No . . No . . No . .
What though your act is beneficial?
You must not do it — that's official.
The public want it? What care we
As long as we've the public's fee?
We charge them thirty bob a year
For what we think they ought to hear
And strive to elevate their minds
With classic stuff of varied kinds.
But when you say it's up to us
To please them — Bah . . . Ridiculous.
We won't allow the Tests — that's flat.
So note; You cannot broadcast that.[33]

A 1ZB clash with broadcasting officialdom late in 1934 concerned the plan to broadcast progress reports from Charles Ulm's aircraft as he attempted to fly the Pacific from Canada to Australia and New Zealand. The plan was particularly Scrimgeour's, whose other major interest was aviation. He had arranged for Ulm to send Morse messages from his plane while in flight and for these to be picked up by Auckland radio amateurs and then read from 1ZB. Again a broadcasting innovation was not permitted by the P&T Department, which vetoed the proposal on the grounds that it entailed a broadcaster competing with government commercial services.

The main difficulty for the B stations was the ending, from 1 April 1934, of sponsored programmes. Apart from the subsidised stations, the B stations, with one exception, now had no revenue other than from the generosity of their listeners. Five closed before the 1935 elections. Two of these — 4ZW Dunedin and 1ZH Hamilton — were purchased by the NZBB and then closed down.[34] 1ZH was one of the subsidised stations but the subsidy was stopped after the increase in power of 1YA, at which stage the board considered Hamilton was adequately served by the Auckland station. Hands considered the £500 price well above the station's value but worth it to get a B station off the air and to forestall the attempts of the Friendly Road and the *New Zealand Herald*, both of which wished to be associated with the Hamilton station.[35]

What is surprising is not that these stations closed, but that so many continued. Their continuation is a tribute to the dedication of their staff and a recognition of the public support they received. Most of the B stations that remained were small concerns existing on voluntary labour and broadcasting for only a few hours weekly. The policy that denied financial self-support to the B stations ensured that they remained undeveloped.

The one B station to develop alternative funding was the Manurewa station, 1ZM, which came to an arrangement with the *Auckland Star*, the details of which were not made public. The newspaper was reported as seeing

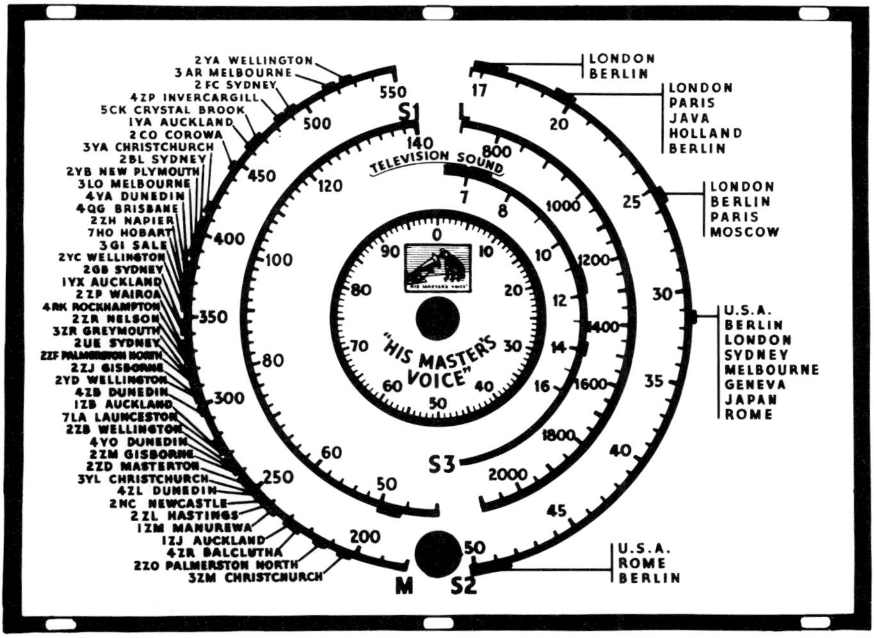

An HMV radio dial in 1935. ALEXANDER TURNBULL LIBRARY

'no question of financial benefit' but Macfarlane was given a different understanding of the *Auckland Star* proprietors' motives. 'Someday when Labour is in power they think that advertising will again be allowed in the air and they want to be in the running.'[36] William Goodfellow was one of the directors of New Zealand Newspapers, the parent company of the *Auckland Star*, and the NZBB considered it was he who had alerted the paper to the future possibilities of radio.[37] As part of the sponsorship New Zealand Newspapers began publishing a monthly magazine, *The Broadcaster*, which contained the 1ZM programmes, some articles on radio broadcasting, and considerable advertising.

BROADCASTING AND CONTROVERSY

The broadcasting of controversial material became increasingly topical during the first half of the 1930s. The matter came to a dramatic climax at 1ZB, but it was an issue that affected all stations, not just the B stations. In its first months the board stopped a talk on monetary reform in mid-broadcast and pointed out it had, 'to be extremely careful to eliminate from broadcasting anything of a controversial nature'.[38] The board and its Advisory Council soon reached the same conclusion as had the RBC — that the ban on controversy was detrimental to broadcasting. The Advisory Council resolved 'that the regulation prohibiting the broadcasting of controversial matter should be relaxed so as to give the Board similar liberty in this respect to that given the BBC in Great Britain.'[39] The government was slow to reconsider the regulation.

As it had with the RBC, the regulation circumscribed any talks programme the NZBB could offer. This applied particularly to the talks given by the WEA — an organisation which, with some justification, described itself as 'the only body in New Zealand concerned with adult education'.[40] The WEA was keen to extend adult education through broadcasting and in 1931 reached an agreement with the RBC to deliver a lecture each week from the YA stations. The experiment was considered successful.[41] In March 1932 the NZBB arranged with the WEA to present talks from each of the YA stations on two nights a week at 7.30pm.[42] The association was told to ensure that no controversial matter was broadcast, and although the WEA clearly disagreed with the regulation, sincere attempts were made to abide by it. The topics chosen tended to be those that would not offend, and those of contemporary social significance were largely omitted. Inevitably, however, complaints were made about the airing of controversy. At first the board instructed that 'more care be taken both in selection of speakers and in matters broadcast'.[43] Later the board resolved that 'copies of talks on doubtful subjects be obtained and scrutinised before delivery'.[44] Eventually in 1934 the NZBB cancelled its agreement with the WEA and no more WEA lectures were broadcast during the board's term. At the time the board was abolished it was arranging to introduce an 'Adult Education Session' under the guidance of different committees at each of the YA stations and under the general control of its supervisor of talks.[45]

A great problem for the board was the broadcasting of speeches by politicians. Under the regulations it was difficult to allow them to broadcast at all since politics could not be discussed. A typical incident concerned the refusal to broadcast the speech by H. E. Holland, leader of the Opposition, at the opening of an industrial exhibition in Dunedin. Holland's broadcast was cancelled after he refused to give an assurance that his speech would not introduce 'politics or any other controversial matters'.[46] The policy was regularly attacked in the House by the Opposition who saw it as deliberately favouring the government. Government speakers were more able to broadcast for, by virtue of their offices, they could announce new aspects of policy, this being acceptable as information rather than controversy. It was always an unclear distinction, and it led the Labour Party and a good number of the public to regard the board as politically biased.

In 1934 the topic of controversial broadcasts was faced reluctantly by the government. At its January meeting the board had to decide on three potentially controversial broadcasts by visitors to New Zealand: George Bernard Shaw, Krishnamurti and Major Douglas. Shaw was accepted immediately and Hands was instructed to endeavour to arrange for him to broadcast during his visit. It was decided to leave the Krishnamurti talk to

Hands's discretion and, in regard to Major Douglas, to seek the view of the postmaster-general.[47]

The postmaster-general did not allow Major Douglas, the advocate of Social Credit, to broadcast;[48] nor was Krishnamurti allowed. The refusals, particularly that of Krishnamurti, aroused controversy. First he had been invited to broadcast from 1YA. After inspection of his script, permission was withdrawn. Colin Scrimgeour then sought permission to relay over 1ZB Krishnamurti's speech at the Auckland Town Hall. Again permission was refused, 'because the draft contained controversial matter'. While the board had left the matter to Hands's discretion, in practice this meant he took the scripts to the P&T Department and the decision was made by his former employers who, as administrators of the Radio Regulations, were the final authority. 1ZB sought to have several studio talks by Krishnamurti, but was told the ban was total. Krishnamurti would not be allowed to broadcast in New Zealand.[49]

When Parliament resumed, the government came in for much criticism. That the prime minister had never heard of the Indian philosopher did not enhance the Opposition's respect for the government's handling of the case. Members noted the various countries in which Krishnamurti had broadcast, including Australia, and wondered at the need for New Zealand's particular sensitivity. The Opposition used the ban to indicate the discrepancy between the stated modelling of New Zealand's system of broadcasting on the BBC and the non-acceptance of the BBC policy of allowing controversial material to be debated: 'The wireless provides a wonderful means of educating the people, and the very fact that on a controversial subject they are able to listen to both sides of the question is itself part of a splendid educative process . . . This is the practice of the British Broadcasting Corporation, and it is surprising that the Minister should be ignorant of the fact.'[50]

The ban on controversial material was not applied to George Bernard Shaw, a figure of such stature in the English-speaking world that the NZBB accepted he should be exempt from censorship. Shaw broadcast twice in New Zealand. His civic reception in Christchurch was taken by 3YA and relayed to the other YA stations. He also spoke from the 2YA studio with his speech being relayed from several other stations, including B stations, and rebroadcast by various Australian stations. He was not required to submit to the usual procedure for speakers and offer his script in advance for vetting. On being asked if he intended to submit his text for approval, Shaw, who went out of his way not to embarrass his broadcasting hosts, replied, 'The question has not arisen, but I should have no objection if I could foresee what I shall say.'[51] His 2YA address was not the impromptu talk such a remark might suggest but was a well-scripted and entertaining broadcast that was easily the most controversial speech made over New Zealand radio. Nor was the board without warning from Shaw as to the content of his 2YA address. Hands reported on his talk with Shaw. 'I pressed him to talk on "Drama" or some

On 14 April 1934 *New Zealand Herald* and *Auckland Star* cartoonists gave their views of George Bernard Shaw's radio talk. NZ HERALD & AUCKLAND STAR

other literary topic, but he flatly refused saying he would talk on what he chose and warned me that he would be as controversial as H—, we could take it or leave it. He was most cordial but was adamant that he must say what he likes and will not be censored. There was no alternative but to accept his dictum.'[52] Shaw entitled his talk 'Observations Pertinent and Impertinent' and in it he, amongst other matters, argued admiringly that New Zealand was a communist nation, second only to Russia.[53] Scrimgeour regarded the broadcast as 'a popular masterpiece'.[54] Shaw's well-received broadcasts, which did not touch directly on the ban on broadcasting controversial matter, still helped show the ban as improper and unbecoming to a civilised country.

Shaw was not completely immune to the broadcasting regulations. 3ZM broadcast his *Androcles and the Lion* during the playwright's New Zealand visit. The district radio inspector would not allow the inclusion of the epilogue of the play because he deemed it controversial. Shaw greeted the action with laughter and wondered why the play itself was permitted.[55]

At the time of the visits by Shaw, Krishnamurti and Douglas the board asked the postmaster-general to amend the regulation so that the ban on the broadcasting of controversial material could be applied 'with less rigour than at present'[56] Hamilton and his Cabinet colleagues felt the request should be favourably considered. Hamilton's suggestion as to the best procedure was to allow station directors greater discretionary power, essentially the same as that of his predecessor to a similar request from the RBC.[57] Again both the government and the broadcasting executives accepted the need for change but the government was not prepared to take responsibility and alter the regulation. With the increasing head office control that had occurred since the NZBB took office, the station directors' discretionary power had declined and they were not the appropriate officials to oversee a relaxation of the ban on controversy. Nothing was to change.

During the rest of 1934 a succession of incidents kept the matter in public and political attention. There were three major instances. In July a speech over 2ZH by the chairman of the Napier Harbour Board was regarded by the minister as one which should not have been broadcast and requiring 'suitable notice [to] be taken of the licensee's failure'. The Opposition saw the broadcast as acceptable action by a responsible public figure.[58] In the same month the banning of two WEA talks provoked discussion. The speakers, two members of the staff of Auckland University, Professor W. A. Sewell and H. D. Dickinson, were to have discussed 'Religion and Philosophy as Manifestations of the Spirit of the Western Civilization' and 'Marxism and the Ideal Equality' respectively, but had their scripts rejected. Speaking in the debate on the annual report of the NZBB, Robert McKeen charged that the two had been treated in a ridiculous manner, with one speaker altering his talk twice but still not being allowed to broadcast it.[59] Professor Sewell, writing against 'the bleatings of the tender-minded', argued, 'The microphone could be and should be an open forum . . . [or] we shall hear nothing of the social and philosophical problems of our time. This seems to me public timidity of a particular poisonous kind.'[60] In September a series of talks from 1ZM on the topic of British-Israelism was banned as controversial, again prompting parliamentary debate.[61]

These and other incidents provoked a growing discussion, in which various public figures spoke against the restrictions on controversial broadcasting. Change came finally with the Broadcasting Amendment Act 1934–35. When the shape of the bill became clear, the board noted appreciatively that there would be 'a freer hand and more scope in the selection of topics'.[62]

BROADCASTING AMENDMENT ACT 1934–35

The change to the ban on the broadcasting of controversial material came as part of an act which also saw the Broadcasting Board's relationship with the P&T Department and the B stations brought closer to the stated intention of modelling New Zealand broadcasting on the BBC.

The Broadcasting Amendment Act 1934–35 increased the size of the board to seven and abolished the Advisory Council. Two board members were to be appointed from persons nominated by listeners' organisations. The government retained control over technical matters such as the allocation of wavelengths. It also remained the licensing authority. This included determining the hours a station could broadcast. Advertising was prohibited. The board was given the duty of supervising the programmes of all stations, including the B stations, with the power to prohibit or alter in whole or part 'a programme which in its opinion is unsuitable for broadcasting'. The board, in respect of a station which had not complied with its programming directives, could only report the matter to the minister who might 'in his discretion cancel the licence'. Finally the number of permitted B stations was limited to no more than those currently licensed.

Vickery requested from the minister stronger powers for the board over the B stations but his suggestions were not accepted.[63] Even the powers the minister was prepared to give were the focus of opposition. During the passage of the bill, the government reduced the board's proposed duty of 'controlling all programmes' to 'supervising all programmes'.

As part of the changes, the two 1932 Radio Regulations covering advertising and the ban on controversy were revoked. For the first time in New Zealand the supervision of programme content was taken from the minister and the P&T Department, and given to a broadcasting organisation, the NZBB. Depending on the board's will, New Zealand broadcasting could now be developed as an independent medium of news and public commentary as well as entertainment. It was not this feature but the board's authority over the B stations which was the major subject of parliamentary debate and public discussion. The B stations were not obviously the worse for the changes. They were totally denied the right to advertise — but they had existed under such a regime since sponsoring had ended. Their programmes were now under board supervision, but this was not necessarily more difficult than the previous arrangement. Their programmes had always been monitored and controlled, though by the department rather than the board. There was no longer to be the total prohibition on controversy, which presumably meant a far greater scope for all stations, B and board, though it remained to be seen how the board would define acceptable programming.

Three and a half years after the appointment of the first board, there was a keener appreciation of the importance of broadcasting and consequently a greater desire by various organisations to be represented on the board. The National Council of Women, the Newspaper Proprietors' Association and the WEA had previously asked to be represented.[64] But the minister appointed no one from any particular organisation, with the exception of the two representatives from listeners' groups. Vickery remained as chairman and Hutchinson was reappointed. They were joined by five new members: Ernest Palliser, Kenneth Phillips, James L. Passmore, W. H. Cocker and H. G. Livingstone. The *Radio Record* noted that 'The cultural side seems to have been well looked after.' Clearly this was an aspect to which the minister had finally paid attention.[65] Both Palliser and Phillips had strong musical backgrounds. Palliser, who had been the chairman of the 2YA Musical and Dramatic Committee under the RBC, was president of the New Zealand Bands Association and a founder of the Wellington Musical and Elocutionary Competitions Society. Phillips was the first and only Doctor of Music of the University of New Zealand. Passmore was the same James Passmore who had begun wireless telegraphy experiments as a sixteen-year-old in 1901. He and Palliser were the listeners' representatives, though no election was involved and, like the other members, they were appointed by the minister. Once again listener representation was at best tenuous. The *Radio Record* regarded only one of the pair, Passmore, as a listeners' representative.[66] Cocker was a member of the Councils of the Law Society and Auckland University College. Livingstone was a businessman active in Christchurch local and sporting bodies. As the Opposition soon pointed out he was also active in Reform Party and Coalition politics.[67]

THE NZBB AND ITS NEW AUTHORITY, 1935

In his address at the new board's first meeting, the minister said the government intended to give the board a free hand but that the board should consider the government's position in anything affecting the government. Hamilton stressed that the matter of political talks was left entirely to the board.[68] His expectation that the board consider the government's position was no more than the British government had expected of the BBC. The minister and the Cabinet, particularly Gordon Coates, had become convinced that the previous curbs on the use of radio as a medium of public discussion were no longer acceptable. What would develop in practice had yet to be worked through, but a genuine change was offered by the Coalition Government.

The board now had considerably greater scope than it had previously enjoyed. The change also came at an important point in the country's history. It was an election year coming after years of depression that had wrought changes in the allegiance of New Zealand voters. Radio ownership had grown to the point where the medium could, for the first time, play a major part in the nation's life. The board had the opportunity to make radio a matter of real political substance in 1935. Yet the board's actions were timid and were an attempt to withdraw radio from political involvement or significance.

The board did not intend to ban controversial material permanently — and it told B-station proprietors so.[69] Equally it was in no hurry to change past

Alan Mulgan; cartoon by McNamara.
RADIO RECORD

practices, especially in the election year of 1935. Its procedure was to add to its staff a supervisor of talks; to this end Alan Mulgan, a leader-writer and literary editor for the *Auckland Star*, was appointed. He was offered the position in July 1935, joined the NZBB in September, and it was 1936 before the board was prepared to address the topic of controversial material. By that time the board's view was irrelevant.

The NZBB failed to make immediate use of the new freedom it was offered. At the first meeting of the new board there were three applications for broadcast talks. All were refused.[70] Two refusals, of the applications from the Friends of the Soviet Union and from the Theosophical Society, indicated the start of a pattern that continued for the duration of the new board. The previous ban by regulation of the broadcasting of controversial material was continued by administrative fiat. The third application was for broadcast talks by the candidates for the Wellington mayoralty. The decline of this request was the first example of the board's unwillingness to allow radio to be used in political contests. It was an indication that the board, in spite of its new permission, intended to retreat from any political involvement in the election year of 1935.

The B stations were also discussed at the first meeting of the board. All received letters informing them they were now under board supervision and telling them of the board's prohibition of political discussion. All B stations were 'asked to submit information regarding the programmes they broadcast and in particular the nature of the spoken matter transmitted'. They were also told that 'No speaker should be allowed to express an opinion regarding the actions of Parliament or Government, or the Administration of Government Departments, boards, or local bodies set up in pursuance of statutory authority'.[71]

By June the board decided on a complete banning of political election material from radio. All stations, board and B, were issued with a new instruction: 'Candidates for political honours at the next General Election are not to be permitted to broadcast in any capacity. Exception is to be made in cases where a candidate holds a Mayoral position and broadcasts in a Mayoral capacity, and does not deal with politics.' This was accompanied by a further requirement: 'The manuscript of all talks which might infringe Radio Regulation 138 is required to be forwarded to the Board's Head Office in ample time to permit examination of the matter.'[72] Regulation 138 was that gazetted in 1932 prohibiting the broadcasting of controversial material. It had been revoked as part of the changes brought in with the Broadcasting Amendment Act.[73] Not only did the board not take advantage of its revocation, it instructed stations to continue to act within its confines.

After receiving the instruction, Garland, writing as 1ZB station director, was puzzled: 'We understand that the Board now has power under these regulations to use certain discretionary rights and we would like to enquire whether any new principles have been laid down and if so when they are likely to come into operation.'[74] Hands replied, 'You will be advised of any changes made.'[75]

The Broadcasting Amendment Act's total ban on advertising was interpreted particularly strictly. Not only were commercial announcements prohibited, as was required in the legislation, but hitherto acceptable public announcements were no longer permitted. Thus, among other examples, 2ZO was not allowed to announce particulars of the Sanson Methodist Diamond Jubilee, 2ZP and 2ZL could not announce the Sunday concerts arranged by the Hastings branch of the New Zealand Relief Workers' Association in aid of the unemployed, and the Dunedin B stations were refused permission to announce a ball held to raise funds for the local Children's Rest Home.[76]

Significantly, these decisions against the broadcasting of announcements were made by the P&T Department rather than the board. This is indicative of the confusion about what authority the board and the P&T Department held over broadcasting. The board had supervisory authority over all programme content, though its disciplinary action was confined to reporting infringements to the minister through his department. The P&T Department continued to police the radio regulations. Control over other matters in the Act, such as the ban on advertising, was not specified. In spite of the changes made in the Broadcasting Amendment Act, which gave the board considerable supervisory power over broadcasting, the P&T Department was loath to relinquish its power. The board continued to accept the department's authority.

This was most noticeable when the P&T Department continued to wield authority over the NZBB concerning the content of programmes. As an illustration, McNamara raised with Hands the topic of the president's speech at the opening of the Auckland Winter A&P Show, which was broadcast from 1YA. He asserted that the speech 'contained matter of somewhat political bias' and asked, 'Perhaps you would kindly cause enquiry to be made and furnish the comment of your Board upon the matter.'[77] In this, and in similar incidents, rather than reminding the department secretary that the matter was no longer one for his jurisdiction, the board did as requested regarding 'the regretted failure'.[78] Thus the P&T Department continued to monitor board and B stations and exert authority over their activities.

COLIN SCRIMGEOUR AND 1ZB

From the time of the first broadcasting regulations the one exception to the ban on controversial material had been the acceptance of religious broadcasting. This is not to say that religious services had entirely escaped censorship — there had been incidents both before and during the board's tenure. During the years of the RBC a sermon from 4ZP Invercargill, which raised the topic of the use of the Bible in schools while this was a matter under parliamentary discussion, resulted in Parsons Radio Supplies being warned they could lose their licence if the offence were repeated. In 1932 4YA cut for two minutes a sermon by the Rev. E. S. Tuckwell when he introduced the topic of alcohol and the closing time for hotels.[79] There were similar incidents, but nevertheless this area of radio remained the one most open to the airing of controversial material.

At 1ZB Colin Scrimgeour was well aware that it was his privileged position as a minister of religion that allowed the Friendly Road to exist and him to broadcast.[80] He also understood that the public support for 1ZB and himself personally was the crucial element which had both gained the Friendly Road its broadcasting licence and enabled it to continue. During 1934 and 1935 much of his effort was spent on expanding the public knowledge of the Friendly Road. He visited North Island centres with a programme relayed from Rotorua and meetings held in New Plymouth, Pukekohe, Hamilton and Putaruru.[81] The power to call out supporters was demonstrated again at a rally on Carlaw Park on 2 December 1934 celebrating the second birthday of the Friendly Road. Six brass bands were at the rally and two services were held in the Town Hall during the day. The hall was packed on both occasions and special trains were required to bring supporters into Auckland from country areas.[82]

In 1934 Scrimgeour also established links in Australia with 2UE Sydney, a private station controlled by the *Sydney Sun* editor Eric Baume. He was in Sydney for several weeks in April and May and again in September and October. The *Radio Record* reported that he founded a branch of the Friendly Road in Sydney.[83] As his biographer notes, he was also making sure of an Australian broadcasting position should he be banned in New Zealand.[84] As well as having a successful show in Sydney and assuring himself of future employment should he need it, the future controller of Commercial Broadcasting had a first experience of commercial broadcasting, and grew to like its Australian style. While in Australia Scrimgeour also approached Major W. T. Conder, general manager of the Australian Broadcasting Commission, to seek employment. This approach came to nothing after Conder asked Hands for information and was told the board regarded Scrimgeour as dangerous and would not employ him.[85]

In New Zealand in early 1935 Scrimgeour again took the Friendly Road message to North Island centres, addressing 'large and enthusiastic meetings of listeners' as far south as Wellington.[86] The board became increasingly concerned that the 1ZB example was spreading and threatening the government's broadcasting policy. This concern increased when Scrimgeour founded and became president of the New Zealand Federation of B Station Proprietors. This was a successor organisation to the Dominion Alliance of B Class Stations, but was restricted to proprietors, and was therefore smaller and easier to coordinate. There is no record of membership, but the majority of proprietors joined. The federation, with Scrimgeour as president and Ian Mackay of 2ZR as secretary, was energetic in promoting its cause.[87]

1ZB's example was followed in Dunedin in April 1934 when the Rev. Leslie Neale (Uncle Leslie) started New Zealand's second radio church — the Church of the Helping Hand, which operated through 4ZM. A service was broadcast every day the station transmitted. By the time the church celebrated its first birthday there were some 2500 members of the 4ZM radio club. The birthday celebrations, held in the Dunedin Town Hall, attracted well over 3000 people, and several hundred were unable to get into the packed hall.[88]

The conventional churches were unhappy with the public attention the radio churches were receiving. Their dislike of the situation threatened the existing NZBB system of religious broadcasting. The cooperation of the churches on the church broadcasting committees had always been fragile. Many smaller churches were denied admission to the roster and the larger churches, particularly the Church of England, considered themselves entitled to a greater share of the radio time. The Anglicans had always argued that they should conduct the majority of the broadcast services as their adherents made up the majority of the New Zealand population. With the advent of the radio churches the Church of England argued it should have its own licence so as to operate its own station.[89] The fear of the NZBB in 1934 and 1935 was that the number of B stations, which had been shrinking, would again expand with a growth of church-based private stations.

After the NZBB became responsible for supervision of B stations, the YA station directors were instructed to bring to Hands's notice 'any instances of B stations infringing regulations or otherwise acting in a manner likely to cause trouble'.[90] Much of the consequent information concerned advertising from B stations or their charging, against regulations, for announcements. 'With regard to the announcement during 4ZM's transmission that a St Bernard dog named Sandy had been lost we managed to ascertain by the simple process of ringing the telephone number concerned that the B station charged the individual 2/6d for this service.'[91] Most of the information from Alex O'Donoghue, the 1YA director, concerned 1ZB and Colin Scrimgeour and was regarded as indicating a deliberate and increasingly serious disregard for the board's authority.

Scrimgeour again worked for 2UE Sydney for three months from May 1935, with a daily 9am session as well as Sunday evening broadcasts. As in New Zealand, his broadcasts were very well received. The *Radio Record* reported him as installed in an elaborate suite of offices and receiving a daily fan mail greater than that of any other Australian celebrity.[92] He also managed to continue his 1ZB 'Man in the Street' session by using newly available recording equipment. The NZBB saw this use of records as an ominous portent: 'I presume he will endeavour to send them [the records] to other stations, if he can get an arrangement with them. The danger I see is . . . he overcomes the handicap of being on a low power and a short number of hours.'[93] Making records was particularly expensive in the 1930s, so Scrimgeour also sent letters. These were the subject of one of O'Donoghue's complaints: '[Scrimgeour] makes a practice of writing weekly letters to his wife, and these she reads over the air to listeners, and generally they contain a sheaf of propaganda work for B stations, by balancing up the virtues of the commercial stations in Australia against the results accruing to the national service.'[94]

The board's main concern with 1ZB in 1935 was its connection with the political campaign and the Labour Party. This had concerned the board at least since May 1934 when, in an interview in the *Auckland Star* on his return from Sydney, Scrimgeour was 'very disgusted with the regulations under which we have to work in our own station, and with the conditions generally of broadcasting in New Zealand.' He noted that the Australian B stations were so powerful and so popular that any government that attempted to interfere with them 'would go out.'[95] This was seen by the board as a 'political threat' that 1ZB's influence would be felt at the next election.[96] In 1935 the board showed increasing concern with 1ZB's political actions. In July, O'Donoghue complained about Forbes Eady broadcasting from 1ZB.[97] His broadcasts were a series of talks on the voyage of the *Bounty* — but it was the fact rather than the content of the broadcasts that was the issue. Forbes Eady was a candidate for the parliamentary elections and, as such, was not allowed to broadcast. His broadcast came within days of the NZBB instruction that such broadcasts were prohibited. It was the first of a series of offences, the most serious being a broadcast by the Labour MP, John A. Lee, in September. Lee spoke on 'Poetry in the Trenches' — but again the topic was not the point. This 'direct flaunting of the Board's instructions' was seen as leaving 'no alternative but to accept the challenge and take the matter up'. Yet the board was aware that this was what 1ZB wanted: 'They are endeavouring to provoke the Board to action so that they may create and stimulate sympathy for their station.'[98] The board referred the matter to the P&T Department, for the postmaster-general alone had the power to take action against a B station.

Rod Talbot, the 1ZB announcer in whose Diggers' Session John A. Lee had broadcast, went to the newspapers. 'If they [the Government and the Broadcasting Board] attempt to interfere with my effort to entertain the troops in the various hospitals and institutions I will start something they won't be able to stop.' He had powerful backing from the president of the Auckland Returned Servicemen's Association, who stated, 'Every member of my executive and a big majority of the clubs with whom we have so far got in touch are prepared to stand behind Mr Talbot and support him in his fight.'[99]

No action was taken against 1ZB. This was extraordinary. All other stations, board and B, were strictly controlled. The one other major breach of the requirement that political candidates be denied air time occurred in the same week as Lee's 1ZB talk. D. G. Sullivan, the Labour member for Avon and also the mayor of Christchurch, spoke from 3YA. He was permitted to speak as mayor, and had spoken in that capacity. But the speech had been political and included various criticisms of the government. The board officer on duty was dismissed for not stopping the broadcast.[100] 1ZB's freedom from official reprisal indicated the minister's fear that action against the station would again provoke large-scale demonstrations against the government. The lack of action demonstrated that public support for Scrimgeour made him relatively immune to the control exercised over all other broadcasters. It was noted with interest by his colleagues that 'Scrim has been giving the Broadcasting Board and the Government Hell in Auckland but they have not raised one protest.'[101]

Scrimgeour left for Australia again in early October 1935. In an interview at his departure he made it clear he would return for the election, as he had no alternative but to engage in political activity.[102] He had written to Mackay the previous week outlining the political action he was contemplating:

> I spent two hours with Mr Savage this morning . . . he has stated that the first plank in their policy is that the B Stations should not only be allowed to exist, but that they should live up to a standard of which they are worthy. He is in favour of freedom of speech without restriction, within the bounds of decency. He is most emphatically in favour of advertising on the air if no more satisfactory way can be evolved by the B Stations themselves; and he says that if advertising is granted it must be unrestricted. So that sounds pretty promising! I feel more and more that

> the necessity will arise of stating in very definite terms that as far as the whole Dominion is concerned there will be only two parties to be considered and the listeners must be advised to choose the one which makes the clearest promise.[103]

1ZB combined its transmissions with political activity and continued its broadcasts with scant regard for the new requirements to refrain from fundraising on the air. Various publications were issued by 1ZB during 1935. Following the lead of 1ZM, a monthly journal, *The Voice*, was started at the end of July. It gave notice of the station's programmes, carried advertising and was sent free to all members of the 1ZB club. A series of pamphlets were also published. Scrimgeour's life story was published in three parts. The 1ZB Radio Club printed thousands of copies of 'The Scandal of New Zealand Broadcasting'. This called on the government to 'openly acknowledge the necessity for the B stations and permit them to earn revenue from advertising', and promised a survey of the political parties' broadcasting policies. 1ZB, through the Federation of B Station Proprietors, then issued a questionnaire to the various political parties. Only the Labour Party's reply gave full support to the B stations. This was followed in November, the month of the election, by the pamphlet 'More Scandal', in which the questionnaire results were published and which also included a call to support Labour in the election. This was proper political activity in that the information and the call were printed rather than broadcast. But there was little doubt among the public and none in the NZBB and P&T Department that Colin Scrimgeour and 1ZB were Labour advocates. The official apprehension was that Scrimgeour, in the last days of the campaign, would broadcast appeals for the electorate to vote for Labour.

Scrimgeour returned to Auckland in early November 1935. In the final days before the election his attitude to the broadcasting rules was increasingly cavalier. In one incident, he urged people to buy his life story. According to the 1YA station director, Scrimgeour 'stated that owing to the restricted liberty with regard to giving publicity to the book over the air, it has not been selling; but he felt that as money was so urgently needed, he was justified in making a breach of the regulations on this occasion, and putting out a further appeal on behalf of the booklet.'[104]

THE JAMMING OF 1ZB

On Wednesday 20 November 1935 Tom Clarkson, an assistant radio engineer at the Wellington General Post Office, was called into the office of his superior, P. H. Mason. Clarkson was told that, through a confidential source, it had been learned that in his next 'Man in the Street' session — the final one before the election — Scrimgeour intended to tell his listeners how to cast their votes. The radio engineers had been instructed to blot

Scrimgeour's remarks off the air by means of another transmission. Clarkson was ordered to prepare for and carry out the jamming.[105]

The equipment used was 'the roarer', the diathermy oscillator once used in Christchurch by Dr Henry Thacker (see Chapter 6). Its frequency was lowered, it was fitted with fine-tuning apparatus, and sent by train to the P&T Department yards in Newmarket where it was installed in one of the huts. In Auckland, Clarkson and a local department radio engineer, Geoff Gilchrist, completed the preparations to have a strong signal available close to 1ZB's operating frequency.

On Sunday 24 November 1935 Clarkson was in the Newmarket hut listening to the 'Man in the Street' session. He called Mason in Wellington to check that his instructions remained the same. Clarkson's own account reads,

> Scrim worked up his subject, on the disadvantages suffered by some sections of society. He said they should recognise and use any opportunity to improve their conditions, and that the information they got from the usual sources of information about the deciding factors was deceiving. They should understand how their interests could be served. The General Election coming up was an opportunity. Many voters were young, and voting for the first time and . . . I shifted the frequency down by roughly four hundred hertz. The silvery tones of Scrim's voice seemed to increase in volume, but became unintelligible and rather like a chorus of duck quacks.

The jamming was stopped after a few minutes but was restarted when Scrimgeour returned to the microphone and 'picked up his theme about the forthcoming election'. The jamming was maintained until 1ZB closed down, soon after.[106]

THE AFTERMATH

Clarkson presumed he would be required to remove all the equipment rapidly from the Newmarket yards, but Mason ordered him to leave it in place in case it were needed again the following night. Clarkson had been stationed at the Newmarket yards in previous years and was known to many P&T Department workers, among them Percy Humphries, a telephone faultman who was secretary and treasurer of the local Fabian Club. Humphries spoke with Clarkson prior to the jamming and accepted the explanation that Clarkson was in Auckland to replace the faulty Tiri telephone line with a radio connection. After the jamming he realised what was going on, and sent a message to Scrimgeour.[107] Thus, on the Monday, a group of 1ZB supporters — among them Colin Scrimgeour and the Labour MP, John A. Lee — were at the Newmarket yards. There was considerable difficulty in obtaining access to the hut, but eventually the offending apparatus was discovered under a wrapping of hessian. Lee said, as only he could, 'Good God, Scrim blacked out, and the weapon then concealed by a different kind of scrim.'[108] The jamming was a major news item, the *Auckland Star* called it 'the most sensational development in connection with radio broadcasting in New Zealand'.[109] Added to it was now the connection with the P&T Department.

Back in Wellington Clarkson was summoned into the office of the director-general, George McNamara. At this meeting Clarkson was told by McNamara, 'I've made a blunder.'[110] Clarkson understood the decision to jam 1ZB was taken by the director-general, not by the postmaster-general. Clarkson was asked to become a scapegoat and admit to mistakenly and carelessly interfering with the 1ZB signal during experimental work. He refused to do so but offered, as a loyal officer of the department, to keep silent. Clarkson emerges from the whole episode with his honour intact. He had initially advised against the jamming: it was a technically inferior way of doing the job, and he felt the extent of Scrimgeour's public support had not been taken into account. However, he had competently carried out the instructions from his superiors and then resisted the pressure from his department's head to lie about the events. He maintained his silence for forty-five years.

The department released a press statement that the jamming was an accidental consequence of an incorrectly tuned condensor attached to an experimental oscillator, which had been taken to Auckland for power measurement work. From 1ZB Scrimgeour argued that the official explanation was 'utter rubbish'.[111] McNamara insisted, 'We have been perfectly frank in the matter.'

The general election supervened and Labour became the government. It would be idle to argue that the jamming was a dominant factor in the election. Clearly the depression years sweep other experiences aside as an explanation for Labour's ascendency. But the B stations, and particularly 1ZB, were linked with the growing support for Labour. As never before, broadcasting policy was a factor in the 1935 election. The electorate was informed that only under Labour would the B stations continue. They were more and more seen as victims of an antagonistic government. Like the population generally, they were portrayed as oppressed, and the jamming, occurring only days before the election, was a dramatic climax to this suffering. As Lee has noted, the jamming 'kept pressure on the Government right till election day', and 'caused people who might not have voted to be so aroused that few votes were lost through indifference'.[222] Although it had a wider audience, 1ZB was principally an Auckland station. However Auckland was a major area of support for Labour and the party won ten of its twenty-eight seats in the province.

Immediately after the jamming, Coates issued a statement denying any government involvement and, in fact, seeing the government as possibly the intended victim: 'The interference is either a childish rag or an unscrupulous

Savage, at left, and Scrimgeour in congratulatory mood at the 1ZB microphone on the night of the 1935 election. ALEXANDER TURNBULL LIBRARY

attempt to make political capital by throwing suspicion on the Government.'[113] Once the jamming was linked with the P&T Department, Hamilton took the proper constitutional stand as minister and assumed responsibility for his department's actions. However, he denied giving orders for the jamming and acknowledged only ordering that a close watch be kept on Scrimgeour: 'The Inspector found the director [Scrimgeour] was breaking regulations, and met the position in his own way. My instructions were to stop his licence the next morning. I have to take the responsibility, however, because, no doubt, the officer thought he was acting wisely.'[114] In 1936 Frederick Jones, Labour's minister in charge of broadcasting, asked for a report on the episode from the director-general of the P&T Department. He reported McNamara as stating that Hamilton agreed to the jamming, and once it was done, endorsed the action. McNamara also said that the final instructions should have come from him, but that the actual jamming was carried out without his being advised. Hamilton denied this version of the events.[115] It is not likely that Hamilton, who, in spite of 1ZB's previous offences, had accepted the political impossibility of closing the station, would have changed his approach in so politically naïve a fashion as to jam Scrimgeour's final pre-election broadcast. In a June 1936 parliamentary debate he asserted that the jamming was the independent action of a radio inspector. Was the former postmaster-general covering for his director-general, and thus misleading Parliament, or had he still not been told that the radio inspector had been sent from Wellington with specific orders to carry out the jamming?

The NZBB was widely suspected of being involved in the jamming and many complaints were lodged with 1YA on this point.[116] Clarkson was told when first ordered to Auckland that the board was available to help and J. R. Smith, the NZBB chief engineer, was ready to collaborate.[117] In the event no assistance was needed from outside the P&T Department. From their letters to Hands it is clear that neither of the Auckland members of the board had prior knowledge of the event. The NZBB did not take part in the jamming but whether Hands or Smith, two ex-P&T Department officers who kept in close contact with the department, knew of the plan is more difficult to judge.

Within the P&T Department it was assumed that McNamara was responsible. An anonymous judgement on his part in the affair was given three years later in the P&T Department employees' association magazine, *The Katipo*, as one of the verses in a squib suggesting the director-general resign:

> Then pand'ring to the powers that be:
> 'I'll wreck a radio' said he,
> And smiled with hopes of extra hoot
> Until those powers got the boot.
> He wrecked the radio alright
> And his patrons overnight.[118]

THE BROADCASTING BOARD YEARS IN RETROSPECT

That the radio stations controlled by the NZBB would be taken over by the government was not in doubt from the moment of Labour's election victory. It was a policy clearly and consistently stated since Labour first considered the matter. The Broadcasting Act, which brought in these changes, was not introduced until the following June, for broadcasting was not the most urgent of the topics tackled by the new administration. Thus the board was not dismissed immediately. Vickery's appointment was twice extended by the new minister and the board retained control until the end of June 1936. Essentially, however, the old order stopped with the Labour victory. The board, in fact, was to implement, against its own advice, the Labour Party's prize objective for radio; the broadcasting of Parliament.

The NZBB presided over five years of rapid growth in radio ownership. Throughout the board years the country was high, usually fourth, in world rankings of radio licences as a percentage of the population. The statistic that at 31 March 1936 New Zealand had 12.22 licences for every 100 people indicated that more than half the households in the country had a radio.

Commensurate with this growth was a considerable rise in income for the NZBB. Throughout the board years each annual increase was greater than that of the previous year; the 1936 income was more than double that for 1932, in spite of the fact that the licence fee was reduced in March 1935 by 5s to an annual payment of 25s.

The board years saw a substantial increase in hours of transmission, from 10,612 hours per year at the start of 1932 to 27,456 hours per year four years later. Each year the board spent much the same percentage of its income — an average of 31 per cent — on programmes to fill these hours. In its accounts this is a broad category that encompasses payments to artists, gramophone companies and the APRA, salaries and expenses of programme staff, payments for relay lines, records, stationery, and sundry miscellaneous expenses. It was not a large percentage for what was the prime purpose of the organisation, and the NZBB was open to criticism that its expenditure on programmes was low. However, such criticism was muted because, although the percentage of income spent on programming remained stable, there was a continuing rise in the actual amount spent.

The board's largest expenditure was on property and equipment — it built new stations and stronger transmitters — and this was its main contribution to New Zealand broadcasting. By mid 1936 some £150,000 had been spent on upgrading 1YA, 3YA and 4YA and most of a further £100,000 on the still uncompleted 2YA. These items represent the expenditure of some 36 per cent of the board's income, added to which was the establishment of the four auxiliary stations. The businessmen appointed to the board saw their major task as the development of the property and plant for a large-scale, publicly owned, national broadcasting organisation. This they performed competently and to the best contemporary radio standards.

Administrative salaries and expenses were not excessive. Neither were maintenance expenses, which included power and the engineering and operating staff's salaries. The other considerable item in the accounts is depreciation. The board allowed an average of over £20,000 per year for depreciation, equivalent to 13 per cent of its income. In its first report this 'relatively heavy depreciation' was explained by the fact that the equipment was 'comparatively old'. This raised some public and parliamentary criticism. The £59,000 settlement with the RBC had already been criticised, since it meant that the public was paying twice for that company's equipment — through previous licence fees and through the settlement. After the board's report the criticism was that the equipment was worthless. The board was required to pay the £59,000. In June 1936 there was still £27,300 to pay. With the demise of the board, the minister in charge of broadcasting allowed the liquidation of the remaining liability.[119]

The NZBB was not an innovative programmer. Its main new programme category — the breakfast session — was introduced in response to pressure from the B stations; otherwise, programmes did not differ greatly from those that had been introduced by the RBC. It did, however, make an administrative contribution by arranging to keep the various programmes permanently on the air. This is particularly noticeable in educational broadcasting, where it developed the tentative experiments of the RBC into permanent features of New Zealand education; and in sports broadcasting, where it made the necessary payments to resurrect the moribund racing and rugby broadcasts — the central features of New Zealand sports broadcasting. It made mistakes — stopping the American recorded programmes, for example — but it was prepared to accept it had acted in error: by the end of its term it had made arrangements for the importing of American, Australian and British recorded programmes. The long-term perspective of the NZBB can be attributed to the growth of broadcasting and its emergence over the decade from 1925 as a permanent social institution. It is also a reflection of the board's different orientation: the RBC had been a private company on a limited contract, and therefore unable to plan for the future; the NZBB, on the other hand, was founding a major public utility, and was not restricted in its planning.

Programming events of significance during the years of the NZBB were the considerable increase in transmission hours and the centralisation of programme organisation and control in its head office. These, along with the spread of radio, meant that by the mid 1930s many people in New Zealand, whether they lived in cities, towns or country, were regularly listening to the same radio fare. It was a uniformity of experience that added to the shared nature of New Zealand culture. The increase in radio listening is a major point in the change in New Zealand from a regional to a national focus. The NZBB continued a process of nationalisation that had begun in the nineteenth century.[120] The new focus also represented a denial of the importance of the differences among the audience and an understanding that the audience could be treated as one. It was as much a creation as a recognition of an audience. The advantage of the policy was to allow the coordination of programmes nationwide and thus, as transmitters and radio sets increased in power, to allow listeners a choice of programme types from the different stations. Its disadvantages were the turning away of stations from their local communities and the disallowal of broadcasting initiative from local staff. Transmission hours were increased, but the centralisation policy and the removal of the earlier regulation allowing only 25 per cent of broadcasting to be gramophone records meant there was less reliance on local talent and local broadcasting staff. The NZBB argued it had overseen a considerable rise in the quality of the New Zealand talent heard on the radio. This was so, but it was at the cost of the previous easy access to the microphone. The NZBB brought in a more formal relationship between the community and radio.

The NZBB was, of course, not the only broadcaster. The dominant feature of radio in these years was the existence of two distinctive types of stations —

'B' and 'board'. The board stations were well-funded and had government support, whereas the B stations, even those subsidised by the board, were underfunded — often desperately so. The B stations' relationship with government and broadcasting authorities ranged from temporary acceptance while the board grew to take their place — as was the case with the subsidised stations — to antagonism and harassment as the B stations attempted to survive in a regulatory regime which inhibited their growth and hastened their demise. The programmes of the board and B stations were quite distinct. Unlike the board's stations, the B stations were deliberately focused on their local communities.

The B stations' precarious financial status ensured that they remained largely amateur organisations — which is not to suggest that their programmes were devoid of skill or unappreciated. Given their lesser hours and power output, they received a disproportionate amount of attention from listeners. Most B-station staff were volunteers working in their spare time. It was not a type of broadcasting the B-station proprietors, with their agitation for the right to earn an income, wanted to be permanent. In their few short years the B stations developed a record of service, an informality and a rapport with their local communities which earned them real affection. Their actions and escapades also gave rise to a fund of incidents which linger in folklore. These are epitomised in the 1932 incident at 3ZR Greymouth. Unable to afford records and, besides, unable to obtain the latest releases, the station relied on its supporters. Ships' crews were regular providers of new records for the station, and the crew of the *Kaponga*, after delivering the latest hit, requested that it be played for them as the ship left port, so they could listen on the ship's radio. After the record had been played, Mick Spiers, the announcer and proprietor, heard the ship's siren sound. Assuming the crew was appreciative, he played the record again. The sound, however, had been a distress call: the ship had hit the Greymouth bar and was wrecked, fortunately with no lives lost. The record had been 'He Played His Ukulele as the Ship Went Down'.[121]

The programmes of the board and the B stations were often categorised as 'highbrow' and 'lowbrow' respectively. It was not that simple. The board stations' largest programme category was always what was referred to as 'light music', and Hands, like Harris before him, made continual efforts to increase the popular content of programmes. However the board had obligations to present a range of material and was also imbued with the understanding that radio should be used as an educative and uplifting as well as an entertaining medium. Thus it programmed considerable material that did not appeal to a majority audience. The B stations were not completely immune to such thoughts either, and the most deliberate experiment in the transmission of 'highbrow' material in these years was carried out from a B station, 3ZM. The Air Institute, a group of Canterbury intellectuals headed by Dr Frank Birkinshaw and including Professors Arnold Wall and James Shelley and Dr

C. E. Beeby, arranged with the station to broadcast a series of deliberately 'highbrow' programmes — talks, music and plays 'of a type which the New Zealand Broadcasting Board would rightly regard as not satisfactory entertainment for the majority of licence-holders'.[122] This was unusual, however; the B stations' usual emphasis was on popular records.

The board and B stations were increasingly distinguished politically. Originally hampered by the regulation banning controversy, the board continued, after the removal of that regulation, to attempt to keep radio from any involvement in the political affairs of the nation. It was a Canute-like attempt which rebounded, with many seeing the board's ostensible political neutrality as a deliberate political decision to leave the provision of public information to the press which was solidly anti-Labour. The B stations, on the other hand, were gradually drawn into political activity. They had little to lose and much to gain. The government's broadcasting policy did not include the B stations. Led by 1ZB, they turned to the Labour Opposition until, by the 1935 election, the public perceived the Labour Party and the B stations as working in tandem. An election poster in Christchurch typified this view: 'Keep 3ZM on the air by voting Labour'.[123] The board and ultimately the government were unable to stop the B stations' political agitation. Led by that charismatic broadcaster, Colin Scrimgeour, the B stations seized the political initiative; and their actions demonstrated that radio had become a major part of New Zealand society. Its political utility was undoubted. It remained to be seen what use would be made of radio in the post-depression, Labour-led New Zealand.

8
Government Controlled Broadcasting

THE BROADCASTING OF PARLIAMENT

Labour politicians had long argued for the broadcasting of parliamentary proceedings, and began setting it in place soon after their electoral success. Their desire for such broadcasts arose from their antipathy to the New Zealand press. Savage connected the two when he first raised the topic in the House: 'Someday we may even have a microphone on the roof of this chamber to enable the people to know how their representatives are shaping in this House . . . I would sooner put up with publication of my remarks over the air than with the reports which I read in the newspapers.'[1] For Labour, radio was the new means of communication with the public, and not subject to the antagonistic discretion of newspaper editors. Once in power and planning a distinctly new legislative programme, the Labour Government wanted to use radio to present and explain its programme and retain the support of the electorate. Savage argued, once the broadcasts began, 'This Government . . . is going to live, but it cannot do so in the atmosphere of suspicion that is being created by the daily newspapers. By means of broadcasting the people are getting the truth right from Parliament; the pure, unadulterated truth.'[2]

The board was less than enthusiastic about the government's initial proposal to broadcast all parliamentary proceedings from all YA stations. Vickery pointed out that full implementation of that plan would require the regular cancellation or transference to the auxiliary stations of school broadcasts, racing commentaries and other programmes. A considerable proportion of the listeners in rural and semi-rural districts would have no programme available other than parliamentary proceedings. Vickery suggested that either the broadcasts be restricted to speeches on policy and other important matters, or the parliamentary broadcasts be only from 2YA.[3]

Labour's decision was that broadcasting be available for Parliament at any time on instructions from the prime minister.[4] The government heeded the board's advice and when the matter was first presented to Parliament Savage stated, 'It is not proposed to broadcast speeches except those on legislative proposals of considerable importance.' He outlined the planned system for

The debating chamber in Parliament showing the positioning of microphones for the first broadcasts. RADIO NZ

speakers from each party, an allocation which Coates saw as reasonable.[5] The new Parliament opened in late March, and the first broadcast was of the opening of the debate on the second reading of the Reserve Bank Amendment Bill held on the evening of 3 April 1936. The speeches by Nash and Coates, each an hour long, were broadcast, as were the following three half-hour speeches by Savage, Hamilton and H. M. Rushworth — one of the two representatives of the Country Party. This and subsequent broadcasts were from 2YA, with the other YA stations linked with it for major occasions such as the Governor-General's speech and the presentation of the Budget. During the rest of 1936, debates on other pieces of legislation, including the Broadcasting Bill, were broadcast. The confinement of the broadcasts to major pieces of legislation was not continued for long, however. The hours transmitted were gradually increased, with evening sittings regularly broadcast from 1937 and daytime sittings added the following year. In 1938 two thirds of total sitting time was broadcast.

The public recognition and political rewards from speaking directly to the electorate were too great not to be fought over, and the government's control of the parliamentary broadcasts was soon questioned. There was no concerted effort made by the government to deny its parliamentary opponents access to

THE LISTENER'S PREFERENCE

Minhinnick, the *New Zealand Herald* cartoonist, incorrectly assumes listeners will tune to Australian stations once Parliament is broadcast in New Zealand. *NZ HERALD*

the microphone. This would, in any case, have been increasingly difficult as more and more of the debates were broadcast. But both the allocation of speaking time and the determination of the hours to be broadcast were regarded with suspicion by the Opposition. It was also apparent that there was competition among parliamentarians to broadcast, and at hours that attracted a larger audience. Such competition was between members of the same party as well as with their political opponents. Although members were not permitted by the Speaker to address an audience other than the House, the listening electorate was the intended recipient of many speeches. The broadcasting of Parliament changed the parliamentary behaviour of politicians at the same time as it enhanced the public understanding of Parliament.

New Zealand was the first country to make regular broadcasts of the activities of its legislature. The broadcasts were an immediately popular feature. There was no public pressure for Parliament to be broadcast and the newspaper reaction was hostile, with predictions that radio-listening and licence-holding would drop. However the parliamentary broadcasts attracted a large audience and held it for many years. Janet Frame has written of her mother's not unusual reaction to the broadcasts: 'Mother went about her everlasting work with the wireless switched to Parliament while she made her own interjections, and as each speaker finished, she praised the "goodies" and criticized the "baddies", using the first names of the members and talking of them and to them with an intimacy that not even our old friends and neighbours . . . could claim.'[6] The first years of the Labour Government ushered in major changes and the broadcasts were heard with interest and excitement. Many politicians became radio personalities in their own right. In particular this applied to senior Labour figures, none more than the prime minister. Savage, who more than any other pushed for the broadcasts, was resoundingly successful in his plan to circumvent the established means of communication and reach directly to the country.

LABOUR'S BROADCASTING POLICY

While there was little doubt that the government intended to turn the board's stations into a new government department the fate of the B stations was less certain. Labour had promised the B stations would continue with some form of livelihood. Over the following months, in a stunning turnaround, that policy became one whereby the government bought out all but two of the B stations and instituted commercial broadcasting as another state-owned broadcasting system.

After the election, B-station proprietors prepared for the new advertising regime they assumed would soon begin. In Auckland this was particularly the case with Colin Scrimgeour. Not only was he the most prominent private broadcaster; he had also been, without ever becoming a party member, an important figure in the Labour victory. Furthermore he was a close ally to the new prime minister. Savage, an Auckland-based politician, had known Scrimgeour since soon after Scrimgeour's arrival in Auckland and had worked with him since 1930, when Savage accepted an invitation to be chairman and spokesman for the Radio Listeners' League. The two shared a similar appreciation for radio and its programmes. Savage was a radio enthusiast from the first days of Auckland radio and favoured the popular entertainment that was the hallmark of the B stations. His favourite programme was the wrestling commentaries, to which he listened while lying on the floor with his ear close to the receiver.[7] Unlike most of his political colleagues and contrary to contemporary educated opinion, Savage enjoyed 'lowbrow' radio programmes which only in later decades would be graced with the title of popular culture. He was also one of the first politicians to see political significance in broadcasting and planned for its incorporation into his party's programme. Many politicians regarded broadcasting as a novelty and a nuisance, but for Savage it was an opportunity which he well appreciated. He was keen for radio ownership to spread: 'We are going to have sets in every home in the country . . . The day is gone when private individuals could enrich themselves while others were denied the important things of life — and radio has become one of the important things.'[8]

When he became prime minister, Savage assured Scrimgeour of the licence for Auckland and told him to get ready to operate 1ZB. Later he told him that as Wellington had no B station, Scrimgeour should also operate the commercial station in that city.[9] A draft company prospectus was prepared.[10] Scrimgeour began gathering a group of future employees for New Zealand commercial broadcasting. These included people such as Harry Bell from 1ZB, Ian Mackay from 2ZR and Merton Bullivant from 3ZM, with whom he had been associated in the B stations' agitation. Scrimgeour's group also included aviation associates, in particular two comrades of Sir Charles Kingsford-Smith, John Stannage and Beaumont Shiel.

In April 1936, when the NZBB annual report was debated, the

Harry Bell; cartoon by McNamara.
RADIO RECORD

government was still announcing that it would turn the board's stations into a state broadcasting service and leave the B stations as private stations. Lee noted the choices were to allow the B stations to advertise, or to assist them out of state revenue — in other words, give them a portion of the licence fee. Although he made it clear it was a possibility rather than a decision, Lee favoured allowing the B stations to advertise, as that policy 'would be at the expense of the newspapers only' — a prospect that did not perturb Lee or his colleagues. In the same debate Savage reiterated his pre-election promise that the B stations would live. He also saw the choices as being subsidies or the right to advertise. Like Lee he was in favour of advertising: 'If broadcasting is the most up-to-date means of publicity in other respects, why is it not the most up-to date means of publicity in the advertising sense? If people think it is why should they not be allowed to pay for it? I say all that without saying whether the B class stations are going to get advertising rights or not.'[11] No decision had been made, but the government seemed to be signalling that it would allow the private stations to advertise — a prospect which had been assumed since the election.

However, although it had not publicly announced it, the government had by this time changed its policy on the more basic issue of B-station continuation, away from allowing private ownership. The new decision was to purchase the B stations and promote commercial radio as part of a government broadcasting service. The B stations had been an issue of caucus and Cabinet dissension since the election. The party had long advocated state ownership of radio. Its public promises to the B stations were in opposition to its own broadcasting policy and socialist philosophy. After the election there was a split approximately between the Auckland Labour members, led by the prime minister, and the rest, led by Peter Fraser and Walter Nash. In a letter to Savage asking for a rapid decision on broadcasting policy, Scrimgeour 'appreciated the fact that the southern chaps do not fully understand' — but it was the southern group who were to win the Cabinet debate.[12]

The new policy came into effect the day before the debate on the NZBB report, when Cabinet approved the purchase of 1ZB.[13] This turn of events was precipitated by an approach Scrimgeour made to Savage requesting assistance for 1ZB.[14] Harry Bell, the 1ZB secretary, told Scrimgeour the station was insolvent and near to closing down. 'You cannot blame the folk who have helped us so well in the past from feeling that their responsibility has been taken over by the Government . . . For the life of me I cannot see how we are going to carry on.'[15] Scrimgeour asked Savage for assistance and pressed for a rapid decision on broadcasting policy generally. His request for haste was occasioned by a forthcoming meeting of the B station proprietors: 'Candidly I am in a fix all ways. I must call a conference of B Station owners as promised. I hate the thought of meeting these interests without being clear as to which way I am to influence them. And I am hoping that it might just be possible for us to get a fixed idea acceptable to Cabinet before I meet the Conference on Thursday afternoon of next week.'[16]

It was obvious to Savage and Scrimgeour that private commercial radio stations were no longer acceptable to the government. They agreed on the purchase of 1ZB by the government and on payment of a £750 deposit, which enabled the station to remain on the air. The sale was not made public immediately. As late as the end of July 1936, Bell continued to deny even that 1ZB had been offered to the government.[17]

The reason given for the change in policy was the government's fear of its old enemy, the newspaper proprietors, who, the government argued, were poised to take over the B stations.[18] However, the newspaper proprietors' desires were known to Labour long before it became the government. Newspaper ownership was part of Australian B radio and was expected in New Zealand. The *Auckland Star* had shown its intentions with respect to 1ZM well before the 1935 election. The government was shown to have exaggerated the post-election newspaper attempts to move into commercial broadcasting;[19] few had acted since the election, though many were prepared. Mackay wrote of what might have been: 'Today we might have had a chain of privately owned commercial stations similar in pattern to Australia, the United States and Canada competing with a state-owned non-commercial service.'[20] Labour saw the most likely New Zealand form of this as a malevolent press also in control of a private chain of stations. The vision was

of commercial broadcasting's remuneration was so in excess of that of the director of broadcasting or the heads of other government departments.[41] The issue would have been a continuing embarrassment had the arrangements not been changed. The Opposition held that its agitation led to the controller's salary being adjusted.[42] Scrimgeour later argued that he instigated the change, telling Savage, 'If at the end of three months my commission should exceed the National Director's salary I would be happy to have the matter adjusted and limited to the same figure. Mr Savage was very grateful for this and said that I had now in him a friend who would never let me down.'[43] The salary changes still left three broadcasting officials each receiving £1500 a year, a matter the Opposition continued to regard as wasteful.

Labour's opponents raised other matters. Rumours of financial impropriety dogged Scrimgeour. He was never charged, much less found guilty of any offence. But he and other senior executives in the NCBS were questioned regularly in the press and in Parliament.[44] The government and civil service also questioned Scrimgeour. In 1937, when Savage was out of the country, Cabinet approved an inquiry into the activities of the commercial broadcasting stations and listed thirteen points on which it wanted information.[45] Lee later complained to Savage, 'The moment you left we put a couple of sleuths on to try and trap Scrim.'[46] Scrimgeour was viewed with suspicion from within both the civil service and the Labour Party ranks before he became controller. That he now headed a government department rekindled rather than dampened the antagonism. Scrimgeour wrote to Lee, 'He [Savage] quietly informed me that I did not have a friend in the whole of the Civil Service and advised me to walk cautiously . . . I believe there has been a lot of criticism within the Party and I would be pleased if you would advise me of anything you hear.'[47]

What was of greater significance than Scrimgeour's financial arrangements was the broadcasting freedom he was given by Savage. The 'Man in the Street' session was transferred to the NCBS and Scrimgeour was to run it as he wished. Savage's successor as minister in charge of broadcasting noted, 'Mr Savage gave Mr Scrimgeour the right to say whatever he liked and to talk as often as he liked.'[48] It was a permission that contrasted starkly with the rights of other broadcasters within New Zealand and in other countries. The Labour Government did not continue the Coalition decision to allow the broadcasting of controversial matter and only Scrimgeour was beyond day-to-day ministerial censorship. The BBC, the model for the NBS, was allowed to broadcast controversial material but not to editorialise. The New Zealand services could do neither — except for Scrimgeour, who could broadcast as he saw fit.

THE PURCHASE OF THE B STATIONS

The government's policy change surprised all B-station proprietors except

The *Auckland Star* cartoon of 29 May 1937 commenting on Labour's changing attitude to the B stations. AUCKLAND STAR

Scrimgeour. As late as three days before the first reading of the Broadcasting Bill, R. Wells, the 4ZM owner, remained 'most sanguine' about the stations' future and considered they had 'nothing to fear from the present Government'.[49] Even after the bill's introduction the *Auckland Star* — in effect the 1ZM proprietor — described the measure as showing the government's desire 'to give effect to its election pledge that it would enable the B stations to carry on'.[50] When the intention to purchase the B stations was finally clear, the proprietors did not happily or immediately accept that they were no longer a part of New Zealand broadcasting. It was a public issue that provoked much discussion and protest.[51]

The government's response to the B stations' protests was to declare discussion of their future status unsuitable for broadcasting, and to prohibit it under the provisions of the 1936 Broadcasting Act. This aroused further protest.[52] The government was unmoved, and the proprietors eventually capitulated. The majority of stations were purchased by the government in 1937.

The prices paid for the stations were not considered generous, and most stations complained. As Frank Barnett of 4ZO noted, 'The offer received from the Government is far from satisfactory considering the promises made before and since the election. It does not cover even the wholesale cost of our equipment, programmes and stock, let alone any credit for the 8½ years service we have given to the public.'[53] Savage said that in the event of arbitration he would brook no delay: 'We do not want any stonewalling.'[54] Although many cases went to arbitration, the government moved with haste and without generosity. Financially, even 1ZB was treated harshly though the exact price paid is difficult to determine. The price agreed by Savage was £5000.[55] This foundered on a Treasury reluctance to include goodwill in the assets of the station. The matter went to arbitration and the postmaster-general stated that £2100 was paid.[56] Walter Nash, minister of finance, put the price at £2183.[57] The NCBS acknowledged paying £1898 16s 5d for 1ZB.[58] Whatever the exact figure, it was one of the higher prices paid by the government for the B stations. The highest price acknowledged was £2555 for 1ZM, but this included the purchase of a house. The lowest was £180 for 4ZR. The postmaster-general listed eighteen stations that had been purchased for a total of £20,732.[59] This did not include two purchases made in 1938.

Although Savage announced that stations would not be forced to sell and that subsidies would be available for those that continued broadcasting, the government did not make it easy for them to carry on. It refused to open negotiations on the number or size of subsidies until a final decision was available from the stations as to the number that would sell. It also transpired that subsidies were not automatically available to all stations, only to those the government considered necessary. There was no understanding made public as to which stations were necessary. Thus proprietors were forced to decide whether or not to continue, without full knowledge of the financial consequences of continuing.[60] Dispirited after years of non-acceptance from officialdom and now repudiated by the government which originally had appeared as their saviour, most who were involved with the B stations were left with neither the desire nor the energy to continue.

By the end of 1937 only four stations had not been sold to the government. Two of these, 1ZJ and 4ZM, sold in 1938 after deciding initially to accept the government subsidy but returning to the bargaining table disappointed at the size of the subsidies offered. Reverend Leslie Neale, giving the final broadcast from 4ZM, declared, 'The greatest tragedy is the termination of individual and community freedom by the domination of controlling authorities.'[61]

THE LAST OF THE B STATIONS

The two that held out were the Otago Radio Association, licensed for 4ZB, and Percy Stevens in Gisborne, who had the licence for 2ZM. The government offered the Otago Radio Association £400 for its station. The association considered the offer generous, but did not accept it. Led by D. G. (Toots) Mitchell, who had first crossed swords with radio officialdom thirty years earlier, the Otago amateur enthusiasts had no intention of closing their station.[62] The call-sign was changed to 4ZD, and the Rev. Leslie Neale with his Church of the Helping Hand transferred to the station from the defunct 4ZM. Otherwise the programmes carried on as previously.

When the association declined to sell, it asked for a subsidy which, in terms of the government's announced policy, was the alternative to sale. In the negotiations for the subsidy Mitchell estimated the 1936 cost of running the station to be £190; and in 1937, as a new studio was involved, the cost would be £250. The government decided to give an annual subsidy of £75 and to include 4ZD among its own stations for the purposes of protection against copyright claims by APRA and the record companies.[63] The association was happy with the subsidy — 'the first real outside assistance we have received during the fifteen years of our existence'.[64] The annual subsidy was raised to £85 in 1939 and £100 in 1944. These amounts were of the same relative order as those paid by the NZBB to its subsidised stations, and indicated that the government's policy for those B stations which would not sell was that of its predecessor: private stations could live but not grow and flourish.

In 1942 the association was again asked to sell and again refused. Mitchell stated their case: 'The members of my Association do not subscribe to the opinion that it would be in the best interests that all stations be State owned . . . listeners like to know that . . . they can get a change from the National and Commercial Services.'[65] The point was conceded two years later by Shelley to his minister. After agreeing that it was clearly against government policy for 4ZD to exist, he noted, 'There would be considerable local objection if the station through any act of the Government were unable to carry on.'[66]

In 1959 the station (renamed 4XD) was given permission, against the advice of the director of broadcasting, to increase its power output from 60 to 250 watts.[67] Otherwise it received no concessions and its £100 annual subsidy continued as $200 well into the decimal era, constantly eroded by inflation. At the time of writing 4XD is still broadcasting.

The Gisborne B station 2ZM, owned by Percy Stevens, was not for sale either. Like 4ZD, it received a subsidy — £125 a year — and the amount remained unchanged for the rest of the station's existence. Unlike 4XD, which had always been attentive to the broadcasting regulations, Stevens continued the B-station tradition of disdain for the requirements of officialdom. He was regularly cited for such broadcasting lapses as references to local businesses, use of locally sponsored programmes, appeals to join the 2ZM club at an annual subscription of 5s, and rebroadcasts without permission of NBS station programmes, particularly sports broadcasts.[68] However there was always some reason for not taking action against him. As Shelley noted in 1948, 'Mr Stevens is very much in public favour in Gisborne, owing to the manner in which he made use of his station in the interests of the district during the recent flood emergency . . . [His action] has rather tended to place in a poor light . . . our own station . . . Mr Stevens extended his hours and gave almost a continuous service during the emergency . . . Any disciplinary action against Mr Stevens at the present time would be most unpopular in the Gisborne district.'[69] Percy Stevens died on 31 May 1963. The licence for the station (renamed 2XM) was in his name and lapsed at that date.

4XD is the last link with the B stations. The stations are not to be confused with either the government commercial stations that started in 1936, or the private commercial stations that became accepted in the 1960s. The B stations were those run by amateur enthusiasts who started New Zealand

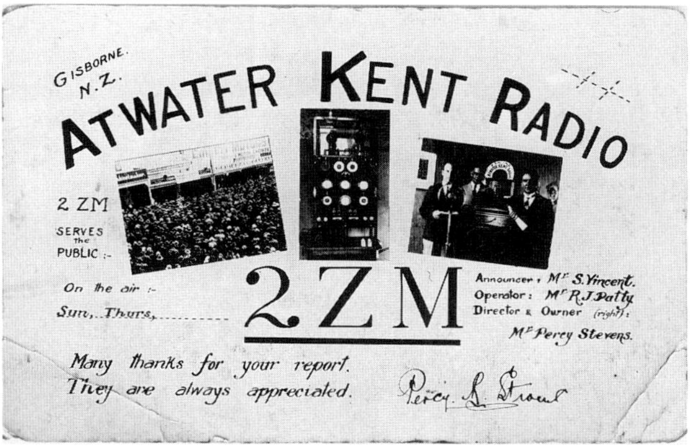

Stations gauged the distance and size of their audience from listener reports. Report acknowledgements such as this one from 2ZM Gisborne became collectors' items.

broadcasting and carried on as successive governments made it plain they were not part of official plans for broadcasting. This opposition was always from a central officialdom, rather than local. The stations' continuing amateur status was forced on them, but they enjoyed the support and affection of their communities as they developed a popular style of local and unashamedly lowbrow broadcasting.

STARTING COMMERCIAL BROADCASTING

1ZB Auckland, the first of the NCBS stations, opened on the evening of 29 October 1936. It had a power output of 1 kW, and was joined within a year by similarly powered stations in Wellington, Christchurch and Dunedin. In October 1938, 2ZA Palmerston North became the fifth NCBS station. The head office of the service was in Wellington.

The NBS acted as the banker for the NCBS in the establishment of the new enterprise, by advancing it funds from the licence revenue: £11,270 in the months to 31 March 1937, and a further £98,975 the following year. Interest at commercial rates was paid — a practice about which Scrimgeour complained: 'We are contributing not only to the licences fund for money which has been used for the benefit of listeners but we are also contributing to a Service which is, in effect, "active opposition".'[70] The debt reached £101,926 at 31 March 1939 and began to be reduced from this time. Labour's financial arrangements for the two broadcasting services was designed to allow this borrowing. It was also designed to keep the two services separate for taxation purposes. The NCBS payed usual taxation rates on its advertising revenue with no possibility of this revenue offsetting losses in a larger broadcasting concern.

The reason the NCBS was able to repay its debt was, of course, its income from advertising. Growth in advertising revenue was slow at first, for the new service had to overcome the opposition of established means of advertising.

Carrick, in *Jack Blunt* on 23 February 1937, gives his view of the press and broadcasting relationship.

Once 1ZB began as the first of the commercial stations it was confronted by the Auckland newspapers. In spite of their running weekly radio columns, both the *New Zealand Herald* and the *Auckland Star* failed to mention the new station. The first time the NCBS 1ZB was mentioned in the Auckland press was in the *New Zealand Herald* on 12 November 1936 — and that was an unfriendly story referring back to the privately run 1ZB and questioning where the funds paid by the government to purchase the Friendly Road station had gone. Scrimgeour was informed that the NPA had instructed all newspapers not to mention the NCBS. The newspapers obstructed radio advertising as much as possible. Retailers who referred in newspaper advertisements to their sponsored programmes on radio were charged the full casual rates for their advertisement, rather than the lower contract rates.[71] The newspapers also used their influence over the accredited advertising agents to hinder radio advertising. In the early months of radio advertising Scrimgeour and his advertising manager saw this campaign as particularly successful in denying the NCBS national advertisers — a matter which was important for the programming schedule as well as financially: 'We have only one genuine National Advertiser . . . We need National Advertisers from the programme point of view because [they] . . . are the only ones who can afford to buy expensive overseas features.'[72] Such difficulties were considerable but temporary irritants. Radio advertising grew, and eventually forced a grudging acceptance from the press.

The early difficulties for radio advertising were not entirely due to newspaper antagonism. For the first six months there was only one commercial station. This meant the NCBS was by no means a national service, which lessened its attraction to national advertisers. Also the NCBS's own policy aroused antagonism. The stations were deliberately modelled on the

Australian commercial stations and the abrupt introduction of the new style was difficult for many to accept. Merton Bullivant, on the staff at 1ZB, described it as 'quite a shock. 75% hated it. They thought they were being pressurised.' It did not follow that the new station did not attract an audience, 'But it was a love–hate relationship. The early surveys within the first three to four weeks showed 1ZB was holding around 80–89% of the available audience and that sort of pattern followed right through until I left the service in 1946.' The NCBS stations inherited the greater popularity of the B stations but their style did deter many advertisers, with those in the southern cities being the slowest to accept commercial radio.[73]

Advertising revenue for the NCBS averaged £570 a week by 31 March 1937.[74] As the NCBS was accepted by advertisers and as more stations were started this income grew correspondingly. Total NCBS advertising revenue was £98,407 for the 1937–38 financial year and £161,305 for 1938–39.[75]

The NCBS calculated its ratio of entertainment to advertising as 93.4 to 6.6.[76] The 93.4 per cent was also greatly influenced by advertisers — a matter to which Scrimgeour proudly drew attention: 'To a large extent the programmes offered are selected by the advertisers, who are constantly seeking for better programmes which will give them greater sales results. Consequently, the programmes include subjects of infinite variety planned and purchased by the best advertising brains of New Zealand and overseas.'[77] This view was shared by the government. The postmaster-general noted, 'As the value of radio advertising will diminish or increase according to the standard of programmes presented by the advertiser, I feel sure that the public will be able to rely upon a high standard of entertainment being provided by commercial stations.' The postmaster-general also held that the ultimate government control would ensure a high advertising standard: profits would be retained by the state rather than distributed among shareholders, and government control would 'ensure that the public will not be exploited by the radio advertising of useless or harmful products and unscrupulous traders will not be allowed to hoodwink or deceive the public'.[78]

DIRECTOR OF BROADCASTING

While the National Broadcasting Service begins with James Shelley, in reality he was appointed to head a group of stations that had been broadcasting for some years, that were well advanced in a programme of technical improvement and that were staffed by people of considerable broadcasting experience. It could have been a difficult environment for a non-broadcaster, but Shelley was soon accepted by his staff, not universally but widely, as an inspirational leader. Collins notes,

> The first day Professor Shelley took up his appointment we knew that here was someone different. During that evening he was, as he was afterwards often to be, in

James Shelley; cartoon by McNamara.
RADIO RECORD

> the control room of 2YA, watching, questioning. Later that evening when the station had closed down, and until an early hour, he held us enthralled as he roamed over many subjects — Greek art, his student days, how to relax (few people, he said, knew the fine art of completely relaxing), painting, drama, music, broadcasting.[79]

Shelley was intelligent, educated and entertaining and was to imbue his staff with an understanding of broadcasting's purpose that long remained the hallmark of the non-commercial stations. However he was not an original thinker and his philosophy of broadcasting was essentially an application of Reithian thought as developed at the BBC. He saw broadcasting as firmly allied with the traditional arts and accorded to this alliance a social purpose of basic importance, 'to the understanding of life values, and . . . the emotional discipline of a community.'[80]

Shelley was much involved in programme production and presentation along with other tasks within the organisation. These were not usual pursuits for the chief executive officer but were part and parcel of the man's nature and were both accomplishments and failures. In his first week as director, Shelley read a special news bulletin, that announcing the abdication of King Edward VIII. He made a practice of personally reading important news bulletins. He was an experienced actor and had been much involved in drama production in Canterbury. This continued in his new occupation. As Collins notes, 'We did a whole series of Shakespearean plays with Professor Shelley taking the principal parts, and assisting in production. Perhaps it would be more correct to say the professor produced, with the assistance of Bernard Beeby, the producer!'[81] Shelley was also a skilled craftsman, and he exercised this ability within the NBS. He designed the masthead for the *Listener*, the label for the

Recording a play at the NBS studios in 1937 with a cast including one current and one future director of broadcasting. From left, Walter Plinge, James Shelley, Karl Atkinson and William Yates. ALEXANDER TURNBULL LIBRARY

service's recording discs and even made a highly regarded cutting stylus for disc recordings.[82]

The continuing complaint about Shelley, though it was made as much in appreciation as annoyance, was that the man could never be found in his office. He was always elsewhere; in a studio, with technicians, rehearsing or participating in any of the myriad tasks of broadcasting. Although the well-paid professional director, Shelley exhibited throughout his years of leadership of the NBS the best characteristics of amateurism. His wide-ranging artistic accomplishments were a constant message to his staff of the primacy of production over administration, and of the ever-present need to strive for excellence. For many within the NBS Shelley's actions were inspirational. Some regarded them less kindly. The director of broadcasting is not an official to be denied, should he decide to participate in the work of actors or technicians or announcers. While his attention was initially a flattering recognition of the importance of the work of his staff, Shelley's presence was regarded by some as interference backed by improper use of an authority that could not be refused. Nowhere in the NBS was this more noticeable than in the relationship between Shelley and his chief announcer, Clive Drummond. After his introduction to announcing, Shelley was captivated by the task and increasingly took for himself the presentation of major announcements. Drummond was unappreciative of his usurpation. Eventually Shelley was the *de facto* chief announcer to the point that during World War II he made a habit of always reading the main 9pm news bulletin. Drummond left the NBS in 1942, at the age of forty-eight. Although neither

Clive Drummond, radio announcer.
LISTENER

he nor Shelley would comment on the apparently early retirement, *Truth* noted, 'The more the Chief Announcer was pushed into the background the more did the Director's sonorous accents fill the air . . . The navigation of a ship is not improved by the captain leaving the bridge and taking upon himself the duties of the crew.'[83]

Shelley's great failing was not that he continued to follow his interests in broadcasting, but that he neglected matters in which he was not interested. The director's concerns should not be limited to personal interests, no matter how wide-ranging. This was of real significance for broadcasting, since the most important of the topics Shelley neglected was news and current affairs. Shelley's reading of news bulletins indicated a passion for public performance rather than a concern for news broadcasting itself.[84] While it is unlikely that Shelley could have altered the government's resolve to control and use radio's news potential, he acquiesced in it and confined his involvement to other aspects of broadcasting. As a consequence there was no counter, within the broadcasting service, to the developing political use of broadcasting.

OFFICIAL NEWS SERVICE

The broadcasting news services inherited from the RBC and NZBB were organised to produce the daily 7pm news hour, a misnomer for a programme which included material such as lectures and book reviews. The NBS continued this programme, with the first significant change being the use of one major news source ignored by the NZBB — the British Official Wireless News, which was supplied to governments within the empire.[85] While the United Press Association paid the New Zealand government £250 a year to use this material, the NZBB had made no such approach. The British Official Wireless News was not broadcast as received from Daventry, the BBC shortwave transmitting site, but transcribed and read by local announcers. The NBS augmented this material with further news material from UPA cables and direct from local newspapers.

The Commercial Service made no significant attempt to begin a news service. The topic was mentioned by Scrimgeour in his 'Man in the Street' session on Sunday 4 April 1937 when he argued that the real reason for the antagonism of the press against radio was radio 'could tell the truth about events, rather than suppress it as the papers do.' He stated, 'We shall have a shot at giving a proper news service, and one that will give you the truth, and not opinion backed by interests, not the truth coloured by . . . well, we will let that go.'[86] Whether Scrimgeour was musing aloud or announcing a definite intention is unknown, but his statement was not followed by the establishment of a NCBS news service. Savage was out of the country at the time. The press reacted angrily to Scrimgeour's suggestion and Fraser, the acting prime minister, was also unsympathetic.[87] Scrimgeour's biographer argues that from this point Scrimgeour was unable to continue his attacks on the press or plan for a news service for fear of placing his ally, Savage, in the position of publicly having to choose between Fraser and Scrimgeour.[88] This is possible, but it is more likely that after these events Scrimgeour accepted Savage's decision that the provision of New Zealand radio news was to be Savage's responsibility.

From 20 September 1937 the NBS began broadcasting an Official News Service. While this service included non-New Zealand news from the UPA cables and the British Official Wireless News, it began a new presentation of New Zealand news. This was produced in the Prime Minister's Department and was sent to the NBS to be broadcast as received without alteration or omission. Thus the government, and in particular the prime minister, became responsible for national radio news. The compilation of the news was largely from press releases of the various government departments. The aim was to communicate the government's actions positively to the country. With the parliamentary broadcasts, the Official News Service was the second of the Labour Government's attempts 'to take its message to the people.'

As with the parliamentary broadcasts, Savage's inauguration of the Official News Service should be seen as an expression of his government's antipathy to the New Zealand newspapers. While Savage was in London he discussed broadcasting with Reith. When Reith asked whether a government-run news service would not be one-sided, Savage's replied that the New Zealand newspapers were already one-sided and the Labour government could not make a news service worse.[89] Labour, convinced that it was doomed to biased, unfavourable treatment from the newspapers, set up the Official News Service as a corrective to and in competition with the press. There was no attempt by Savage to disguise his part in the preparation of the Official News Service bulletins. He acknowledged on various occasions that nothing went out without his authority. The daily bulletins were brought to him for approval before they were broadcast.[90] The Opposition regarded the system as improper. Walter Broadfoot's comments reflected this: 'Every night now at seven o'clock we have a news-reel from the Prime Minister's office. If that is not an abuse of power I do not know what is.'[91]

The Official News Service continued for twenty-five years. For broadcasting it meant that for those twenty-five years the involvement in news was limited to the reading of bulletins prepared in another government department. New Zealand broadcasting remained without a news staff — a section regarded as essential in broadcasting services in other countries.

POLITICAL BROADCASTING

The Opposition regarded the Official News Service as a form of government propaganda. It had a wider concern — that not just the news service but the entire broadcasting organisation was a political prop for the government. Colin Scrimgeour was regarded as particularly at fault. George Forbes made the Opposition's view of him plain when he remarked, 'He put over political propaganda to aid the Labour party, and now he is the Labour Government's man in charge of the National Commercial Broadcasting Service. What confidence can we have in him when he has a friendly Government behind him.'[92]

Shelley was not included in this criticism. However, the NBS stations he controlled, along with the NCBS stations, were regarded as under the day-to-day control of a government determined to use them for its advantage. In a typical complaint, Opposition member Colonel Hargest asked in 1937, 'Whether two broadcasting stations, namely 2YA and 2ZB, were compelled to sacrifice their ordinary programmes on Saturday 18th September in order to broadcast political speeches of Ministers on the occasion of the occupation of the first of the State-built cottages at Miramar?'[93] The occupation of its first state houses was a major occasion for the new government, but the broadcasting publicity given was not unusual. Ministers and other government speakers had frequent access to the microphone. Sydney Smith, the Opposition member for New Plymouth, complained about being unable to hear the commentary of the 1938 Taranaki v. Auckland rugby match, as it had been relegated to the weaker 1ZM while speeches from no less than five government ministers were broadcast that Saturday afternoon from 1YA and 2YA.[94] The importance of and need for such broadcasts was decided by the minister in charge of broadcasting, rather than within either of the broadcasting services.

The Labour Government reinstituted the broadcasting policy whereby controversial material was prohibited. It broadcast Parliament but, unlike its predecessor, did not permit the director of broadcasting to extend the range of radio discussion generally. Only members of Parliament enjoyed a freedom of expression over the air, and that only during parliamentary debates. The broadcasting of Parliament was used by the government as a rationale for its failure to allow other radio discussion of controversial subjects: 'It should be remembered that in broadcasting Parliamentary debates the New Zealand Broadcasting Service has provided for listeners far greater opportunity of hearing controversial topics discussed over the air than has been the case in other countries where controversial talks and discussions are permitted only under considerably more restricted conditions.'[95]

While parliamentary broadcasts were apportioned relatively evenly between the parties, the Opposition's radio access was otherwise limited. One Opposition member, William Endean, complained in mid 1938 that on two occasions only had the Opposition had the privilege of speaking over the air outside the House. The two occasions were both broadcasts of addresses by the leader of the Opposition.[96] With the government reverting to the policy that controversial material should not be broadcast, the Opposition had little opportunity to broadcast.

Furthermore the government moved, in its first years in office, to ensure that opponents to its policies generally were not heard over the air. The case of A. E. Robinson was one of the more celebrated but is indicative of the changes made. Robinson was the secretary of the Auckland branch of the New Zealand Farmers' Union and had a weekly half-hour session from 1ZM and 1ZJ in which he discussed items of interest to farmers. Soon after the purchase of 1ZM by the NBS, Robinson was informed that he could talk on general farming matters such as cropping and haymaking but not on topics such as guaranteed or compensated prices or the organisation of the union.[97] Robinson took little notice of the directive, and in mid 1938 was effectively muzzled. He was told his material was too controversial and in future his sessions would not be broadcast unless he submitted each script for vetting. The minister explained to one of his colleagues, 'The amount of anti-Government propaganda in recent months has caused some concern . . . Mr Robinson's talks were supposed to be on farming subjects but much of the

A cartoon in *Progress* on 1 June 1938 has Robinson of the Farmers' Union stopped from broadcasting while Savage and Nash continue. *PROGRESS*, ALEXANDER TURNBULL LIBRARY

matter bore little relation to farming and was of a controversial nature unsuitable for broadcasting.'[98] The government's stance was in contrast to its stated intentions when in opposition that it would encourage a lively broadcasting of differing points of view. Its new policy was also applied so distinctly to its political opponents as to promote an understanding that the rules against controversial broadcasting were designed to eliminate opposition rather than controversy as such.

THE 1938 ELECTION

The 1938 election which saw the Labour Government returned with a resounding majority was the first for which Labour had to determine the role of broadcasting. Savage decided the campaign on radio would consist of six broadcasts from government speakers, four from the Opposition and two from Independents. On the final night of the campaign the prime minister and leader of the Opposition were each to speak for one hour over the air. Savage later changed this to fifteen minutes. The speeches were from 2YA, plus one other station to be designated by the speaker. Thus each speaker could ensure his speech was heard in his local area. The allocation of times and speakers was not greeted with universal satisfaction, but Savage's actions were accepted as fair and reasonable.

Savage was concerned not only about the speeches from his opponents but also about those from his allies. He moved to ensure he remained in overall control of his party's electoral broadcasts by instructing Shelley that ministerial statements were only to be broadcast subject to Savage's personal consent.[99]

Within the Labour caucus many wanted the NBS to show enthusiastic support during political broadcasts. As Lee put it, 'The man at the microphone must keep up a running commentary and it should be his duty to produce an atmosphere of expectancy prior to the Chairman's remarks and the commencement of the speaker's speech.' Shelley argued successfully against this view. 'This would . . . undoubtedly lead to severe criticism. It is

A 1938 election poster with Scrimgeour portrayed as one of the non-elected dominators.

my view that political broadcasts should be dealt with in a dignified manner by announcers who should obtrude as little as possible.'[100] This was one of many such instances. Both at election time and generally, many within the Labour Party argued that the broadcasting services should be required to do much more to support the Labour cause.

Within the NCBS Scrimgeour also argued that 'it would be highly undesirable that the Commercial Service should be used for political propaganda, and I would seriously suggest that our policy should be to advise that we will only accept announcements concerning meetings at standard rates'. Savage's decision introduced a format suggested by Scrimgeour. The time from 6.30pm to 6.45pm was set aside for the broadcasting of announcements of political meetings. Each speaker was allowed a maximum of one twenty-five word announcement per day. The announcements were confined to advertising the speaker, the date, the time and the place of the meeting. A special rate of 10s per announcement was charged.[101]

Scrimgeour's political neutrality for the NCBS was one thing, but his own programme, the 'Man in the Street', marched to a different beat. The programme's major contribution to the 1938 election campaign came when Scrimgeour chose to answer accusations by Justin Power, a National Party organiser, that 'poisonous political propaganda' was being spread on Sunday evenings 'in the name of religion'. It was an obvious reference to the 'Man in the Street' programme. Scrimgeour's reply to Power the following Sunday evening was a sustained attack on the Opposition and the press and included the declaration, 'You and the gang who employ you are unmitigated and malicious liars'. Forbes led the counter-attack to Scrimgeour's 'breach of the privileges and duties of a public servant' and an extensive parliamentary and public debate followed. Savage stopped short of endorsing Scrimgeour's tirade but stated the broadcast was understandable and made it clear he would not be accepting the resignation of Scrimgeour, who remained his personal friend.[102]

NATIONAL COMMERCIAL BROADCASTING SERVICE

A concern for Colin Scrimgeour in his negotiations with the prime minister was that the Friendly Road should continue. Savage accepted this and when 1ZB began as the first of the commercial stations, it was recognised as a continuation of the Friendly Road. The call-sign was the same and the first programmes were previously heard on the private 1ZB. The station continued with daily morning devotional services, usually conducted by Scrimgeour until he moved to Wellington. This identification with the B stations was also notable in the three other ZB stations, all opened in 1937. 2ZB used the wavelength of the Wellington B station 2ZW, and L. E. Strachan from 2ZW was the foundation programme organiser. The 2ZB studios were in the Hope Gibbons building, thus returning Wellington radio to the premises occupied in 1922 by Charles Forrest. The 5000-square-foot ground floor was leased by the NCBS, with the front half being used as the head office and the rear as the 2ZB station.[103] This increased the Friendly Road influence as various of the 1ZB staff moved to Head Office and to 2ZB. In Christchurch 3ZB was a continuation of 3ZM, using its transmitter, wavelength and premises and employing most of its staff.[104] Dunedin was the most difficult city for the NCBS. 4ZB competed there with the NBS stations and with two surviving B stations, 4ZD and 4ZM. 4ZM in particular was strong competition. It had a popularity approaching that of the Friendly Road itself and, with the Rev. Leslie Neale's Radio Church of the Helping Hand, had a similar religious orientation. The fifth NCBS station was 2ZA Palmerston North. With personnel and equipment from both the local ex-B stations, it opened for the NCBS in 1938 with a power output of 200 watts.

The NCBS stations continued the B stations' concentration on contemporary light music, with an emphasis in the evening on serials and in the daytime, particularly in the morning, on women's sessions. Both the serials and the women's sessions had been introduced to New Zealand radio earlier. In the new commercial regime they became programming mainstays. The evening programmes were based around serials played one after another, generally in quarter-hour units. Initially most were American programmes, but these were soon exceeded by Australian productions.

The daytime programmes and radio's development as home-based entertainment were of considerable influence on women in particular and offered a recognition and a definition of their lives. In the 1920s women were not employed as announcers, on the assumption that they would not be accepted by listeners as authoritative enough. With daytime broadcasting,

AUNT DAISY

Daisy Basham; cartoon by McNamara.
RADIO RECORD

women became a large part of the audience, and announcing and programming practices were altered accordingly. Serials were played during the daytime, women announcers were engaged, and in commercial radio greater emphasis was placed on selling to the daytime female audience. Women were acknowledged as being in the home and in charge of its provision and maintenance. Radio's appeal to women both recognised and upheld this view of women's role.

The women's sessions under the new advertising regime were the vehicle to national stardom for Daisy Basham. Her first commercial broadcasts were from 1ZB, but in 1937 she moved to Wellington and 2ZB. Her broadcasts from there were heard from all the NCBS stations. While continuing to broadcast, she soon left the Broadcasting Service and joined the advertising

Left: An advertisement from the *Listener*, 18 April 1952.

Below: Paterson in the *Dominion* of 14 June 1937 gives an opinion of the extension of daytime broadcasting. The cartoon was reprinted in the *Radio Record* which took the opportunity to add the reference to its star, Aunt Daisy. DOMINION

sector, where she was employed first by Carlton Carruthers and eventually by her own agency, Aunt Daisy Radio Advertising. She acquired the reputation for advertising only products in which she personally believed. This was a

The 2ZB announcing staff when the station began in 1937. RADIO NZ SOUND ARCHIVES

policy developed by herself and Beaumont Shiel and, while the reputation was not without foundation, it clouded the fact that Basham believed in and advertised those products for which she had an individual contract. As well as being the first and most successful salesperson on radio, she was among the most astute in obtaining rewards for her skill. She was the first of a group of women announcers who developed daytime commercial radio as one focused on a home-based women audience.

Daisy Basham epitomised a style of announcing that characterised the commercial service and differentiated it from the NBS. Mackay exaggerates both the nature of the non-commercial broadcasting and the speed with which the new style emerged but he does point to the change: 'Previously the radio announcer had merely been a voice timidly intruding at frequent intervals and scrambling out of the programme with the least possible delay but in the ZB era the voice was to occupy a different role. The station and programme revolved around the announcer who, overnight, became a radio personality.'[105] The change in announcing styles was made deliberately to distinguish the commercial stations from those of the NBS and claim the majority audience for the NCBS. As M. D. (Doug) Laurenson, who joined 1ZB early in 1937 noted, 'The more the audience knew of the person behind the voice, the more he was regarded as a friend and welcome guest. So people would listen to us — that was the idea — in preference to the YAs or the YA announcers, who were most strictly anonymous.'[106]

It was Scrimgeour who chose the majority of the NCBS staff. Savage accepted Scrimgeour's argument that some Australians, experienced in commercial radio, were needed; but he insisted that the majority of positions go to New Zealanders. Most of those chosen had been associated with B stations. B-station work had been part-time and voluntary, and not all involved

As these photographs for John Gordon and Rod Talbot of 1ZB indicate, the NCBS followed contemporary film-star styles in publicity for its announcers. RADIO NZ

in it wished to start a new career. Most of the proprietors had no wish to become employees of a government service. But many B-station broadcasters were keen to continue in radio and received their opportunity in the new commercial regime. There were some exceptions to the predominantly B-station background. Among them was John Gordon, a 4YA announcer who was posted to 1ZB in January 1937. Gordon thought he had been chosen because Scrimgeour wanted another New Zealander to leaven the Australian influence — but he was a fine announcer in his own right. Hands had greeted his audition in 1932 by exclaiming, 'We've found a voice.' Scrimgeour, in spite of considering Gordon a 'Red Hot Tory', chose the man for his professional prowess.[107]

One notable addition made to the staff of the commercial stations was the introduction of Maori announcers. Uramo Paora (Lou Paul) was appointed to 1ZB, Kingi Tahiwi to 2ZB, Te Ari Pitama to 3ZB and Airini Grenell to 4ZB. This was Scrimgeour's innovation. Each station was required to have at least one Maori announcer and other Maori staff were also employed. It was a policy welcomed by Eruera Tirikatene, the member for Southern Maori, who had previously voiced his disapproval of broadcasters' actions with respect to his people.[108] Scrimgeour encouraged Maori to become radio listeners and was eager to do what he could to raise the Maori profile in New Zealand society generally. He felt his department could 'contribute something of permanent benefit to the Maori race'.[109] Of particular concern at the time was the question of the pronunciation of Maori. Many Pakeha believed that the anglicised pronunciation of Maori words was acceptable. Scrimgeour's policy

Colin Scrimgeour;
cartoon by McNamara.
RADIO RECORD

was deliberately opposed to this view. He was not the first to have specific Maori programmes, but he was the first broadcaster in authority who thought beyond the immediate topic of programming and considered a policy in terms of long-term benefits to the Maori.

Much of the success of the commercial stations can be attributed to Scrimgeour. He employed a capable and loyal staff, enjoyed the popular acclaim of much of the population (even during the years when he was falling from political favour), and proved he had commercial ability. As he was being removed from office, it could still be said of him, 'To do him justice many people who have had business transactions with the Commercial Service consider Mr Scrimgeour as Controller has done a first-rate job. Of his ability, cool judgement and general acumen there can be no doubt.'[110]

In his relationships with his staff Scrimgeour accepted their independence and showed an informality that was unfamiliar in the traditional public service. The staff returned his attitude with similar informality and with affection and achievement. Coincidentally arriving in Christchurch on the May morning in 1938 when the 3ZB studios were destroyed by fire, Scrimgeour was greeted amidst the debris by the station director's startled, 'What the bloody hell are you doing down here?' and allowed himself to be firmly escorted from the premises. That evening Scrimgeour was told, 'OK, Chief, you can come around anytime you like.' He discovered the station had managed a full day's broadcasting and had not missed one advertisement.[111]

Scrimgeour's administrative style was not universally appreciated, however. Commercial broadcasters were quick to enthuse about and try new ideas. Action rather than scrutiny was the approach, and the NCBS regularly sought approval for its innovations and expenditure after the fact. This procedure was

The controller, Colin Scrimgeour, is presented with a baton by Theo Walters and the 1ZB Orchestra in appreciation of Scrimgeour's efforts on behalf of local musicians.

not appreciated by Scrimgeour's superiors — as is illustrated by the reaction to the NCBS extension in 1939 of its Saturday night dance music programmes to 1am on Sundays, 'for the convenience of those holding parties'. Asked in the House about this 'desecration of the Sabbath', Jones, then acting minister in charge of broadcasting, made it clear that 'The change was made without my authority, and as soon as it came to my knowledge instructions were immediately given to restore the stations to their normal hours of transmission.'[112]

Scrimgeour argued that commercial broadcasting required rapid action. He was soon regarded as intractable and not amenable to civil service discipline. Furthermore, the NCBS cultivated a business image and pattern of expenditure that were seen as unbecoming to the civil service. The ZB stations and NCBS head office were described as unnecessarily luxurious and Scrimgeour was criticised for his own expenditure. This applied particularly to his travelling. He was reluctant to use a train or boat when a plane was available. He made trips to the United States and to Australia which were attacked as unnecessary. His personal first-class style was defended as part and parcel of the business image of commercial broadcasting, but won him few friends outside the NCBS.

NBS v. NCBS

The NBS and NCBS were separate and competing organisations until their amalgamation in 1943. Their relationship was seldom amicable. That the staff of the NCBS were largely from the B stations did not make for good relationships between the two broadcasting services. The staff of the NBS were those who had worked for the NZBB. Even Hands remained, taking a drop in salary and a position below Shelley as business manager. To these people the B stations had long been not only competitors but an enemy. Over the years an analogous antagonistic attitude had developed within the B stations towards the NZBB stations. Not only did it appear that the B stations had triumphed, but their new government-backed format was headed by Colin Scrimgeour, a man regarded within the NZBB as the epitome of all that was wrong with the B stations. Nor was there a good relationship between the principals of the two organisations. Scrimgeour and Shelley each felt the other was unnecessary and that he should be the head of a combined service. They came from the opposite ends of the broadcasting spectrum, with Shelley's Reithian and Scrimgeour's populist sentiments always at odds — and this was compounded by a mutual personal dislike.[113]

Shelley made his programming preferences known to his staff. He once stormed into a studio and broke a Vera Lynn record over his knee. Informally within the NBS, records were divided into 'BPWU' and 'APWU' categories, according to whether they should be played Before or After Prof Wakes Up. Shelley never made his cultural preferences the exclusive focus of NBS programming, however; and he was ready to compete with the NCBS. For the NCBS, the most difficult competition was for the rights to serials. Serials were the major evening programming for the commercial stations, but they also featured on the NBS. The competitive bidding between the two services meant an increased price had to be paid. The NCBS often lost rights to serials it had previously broadcast. Thus, after the first series of *Dad & Dave* was broadcast by the ZB stations, the NCBS was outbid by the NBS for the second series. The NCBS saw such practices as particularly harsh on its advertisers. Mackay, an NCBS employee, commented on another incident when, after signing an advertiser to sponsor the first series, twenty-six episodes in all, of *Coronets of England*, the NCBS saw the NBS purchase the second series: 'The sponsor was faced with competition on his own programme, and while the first series was playing on ZB, National stations programmed the second series thus robbing the advertiser of some of the popularity of the feature which his money had been spent to create.'[114]

Along with the contests for serial purchases the NBS competition with the NCBS was most notable in the inauguration of a new type of national station. On 25 April 1937, four days before the opening of 2ZB, the NBS opened 2YD, which broadcast nightly from 7pm to 10pm. This was 'Wellington's new entertainment station, specialising in light drama, variety, thrills, rhythm, new features and new personalities . . . 2YD has adopted a completely new style of presentation. The announcements will not be of the usual dignified style. They will be snappy.'[115] The opening of 2YD had not been previously announced and was a surprise move to counter 2ZB. Late in 1937 2YD was joined by 1ZM broadcasting a similar fare. 1ZM was the previous Manurewa station which after purchase was raised in power to 750 watts.

NATIONAL BROADCASTING SERVICE

The NBS incorporated B stations on a larger scale than the NCBS. In 1937, along with the opening of 1ZM, 2ZR Nelson was also taken over by the NZBS. The station was renamed 2YN and broadcast on 30 watts. 4ZP Invercargill was renamed 4YZ and reopened for the NBS with a raised power output of 5 kW in 1938. Five other B stations that had been purchased were continued by their previous owners under contract to maintain broadcasts until new stations were opened. Of the five, by the end of 1938, 2ZA of Palmerston North was an NCBS station and 2YH Napier and 3ZR Greymouth were NBS stations. This left two B stations, 2YB New Plymouth and 2ZJ Gisborne, broadcasting under contract. They did so until 1944, at which time they also became NBS stations.

Thus by the end of 1938 the NBS stations, which at Shelley's appointment had totalled eight — the four YA and the four auxiliary stations, had been increased to fourteen, plus two stations on contract. It was the end of expansion for slightly more than a decade — the next permanent station was not opened until 1949.

By opening 2YD and 1ZM as light entertainment stations, Shelley showed he was ready to look at radio as more than a medium for cultural improvement. But these two stations were relatively low-powered and were very much the lesser of the NBS stations. The principal stations remained the YA stations, which transmitted a nationally coordinated programming designed to appeal to an educated élite.

In line with the desire to raise standards generally, one of Shelley's first actions on appointment was to assess the speech of announcers. Arnold Wall, then the recently retired professor of English at Canterbury University College, was employed to spend at least a fortnight at each station analysing the speech of the announcers and instructing them in improved pronunciation. This improvement and standardisation of announcers' speech was a continuing feature of Shelley's term of office.[116] It was a major aspect of his approach to broadcasting, and continued the NZBB-introduced system of a centrally coordinated and uniform pattern of broadcasting. It contrasted strongly with the ZB stations' popular and often locally oriented programming, and offered a real choice to listeners, or at least to that majority of listeners who were able to receive both types of station. There was no doubt about listeners' preferences. The ZB stations tended to exaggerate their popularity, but clearly a large majority of listeners tuned habitually to the commercial programme.

The cover of a book by Professor Wall, published to accompany his and Shelley's efforts to improve the pronunciation of New Zealand announcers. WHITCOMBE & TOMBS

THE BROADCASTING AUDIENCE

By 1939 an estimated 84 per cent of households had radio sets. Since the 1860s the daily newspaper readership had been larger than any other grouping, but now it had been surpassed in size and type by the radio audience. Not only were stations broadcasting for many hours daily, unlike the newspapers' once-a-day production, but radio had none of the difficulties newspaper proprietors experienced in distributing their product.

Radio communication was profoundly one-way. Many could listen but few could transmit. This is not part of the nature of the medium — it is inherently two-way — but in New Zealand, as in other countries, there had been a deliberate restriction of access to radio transmitters. That announcers spoke and listeners listened was part of the tradition of entertainment, where a few artists performed for an eager audience. With radio, the scale was so increased as to change the nature of the experience. Eventually, one cast could entertain not just a theatre but a whole country. One speaker could have the entire population for an audience.

Broadcasting was at first seen as a medium particularly appropriate for a rural audience. Jim Henderson, writer and broadcaster, relates his mother's comment as she switched on the radio for the first time at their home on Takaka hill: '"We'll never be isolated again."'[117] Not only was isolation alleviated for rural dwellers, but radio became a source of entertainment and information for the entire population, rural and urban. The two previously disparate groups now shared new experiences, interests and understandings.

The widespread acceptance of radio changed people's lives. Radio was developed for a home-based audience. It became the main form of family entertainment and leisure-time activity. The home, previously a relative sanctuary from external influence, was now the reception point for a centrally controlled means of communication beamed directly into each dwelling. By making the same message simultaneously available everywhere, radio allowed and encouraged a transformation of the society itself. The implications of these changes were not all obvious in the late 1930s. What was obvious was that radio had become a normal part of daily life in New Zealand.

9
Broadcasting in the War Years

1939 PLANS

By 1939 the Labour Government had been reelected resoundingly and the two broadcasting services were well established. Both services were eager to expand and, with the increasing revenues from licence fees and advertising, were in a position to do so. Each service had a particular building project at the top of its priorities. The NCBS wanted new studios and offices for 1ZB and had selected a Durham Street site on which to build them. Savage also forecast the opening of more NCBS stations: 'The very nature of the commercial broadcasting service in New Zealand,' he declared, 'means that it cannot and must not stand still and extension is very likely in the future'. The *Radio Record* interpreted this as meaning that there would be new commercial stations in provincial centres.[1] With the Act requiring a non-commercial station wherever there was an NCBS station, it could be assumed there would also be a provincial growth of NBS stations.

The NBS priority was for a broadcasting centre on its Boulcott Street site next to Parliament. In October 1938 Savage stated that the Wellington broadcasting centre would include a conservatorium for the study of the arts, and of music in particular.[2] Plans for a five-storey building were made public in December.[3]

NBS STRING ORCHESTRA

The establishment of a music conservatorium and ultimately of a symphony orchestra was part of Shelley's wish for broadcasting. He made the first public mention of a National Conservatorium of Music and the Spoken Arts in January 1937 at the opening of the Titahi Bay 2YA transmitter.[4] The Broadcasting Centre and its conservatorium was never built, but the first steps towards a symphony orchestra were taken in 1939. With the New Zealand centennial little more than a year away, a decision to form a string orchestra was taken to ensure New Zealanders provided the music for the forthcoming celebrations.

The YA stations' orchestras provided part-time employment but the

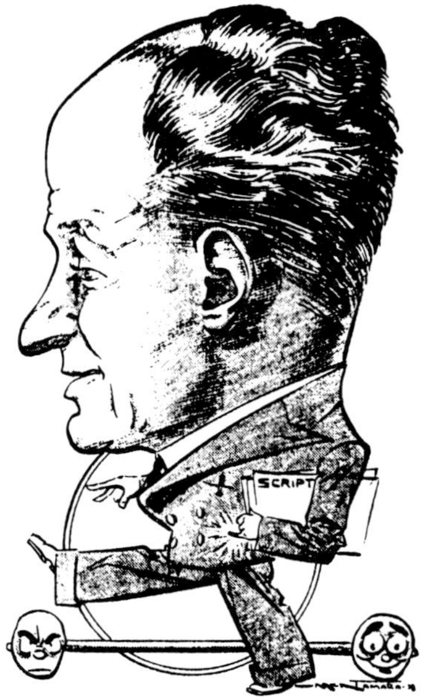

Bernard Beeby . . . Play Producer to the NBS; cartoon by McNamara. *RADIO RECORD*

formation of a string orchestra opened up permanent employment opportunities for some of the country's best classical musicians. Haydn Murray, appointed leader of the second violins, saw the importance of the orchestra as 'the fact that it started. It gathered together the last of the "dyed-in-the-wool" professional musicians in New Zealand, those that had survived the Depression, and hadn't gone away doing other things.'[5] The orchestra was established as the NBS String Orchestra with thirteen players, under the direction of the English violinist Maurice Clare.

RECORDING

Along with the establishment of a symphony orchestra, Shelley also had a passionate interest in drama. He was able to increase its production because of the availability of recording technology, the innovation of the 1930s that brought the greatest change to broadcasting. Previously all broadcasts other than the playing of gramophone records had been live; there was no way to store and broadcast them at a later date. This began to change in the mid 1930s in New Zealand, with the introduction of acetate discs. Commercial recording had improved enormously since its introduction in 1890, but required a pressing plant and was suitable only for the production of large numbers of each record. In countries such as New Zealand, which did not have a pressing plant, and in circumstances such as the recording of a speech or a play, where only one copy was required, acetate discs came into their own.

They were lacquer-coated metal discs, usually made of aluminium. The recording was made by cutting the disc with a steel needle with a cutting mechanism rather like an ordinary pick-up. The discs, played at 78 revolutions per minute, were 10 inches in diameter. Twelve-inch discs were later used, which gave a total of around four minutes' recording.

Recording equipment was introduced to New Zealand in 1935. Both 1ZB and the NZBB began using recorders as soon as they were available.[6] However it was with the NBS and NCBS that recording changed the nature of radio broadcasting. Portable recorders existed but were not satisfactory and most recording was done in studios with more substantial equipment. NBS staff, faced with problems in obtaining equipment from overseas and not content with the quality of the purchased recorders, designed and built many machines locally.[7]

Important speeches were recorded, along with projects such as the NCBS recording of historical Maori material. The development of sound archives begins with the advent of acetate discs. However their principal use was in day-to-day programme production, with talks, commercials, serials and plays being the major items recorded. Serials and plays were recorded in three to four minute sections, thus ending with a stack of discs. The excitement and tension that had been present during live broadcasts were lost when recording began, but were replaced by a new level of professionalism in production standards.

An acetate disc recorded only a small amount, was cumbersome and subject to deterioration over time. But these are judgements of later generations used to improved recording methods. With the introduction of the discs, broadcasting could begin to escape from the confines of impermanence. The new technology was welcomed with the same enthusiasm as radio's own introduction. 'That the voice should carry through the air across the world seems to me', wrote Alan Mulgan, then starting his broadcasting career, 'not more wonderful than that a shallow track on a disc should reproduce the voice faithfully, and also all the varieties of tone in an orchestra.'[8] No longer was a broadcast necessarily something ephemeral, immediately lost upon the air.

DRAMA PRODUCTION

Earlier plays that had been broadcast were produced on contract separately in each centre. With the new recording equipment, Shelley greatly increased the number of plays broadcast, and centralised their production in Wellington. This was necessary, since the recording equipment was available only in Wellington; and at the same time, it reflected a deliberate policy to concentrate the available talent in one place. Bernard Beeby, a contract producer from 3YA who had been associated with Shelley in Canterbury drama productions, was appointed in 1938 to supervise drama production.

Bernard Beeby directing a play in the NBS studios in the early 1940s.

An increase in drama production was a hallmark of Shelley's NBS. In the 1938-39 year 1157 plays were presented, and the two recording plants owned by the service were in almost constant use.[9] Importantly, much of this production was of New Zealand plays. New Zealand playwrights were encouraged to submit scripts, and new writers were discovered by the promotion of play-writing competitions. The NBS greatly influenced the development of New Zealand drama by enabling it to exist on a permanent and professional footing. For many years drama was oriented to radio rather than to live theatre — there existed the majority of work opportunities. But for the first time a cadre of writers and actors found reasonably regular employment in New Zealand.

The developments in drama production were slower in commercial radio. Serials were not so much produced by the NCBS as chosen and purchased for it by sponsors and advertising agencies. The first New Zealand radio serial was *One Man's Family*, first broadcast in 1937. It originated in the United States. The rights to produce it in New Zealand were obtained by Carlton Carruthers, a local advertising agency. The serial, which ran for eighteen months, was produced in their recording studio, Staradio, the main function of which was to record radio commercials.[10] Australian production of radio serials was also established at this time and became a flourishing industry. In

Station 5ZB broadcasting on North Island provincial location in 1939. RADIO NZ SOUND ARCHIVES

New Zealand it was realised that serials could be bought from Australia much more cheaply than they could be produced locally. As a consequence few serials broadcast in New Zealand were produced locally and Australia became the major supplier. NCBS drama production increased significantly from 1940. Wartime circumstances made the importing of programmes more difficult, and an NCBS productions department was started.

5ZB

The NCBS wished to expand commercial broadcasting and, in 1939, tested the public reception by introducing a mobile broadcasting station which toured many North Island centres. The aim was to generate 'publicity and goodwill for our main stations . . . [and] to investigate the sales possibilities of commercial broadcasting in a number of major provincial centres.'[11] The call-sign was 5ZB and the station, which had a power output of 250 watts, was built within a railway carriage. The tour commenced in Rotorua on 4 April 1939. In the next seventy-nine days the station covered 1650 miles and broadcast from Rotorua, Hamilton, Whangarei, Te Kuiti, Taumarunui, New

Ian Mackay on the steps of 5ZB. RADIO NZ

Plymouth, Hawera, Wanganui, Palmerston North, Dannevirke, Napier, Hastings and Masterton.

The station had a three-man crew, with Ian Mackay as director and announcer. The other two were a sales manager and an operator. The travelling station was a novelty and attracted considerable attention. Visitors were welcome: Mackay reported that 'It is safe to assume that more than 15,000 people have seen the interior of 5ZB.'[12] Financially the tour was a success with returns more than doubling expenditure. M. D. Laurenson, the 5ZB sales manager, recommended the tour become a permanent feature of broadcasting. His suggestion was that, allowing for four weeks annual refitting and four weeks travelling, 5ZB could transmit for forty-four weeks a year from various centres.[13] The counter view was that, instead of a travelling station, new and permanent stations should be located in many provincial stations. Neither view was to prevail before the outbreak of war led to both being abandoned.

1940 CENTENNIAL

The 1939 tour was not the only appearance of 5ZB. It was dismantled in 1940 as the rolling stock was required by the Railways.[14] Before this 5ZB made a

Anderson Tyrer and the Centennial Orchestra. RADIO NZ

six-month appearance at the Centennial Exhibition in Wellington, opening on 8 November 1939.

In spite of the war it was decided to go ahead with the centennial celebrations. Most of the work was completed by the time war was declared, and Fraser considered that 'The Centennial efforts themselves will be a most valuable aid to cementing the national spirit which is so necessary now.' The Exhibition was on a 22-hectare site at Rongotai. The crowds flocked to it: at a time when New Zealand's population was less than 2 million, there were 2.6 million visitors to the Exhibition during the six months it was open.

The broadcasting services made two major contributions. One was 5ZB, which served as the official voice for the Exhibition. Ian Mackay, the resident manager, lived aboard the railway carriage that was 5ZB for the six months of the Exhibition.

Broadcasting's other contribution was musical. The NBS String Orchestra played separately and as part of the thirty-four member National Centennial Symphony Orchestra. Maurice Clare was leader of the orchestra and the conductor was the Englishman, Andersen Tyrer, a composer and well-regarded pianist. He had previously toured New Zealand and his *Dr Faustus*, a symphonic setting of Marlowe's play for orchestra, chorus and narrator, which was dedicated to Shelley, had been played over the YA stations in 1938 with Shelley as narrator. The Centennial Orchestra played at the Exhibition and toured the main centres with English and Australian soloists and one New Zealander, the bass Oscar Natzke. Symphony concerts, oratorios and two versions of the Faust story — Gounod's opera *Faust* and Tyrer's own *Dr Faustus*, again with Shelley as narrator — were played.

The orchestra's playing was well received. Fraser noted that the tour was something the country had never experienced before. The orchestra would not have been possible without the backing given by the NBS. Its formation and success were due in no small measure to Shelley's backing and to the support given by Fraser. Shelley's enthusiasm was joined by that of J. W. Heenan, under-secretary of Internal Affairs — the department which shared the cost of the orchestra with the NBS. However, in wartime conditions the orchestra could only be shortlived, and it was disbanded after its tour. During the war years the NBS String Orchestra continued with various guest conductors including, notably, the composer Douglas Lilburn, who had won three of the four Centennial music competitions. Fraser, Shelley and Heenan retained a dream of a national orchestra supported within the broadcasting service.

THE *NEW ZEALAND LISTENER*

A further 1939 addition to New Zealand broadcasting was the weekly journal, the *New Zealand Listener*. Savage advised the publishers of the *Radio Record* in January 1937 of the decision to begin a new journal, but it was two and a half years before the *Listener* was first issued.

The *Listener* began publication on 30 June 1939 with 380,000 copies given away free, one to each household in the country. After the first week it was sold at 3d per copy, one penny less than the *Radio Record*, and settled down to a circulation of 30,000. By the end of its first year of publication the circulation had risen to 40,000.[15]

Publication of the *Listener* was the death knell for the *Radio Record*. Since the mid 1930s and the loss of its subsidy from the NZBB, the *Radio Record* had taken the character of a general magazine with, in particular, an increasing amount of material devoted to films and film stars. Its character had progressively become less august. Its previous support for the RBC and NZBB (and opposition to Labour's broadcasting policy) meant that by the time Savage and Shelley were in control it was inevitable that a state-owned broadcasting journal would be a new venture rather than a renewal of the *Radio Record*. Charles Earle and the Wellington Publishing Company severed their connections with the *Radio Record* in October 1938 leaving Arthur Heighway in sole control. The *Radio Record* did not long survive its loss of the right to print programmes in advance. It was bought by the NBS and incorporated into the *Listener*.[16] The parent company, National Magazines Ltd, collapsed and Heighway was bankrupted. After clearing his commitments in New Zealand, he moved to London in August 1940 and became editor of the *World's Press News Magazine*, a position he held until his retirement in 1953.[17]

Although the Opposition wondered whether the new journal would be used for political propaganda, the *Listener* did not become a government apologist.[18] In line with Shelley's attitude to broadcasting generally, its first editor accepted that independent political commentary was impossible. The

Listener instead gained independent status in New Zealand as a journal of literary and cultural discussion.

Oliver Duff, previously editor of the *Timaru Herald* and the *Press*, was appointed inaugural editor of the *Listener*. Shelley had a particular vision of the new journal and gave detailed instructions as to its purpose and the nature of the material that should be included. He made it clear to Duff that the *Listener* was a Broadcasting Service publication and that he intended to give it close scrutiny, at least in its early period. 'In general any contemplated material which involves politics, religion, or foreign countries should be submitted to me before publication. The whole contents of the journal should be submitted to me before printing until further notice in order to ensure that matters of broadcasting policy are checked.' Shelley had definite views as to the nature of the *Listener* and intended it should not be the general publication the *Radio Record* had become: 'The journal is a *broadcasting* publication and everything in it (apart from advertisements) should have a fairly direct connection with broadcasting — the journal should not attempt to be a general family magazine.' He intended it to be a tasteful but not an intellectual publication. 'Beware of adopting too high an intellectual standard — the journal must not sacrifice "life" for "learning". In matters of taste — if in doubt, cut out. There must be nothing "questionable" in articles or advertisements.'[19]

There was disagreement between Duff and Shelley on the nature and control of the new publication. Duff regarded the *Radio Record* as 'objectionable and even nauseating' and concurred with Shelley's desire to start a quite different weekly broadcasting publication. But on the editor's part, there was a mounting frustration. He complained about his lack of power to make prompt decisions: 'If I have to discuss every step with you, and you in turn discuss it with someone else, *before I take it*, action will always be too late . . . If I am to have no freedom I can show no initiative, and the *Listener* will become a dull official journal which the public will not buy.'[20]

Duff was of a similar age to Shelley, had his own distinguished background in New Zealand intellectual circles, and, in spite of the early instructions he received, did not surrender control of his journal to the director. In particular he disagreed with Shelley about the policy of the *Listener* and determined both to set a high intellectual standard for the journal and to ensure that its character would reach beyond broadcasting. Through his efforts, which were continued by his successor, M. H. Holcroft, the *Listener* became well regarded in New Zealand literary and intellectual circles, while retaining the large circulation that flowed from its sole right to print radio programmes in advance. The priorities and the importance of the first editors are encased in Mulgan's assessment of their publication as 'a first class critical journal and a keen encourager of new writing-talent'.[21]

While its contribution to New Zealand literature is undoubted, the *Listener* during Duff's editorship paid little attention to broadcasting. Leading and general articles on the topic were few, and broadcasters, both commercial and non-commercial, complained that the loyalty of the *Listener* was to New Zealand cultural activities other than radio. Holcroft published more articles on broadcasting than his predecessor, but the complaints continued. In particular, the *Listener* was seen as unsympathetic to commercial broadcasting. Tackled on the matter soon after he became editor, Holcroft made it clear commercial radio would receive no more space until its material conformed to the high standards of journalism expected in the *Listener*. This could not be achieved, for it ran counter to the philosophy of the NCBS which was to accept and cater to popular taste. Commercial broadcasters for years regarded the official broadcasting publication as unhelpful to their efforts and even as antagonistic to their aims.[22]

OUTBREAK OF WAR

World War II ushered in enormous changes to broadcasting. Many peacetime priorities were discounted and the broadcasting services took on new responsibilities and a new importance. The first priority was security of broadcasting facilities. Within stations, steps were taken to lessen the chances of unauthorised people gaining access to the microphones. The general public was no longer allowed access to broadcasting buildings. A major consequence of this was the banning of studio audiences — radio stations were no longer community facilities that could be watched in operation.

From the declaration of war, at least two persons were in constant attendance at all studios and transmitters. When war began in the Pacific, measures were taken to protect broadcasting premises. All windows in all transmitting stations were blocked by concrete slabs to deaden any vibrations which might break glass valves. Low walls were built around transmitter masts to protect them from the effects of bombing. The NBS head office and 2YA premises in Waring Taylor Street were considered unsatisfactory as they were close to the wharf area and were structurally unsuitable for the addition of air-raid shelters. New premises were eventually found in the Masonic Buildings at 38 The Terrace and, from October 1942, they were occupied by the head office administrative staff, the programme section, the record library and, importantly, the 2YA control room.

The expansion plans for both broadcasting services were halted. It was to be a decade before any new station was opened in New Zealand, and all plans for the development of existing facilities were shelved. It was acknowledged in Parliament that the war would affect the continuation of a large number of public works.[23] For broadcasting the major casualty was the proposed broadcasting centre. Work on the foundations was halted. The foundations filled with water and were stocked with goldfish to counter the mosquitoes that soon began to breed. Long after the war when the project was

The illustration on the cover of the *Listener* at the outbreak of war, 15 September 1939. LISTENER

reconsidered, it was modified to a more modest design that did not include a conservatorium. The only planned building that was constructed was the one in Durham Street, Auckland, to house the 1ZB studios. It was completed at a cost of £85,000 and occupied on 6 October 1941. Walter Nash contended that, at the outbreak of war, the construction was at too advanced a stage to stop and that, besides, the Department of Health evidence was that conditions were 'unhealthy and unfair to everybody there' and a new building was 'justifiable'.[24] The expense was not entirely for broadcasting purposes — the NCBS occupied the two lower floors of the building, but the top two were used by other government departments.

2YA's status as the main broadcasting station was enhanced by the wartime conditions. All stations were required to monitor 2YA constantly and to rebroadcast emergency transmissions. The first changes to wartime programmes were to news broadcasts and the length of daily transmissions. From the Munich crisis in September 1938, BBC shortwave transmissions were regularly rebroadcast by the YA stations, and during the crisis the ZB stations also used BBC news and commentaries. At the outbreak of war NBS stations broadcast twenty-four hours a day for several days — such was the public interest and anxiety. This practice was stopped when it became clear that it was provoking rather than alleviating anxiety. The NBS stations began broadcasting for eighteen hours a day, from 6am to midnight. This was an increase on their pre-war hours and brought them up to the same transmission hours as the NCBS stations. The number of transmission hours a day was decreased during the final eighteen months of the war so as to reduce electricity consumption.

Disc recording in progress during the Second World War, with technician Basil Clarke recording a news item.

During the six months of the 'phoney war' the number of news bulletins was reduced, but these were stepped up again with Germany's attacks and Italy's entry into the combat. For long periods the national stations broadcast daily up to three hours of news, commentaries and talks, mainly bearing on the war. All regular BBC news bulletins were recorded, and officers were always on duty at the 2YA control room and the shortwave receiving station ready to drop the cutters of their recording equipment should there be news flashes or emergency announcements. From September 1940, a year after the start of the war, NBS programmes were changed to allow for the increase in news bulletins. For the many who were not interested in hearing more news, the alternative stations offered musical programmes at the same time as the YA stations gave news broadcasts. The 7pm news hour was replaced by a normal news bulletin and 9pm became the time for the major daily news programme. As with the previous 7pm programme, it included bulletins from the Prime Minister's Department, UPA cables and the British Official Wireless News. To these were added the main items from the BBC news. So as to overcome the often poor reception from the BBC shortwave station, these items were re-read from the 2YA studios.

At the September 1940 reorganisation of programmes, the opportunity was taken to introduce a system to the daytime NBS programming. As explained in the *Listener*, each session of music was to be continuous in its appeal; it would not change from one style to another: 'If [the listener] does not like it, he or she can know to switch off or tune elsewhere.'[25] The coordinated programming schedule, first introduced by the RBC, had applied only to evening programmes. The 1940 changes were a first attempt at systematic coordination of the daytime programmes.

Ken Collins, long-time broadcaster and wartime announcer technician at Parliament.
LISTENER

WARTIME BROADCASTING FROM PARLIAMENT

The broadcasting of Parliament continued during the war, but a new measure of censorship was introduced. The broadcasts had been under the control of the prime minister. During the war the Speaker was given authority to censor by stopping any broadcast from the House. Wartime censorship had already started; Fraser reported an early instance: 'A few days after the ships carrying our boys [of the First Echelon] were going away they were being mentioned specifically and in detail in a religious service and the operator had to cut in immediately, because if he had not done so the information would have been broadcast, possibly with disastrous consequences.'[26] Fraser wanted a similar censorship of parliamentary broadcasts if and when needed: 'I do not think that any member can trust himself, when speaking in the heat of a debate, to be absolutely discreet, and we would never forgive ourselves if we uttered a single word that would endanger the lives of our boys.'[27] Mr Speaker Barnard received the authority and promptly assured Ken Collins, the announcer—technician at Parliament, of his backing if Collins were to cut Parliament off the air without waiting for the Speaker's signal. Thus over the war years the NBS, in particular Collins, was responsible not only for the technical quality of parliamentary broadcasts but for exercising a censoring vigilance over their contents for which parliamentarians themselves declined to assume responsibility. Collins often cut broadcasts a split second before potentially damaging information was spoken.[28]

MOBILE BROADCASTING UNIT

World War II was the first time broadcasters accompanied New Zealand armed forces. The broadcasters were both providers of entertainment and a new type of war correspondent. They and their equivalents from other countries altered the general understanding of warfare and changed broadcasting itself.

In February 1940 Shelley proposed the NBS send a team to accompany the Second New Zealand Expeditionary Force (2NZEF). The idea was accepted in March and the Broadcasting Unit sailed on 27 August 1940 with the third contingent. The unit was a three-man team comprising N. R. Palmer, M. D. Laurenson and N. B. Johnston. Palmer, who commanded the unit, was an assistant engineer from the NBS head office, Laurenson was a 1ZB announcer and Johnston was a 2YA technician. The three were given no rank but were accorded the status and privileges of officers. Command of the unit was taken over by A. L. Curry from August 1943 when Palmer returned to New Zealand.[29]

They took with them a 16-foot-long van body, extensively equipped and spacious enough to be used as a working studio: once in Egypt, this would be mounted on a chassis that had been ordered from England. The unit was operational while en route and was used during the voyage from New Zealand to record talks from members of the ship's crew and programmes dealing with life on board a transport ship. The third contingent disembarked in Egypt at the end of September and the unit joined the New Zealand base camp at Maadi, near Cairo.

In the Middle East it became evident that the unit would not operate as envisaged in New Zealand. The large van was immobile until July 1941 when its Leyland chassis finally arrived from England. Once it could be moved, it was taken on a lengthy journey to Syria, Palestine, Iraq and Trans-Jordan. This experience convinced the broadcasters that the van was too large and too slow to maintain contact with army units. From this time the van was used as a base recording and editing studio. Transport was a continuing problem for the unit; at best, it used the standard 15-cwt army trucks of the New Zealand forces. These were scarce, however, and not always available to the unit.

Much of the equipment was too large and sophisticated for field work. With experience, the unit reduced its field equipment to more compact dimensions. Eventually a set of mobile recording equipment comprised a recorder in a leather cover, a suitcase containing thirty discs, cutters, two microphones, cords, a vibrator–converter, a small microphone stand, a 12-volt battery and sundry odds and ends.[30]

Some of the original proposals for the unit could not be carried out. Public address work proved impracticable, and the equipment for this was disposed of. Attempts were made to supply a regular news service via radio-telephone to London for the BBC to rebroadcast in its Pacific Service. Stringent censorship led to this being abandoned. Conditions during campaigns were

The inaugural members of the Mobile Broadcasting Unit pictured on their departure from New Zealand. From left, N. B. Johnston, N. R. Palmer and M. D. Laurenson. *LISTENER*

naturally difficult, and it was possible to record very little. At such times the emphasis was on news commentaries. At other times the unit developed new priorities and wartime broadcasting became distinct from war correspondent press reporting. By early 1941 Palmer and his team had decided that, rather than presenting commentaries, the essential work of their unit was the personal one of recording the spoken words of members of the 2NZEF. Their method was to ask the men to write down their experiences, and then read them. These recordings were very well received in New Zealand and the policy was followed for the rest of the unit's existence. At the same time the unit began recording messages to be broadcast in New Zealand. Alan Mulgan, the supervisor of talks in Wellington, called this 'our most personal intimate service'.[31] At first it was thought this material would be dull, and to increase interest, the men were asked to nominate their favourite recordings which would be interpolated in the broadcasts. However, demand in New Zealand rose rapidly for more greetings and no music. Units were visited in rotation and recordings taken from around 10 per cent of the personnel.

From March 1941 the second contingent began arriving from England and the whole division commenced its move to Greece and Crete. The unit did not accompany the division, but after the evacuation the campaign was covered in a series of commentaries and a large number of recordings of personal experiences. Rather than concentrating on authoritative officer reports, the unit made efforts to cover the experiences of all ranks. Again this material was well received in New Zealand, and confirmed the unit's belief that their innovative policy should be continued.

Along with visiting the components of the 2NZEF, the broadcasting unit recorded the experiences of other New Zealanders such as naval personnel and members of the Royal Air Force stationed in the Middle East. Many recordings were made of notable events, from the visits of dignitaries to parades, concerts and sports meetings. There was much cooperation with other broadcasters and considerable joint activity with the Middle East Broadcasting Unit, the British Army unit which conducted the forces programme in Cairo.

The unit's greatest contribution to news broadcasting was in the second half of 1942. The unit moved with the New Zealand Division through the Western Desert. Throughout the period of the Alamein defensive front, Curry, who had replaced Laurenson as commentator, made news despatches which secured wide recognition for the unit.

In New Zealand the unit's news despatches were combined with other material in news bulletins. The other recordings were at first broadcast twice weekly in evening programmes from all NBS stations. In 1941 this was changed and a weekly programme entitled 'With the Boys Overseas' was presented on Sundays from 9am to 10am. The programme also included material dealing with New Zealand forces from other organisations such as the BBC, which forwarded recordings from servicemen stationed in Britain. The recorded messages were repeated on Tuesday evenings with the names of the men who were to speak and the people they were calling read out at the beginning of each broadcast so as to alert relatives. The recordings were airmailed to New Zealand, but after the fall of Singapore this was no longer possible and recordings often took up to ten weeks to arrive in New Zealand.

The unit reported to the NBS the reception given by the troops to New Zealand-prepared broadcasts of the 'New Zealand Calling the Boys Overseas' series. These reports were often decidedly uncomplimentary. One extract from a Palmer report forwarded by Shelley to his head of drama included complaints about most of the programmes sent to the troops: 'They had a hot reception . . . [Talks] must be stripped of all colour, bombast and hot air . . . There is no audience here for advertisements of New Zealand's war effort.'[32] Palmer's reports continued to comment on the New Zealand broadcasts:

> Military subjects should be avoided. The things that men over here want to know are strictly censorable and there is no interest in the rest . . . The subject of American Forces should be avoided as far as possible . . . No opportunity should be given for men returned from the Middle East to air their opinions except in connection with such matters as rehab schemes. At all costs the theme of 'It's great to be home! I hope you're all right over there' should be avoided. The reaction of men here to this sort of thing is amazingly complex, and it is only too easy to cause bad feeling in this way.[33]

In late 1943 the New Zealand Division moved to Italy and the broadcasting unit accompanied the troops throughout the Italian campaign. Communications from Italy were superior to those from the Middle East and Curry's war commentaries from Cassino were recorded from the BBC and heard in New Zealand little more than 24 hours after their despatch from Italy. While stationed at Naples and at Rome the unit continued its policy of recording as many New Zealand servicemen as possible. Many hours of recorded greetings and first-hand personal accounts of battle and campaign life were collected and sent to New Zealand. At the end of 1945 the last of the equipment was loaded on board ship and the unit returned to New Zealand.

In April 1943 a second broadcasting unit was sent to the Pacific theatre of war; it was stationed there until August 1944, when the broadcasters returned to New Zealand. This second unit was deliberately modelled on the first. Half of its recordings were personal messages from servicemen and the rest were talks of personal experiences from servicemen, commentaries and recordings of ceremonies. The unit was based in New Caledonia but travelled considerably as it followed the troops north through the Pacific. Mail from the unit to Wellington was never disrupted and the NBS had two weekly 'With the Boys Overseas' programmes devoted to the Pacific. The Pacific theatre had its own tremendous difficulties for wartime broadcasters, but like their Middle East counterparts the Pacific unit members earned a fine reputation. The final act of the Pacific unit was to donate all its equipment as a gift from the New Zealand government to the people of New Caledonia in appreciation of the assistance given to the New Zealand Forces. In practice this benefited Radio Nouméa, the station from which NBS officers had been presenting a daily 'Kiwi Hour' programme.

The broadcasting units made significant contributions both to New Zealand's war effort and to broadcasting. Initially viewed with scepticism within the 2NZEF (the broadcaster's greeting on arrival at the Maadi base camp was 'Good God, what next are they going to send over!'), the first unit won for itself a respect and affection both in the Middle East and in New Zealand. In the less momentous periods of the war the two units' feature broadcasts, and especially their interviews with servicemen (and occasionally servicewomen) of all ranks introduced a new understanding back in New Zealand of the experiences of the country's armed forces. The broadcasting units were an important factor in the transmission of wartime information and in the maintenance of morale among the troops and in New Zealand. For broadcasting itself the units' significance was twofold. They represented the first presence of broadcasters among New Zealand's military forces. The acceptability of broadcasters in a war zone and of their right to report independently was always going to be contested; but the broadcasting units proved that wartime broadcasting was possible. Under the most difficult conditions, they demonstrated what could be done with the newly invented recording equipment. Secondly, the contents of the broadcasts indicated a new path for postwar documentaries concentrating on recording the voices and the stories of everyday men and women.

WARTIME BROADCASTING

Extensive changes were made to broadcasting practices in 1941 following the sinking of the vessels *Holmwood*, *Rangitane*, *Vinni* and *Komata* and the report of the subsequent Commission of Inquiry. There had been speculation that apparently innocent broadcast messages may have been coded instructions to German raiders. Shelley was called to give evidence before the Commission as was L. J. Greenberg of the NCBS. A section of the report details the nature of the changes:

> They [Shelley and Greenberg] have satisfied us that announcements of missing cars or missing relatives are forbidden unless they come from the police. As to birthday and other like announcements, these ceased to be made by some of the commercial stations on 23 January 1941 and by the remainder on 3 March 1941. The question of their discontinuance is under consideration by the national stations. Professor Shelley agrees that they should be discontinued and so do we. The evidence is that the script of all advertisements is altered before the advertisement goes over the air; that the strictest supervision is exercised over all musical items, and the method of announcing them; and that even the jokes made in the sponsored programme *Chuckles with Jerry* are censored before being projected over the air; and that as to all matters there is a time lag, and that the person who evolves the phraseology is not the person who determined *when* it will be broadcast. As to musical numbers we are satisfied that the programmes are built round an advertising unit so far ahead that it is impossible for any announcer to put on a record, such, for instance, as one which could indicate the presence of warships in any of our harbours without prior notification. Furthermore the script of all projected broadcasts is censored, and now most of them are recorded some time before they are put over so that the speaker himself does not know when his record will be on the air.[34]

These changes considerably altered broadcasting. The general trend was to stop the contributions from listeners for fear they might carry some disguised message. Some types of programmes, such as request sessions and community singing, were dropped. Others took different formats for the duration of the war. Children's sessions had little direct contact with children. Their request sessions were cancelled, as was the practice of reading birthday greetings and other personal messages. In all programmes the broadcasting of names and addresses was eliminated.[35] In keeping with the removal of outside contributions, advertising could also have been banned. This was obviously economically difficult, and instead a system, not entirely satisfactory to advertisers, was developed whereby the wording of all advertisements as given to a station was altered before being broadcast.

Programmes which might be of use to the enemy were also banned. The

An NCBS outside broadcast vehicle pictured in 1940.

particular casualties of this policy were weather reports, which were stopped from December 1940. The exceptions were the few cases where it was judged necessary, usually for the benefit of farmers, to give advice of approaching storms. This banning of weather information was carried out to the extent that sports commentators were forbidden to make references to the weather conditions at sporting events.[36]

Certain topics were not permitted to be broadcast. This applied especially to the locations of troops. All references to forces in New Zealand or overseas, including greetings to soldiers or references to camp locations, were banned. Various records were also deemed unacceptable for broadcasting: after 2ZB played *The Fleet's in Port Again* at the time of the arrival of warships in Wellington, the recording was banned. Greenberg, the NCBS secretary, also eliminated other possibly informative titles, including *The Fleet's Not in Port* and *There's a Boy Coming Home on Leave*.[37]

The most difficult form of censorship was the banning of all ad lib material. Nothing was to be spoken over the air other than that which had been scripted and censored. This requirement proved impossible to implement fully — it would have meant the elimination of sports commentaries. It proved particularly difficult for the NCBS, as the deliberately informal nature of the commercial stations' presentation was incompatible with a fully scripted programme. The Organiser for National Security noted to Greenberg that the Aunt Daisy programme was almost entirely ad lib. The eventual decision was that material from outside the station, such as correspondence, hints and recipes, should be changed and expressed in the announcers' own terms. Daisy Basham's programmes were recorded beforehand so that the censors could listen to them before they were broadcast.[38]

Within stations there was considerable enthusiasm for New Zealand's war

A wartime publicity photograph for *Chuckles with Jerry* with Dudley Wrathall, on left, and George Tollerton seen on the steps of Parliament.

effort and much desire to support it. A sheaf of suggestions from 3ZB was supported by Greenberg and sent to all stations as instructions and to the prime minister as information. Thus, in order not to encourage hoarding, phrases such as 'Buy now while stocks last' and 'Lay in your supply while there is yet time' were banned. Phrases such as 'Business as usual' which suggested firms were carrying on in spite of the war situation were avoided, as were other unwanted sentiments.[39]

Wartime shortages affected broadcasting. The size of the *Listener* varied considerably in the number and width of its pages, largely because of the shortage of newsprint but also due to a falling off in advertising sales. From its original fifty-six pages at the start of the war it was gradually reduced until from March 1943 there were twenty-four pages an issue. A major casualty of the wartime shortages was drama production. Blank discs for recording were scarce and were increasingly reserved to record overseas news and commentaries about the war. For eighteen months from early 1942 the NBS virtually suspended drama production. Restrictions were again necessary in 1944 when aluminium, from which disc bases were made, was in short supply. Glass was used as a substitute but, while satisfactory in the studio, it was heavier and less likely to survive handling and transport.[40]

There were some positive additions to broadcasting. The humorous programme *Chuckles with Jerry*, with Dudley Wrathall as Jerry, the ventriloquist's dummy, and George Tollerton as the ventriloquist, was on all the ZB stations from December 1939. Later in the war, general programmes, representing a large amount of the production from the recording studios, were made both to inform the audience and to enhance morale. These

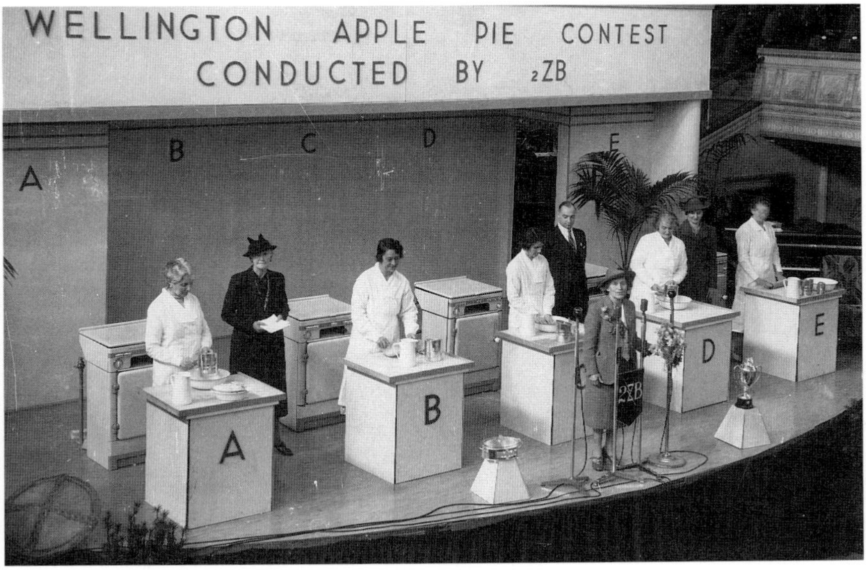

Daisy Basham at the microphone leading the disposal of the 1940 apple harvest.
ALEXANDER TURNBULL LIBRARY

programmes included programmes such as *Tell the People*, *This is War* and *We Work for Victory*. Besides the news broadcasts, perhaps the main use of radio during the war was to exhort the population to action. The stations of both the NBS and NCBS took part in many campaigns in support of the war effort. Scrimgeour, in a 1942 letter, detailed sixteen such campaigns in which the NCBS had been involved, including a drive to dispose of surplus apples, a WAAF recruitment drive and an appeal for binoculars.[41] The apple appeal was especially successful: a million cases that would have been exported but for shipping shortages were sold in New Zealand. Among the appeals, a major addition to the broadcasting repertoire during the war was the radio telephone appeal. The first of these was held by 4ZB in June 1940 for the Red Cross and £1500 was raised in three days. The northern ZB stations greatly exceeded this total in subsequent appeals. The same month 2ZB raised more than £5000 for the Sick, Wounded and Distress Fund. For the same cause 1ZB raised £16,000 on its Sunday night appeal, and carried on the collection when the telephones continued to ring. By midnight on Monday £32,000 had been collected. The NBS and NCBS combined for a nationwide radio telephone appeal on Anzac Day 1941 which raised £45,000 for the National Patriotic Fund. The combination of telephones and broadcasting was a wartime discovery that gave a further indication of the widespread nature of the radio audience and the way it could be aroused to act collectively. This was further exploited during the war in other charitable appeals, and in particular in the sale of war-loan stock.

Both the NCBS and the NBS were involved in camp concerts arranged for the entertainment of troops. These were started by the commercial stations once the large training camps began filling with troops. Later the NBS joined in and the concerts were subject to a more formal organisation and association with the armed services. As only some of the concerts were broadcast it was disputed whether this was a proper activity for a broadcasting service. The dispute was settled by an Order in Council, gazetted on 26 August 1942, which interpreted the minister's power to develop and improve the broadcasting service as including 'power to organise and present entertainments for members of His Majesty's naval, military, or air forces or any other forces for the time being lawfully present in New Zealand'. This was a specific wartime measure, but it was indicative of the wider role the broadcasting services were assuming — and it was extended in later years.

Among the wartime programming changes were those made to accommodate the American troops stationed in New Zealand from the middle of 1942. The first was 'American Hour', a weekly NCBS programme, compered by a marine and featuring the latest records from the United States. Later complete recorded programmes were obtained from the United States and played on Sunday evenings, proving equally popular with American troops and New Zealand listeners. Programmes presented over 1ZB and 2ZB by members of the American forces were recorded and despatched to the United States to be played over the networks there. A KWID San Francisco programme, 'American Commentary', was rebroadcast in New Zealand by arrangement with the United States Office of War Information and played five nights a week. As with the entertainment programmes, it was listened to as much by New Zealand audiences as by the American troops. The broadcasting cooperation with the American troops was at its highest in 1944. In April of that year the control of station 1ZM was handed over to the American army. Shortly after that the number of American troops in New Zealand began to be greatly reduced and the station was returned to NBS control in December 1944.[42]

The 'Talks' programme was another part of broadcasting's war effort. The programme had been under the general supervision of Alan Mulgan since his appointment as Supervisor of Talks by the NZBB in September 1935. He was appointed to coordinate radio talks under the new regime that was to follow from the coalition government's allowing of the broadcasting of controversial material. Mulgan retired in 1946 and never saw his optimism for radio talks reflected in actual broadcasts. Labour's return to a ban on the broadcasting of controversial material was followed by the war which introduced new priorities for the Talks programme. Mulgan administered those new priorities, and under his guidance radio talks became an important part of the war information and commentaries. There were fewer talks on general subjects, but the conflict was a topic of great interest and importance. On the national service time was reserved every night for talks bearing directly on the war effort. While many New Zealand speakers were with the forces and unavailable, there was a continuing parade of military and political visitors who were able to broadcast.

THE DEATH OF SAVAGE AND REMOVAL OF SCRIMGEOUR

Among the most famous of the wartime talks were the twelve Sunday evening 'fireside talks' given very early in the war by Savage. The first was broadcast on 19 November 1939, and they continued to March 1940, the month Savage died. They were scheduled to attract a large audience, being broadcast in the quarter hour on Sunday evenings between the ending of Scrimgeour's 'Man in the Street' session and the 9pm broadcasting of the chimes of Big Ben. In a final indication of his high regard for radio and determination to use it to the full, Savage ordered that his talks be broadcast over all stations, NBS and NCBS.

Savage's death was an event that led to a radio-coordinated national expression of grief. The prime minister's casket was taken from Wellington to Auckland on a special funeral train which also carried broadcasting staff and equipment. Clive Drummond was the travelling announcer and broadcasts were made from twenty-five stops on the journey. The broadcasting of Savage's final journey was an important national event and was especially appropriate for the politician who did so much to further the political impact of radio.

During Savage's time as prime minister much of the administration of the broadcasting portfolio was handled by Frederick Jones, the postmaster-general. But Savage never relinquished his personal control of the radio services. As the *Listener* noted in his obituary, 'Broadcasting was not merely a State function to him, and a public utility. It was a social and political miracle which he never ceased to think about.'[43] Savage was the Labour politician who had been most determined to use radio for political purposes. His lessons were well absorbed by his colleagues.

Within broadcasting, Savage had championed commercial broadcasting and Colin Scrimgeour. With Savage's departure their ascendency did not continue. In spite of his ability, Scrimgeour was very much reliant on the protection afforded by his personal relationship with the prime minister. Savage's death meant the end of that protection; and it came at a time of particular dissension within the Labour caucus culminating in the expulsion from the party of Lee, Scrimgeour's friend and ally. In this turbulent atmosphere Scrimgeour ignored his weakened position and set himself irrevocably at odds with the new prime minister by committing what the *New Zealand Observer* referred to as 'the unforgivable sin'.[44] Broadcasting a panegyric after Savage's funeral, he commented:

> The name of the messenger sent to me by Mr Savage to ask me to accept appointment was — John A. Lee. May I be pardoned for saying that the greatest tribute we can pay to him whose departure we mourn is that the big-heartedness of New Zealanders may rise above the merciless bludgeonings of misunderstanding, and may not lose for us two men who desired to serve the nation — the one by death, the other by dissension.[45]

Scrimgeour was not pardoned by Peter Fraser, the new prime minister, and one of Fraser's first actions as minister in charge of broadcasting was to suspend the 'Man in the Street' session. Fraser was not bound by the personal arrangements made by Savage with Scrimgeour and had no obligation to continue Scrimgeour's unique broadcasting rights. However, in spite of now being prime minister, Fraser was forced to take note of the strong public support and trade union backing for Scrimgeour. The 'Man in the Street' session was resumed after being off the air for only one week. However, Fraser's appointment as prime minister meant that Scrimgeour's power was never the same again.

In January 1941 Fraser appointed David Wilson as minister in charge of broadcasting. It was widely considered that the prime minister remained closely involved in the portfolio, leaving only the day-to-day administration to Wilson. With an understanding typical within the NCBS, Mackay compared the relationship between the two ministers 'in my prejudiced eyes' to that of the 'pilot fish swimming close to a marauding shark.'[46] For the next two and a half years the NCBS was hampered by the appalling relationship between its minister and its controller.

In May 1941 Scrimgeour announced in his 'Man in the Street' session that the topic of one of his future broadcasts would be 'Sex and its Contagious Diseases'. He then proceeded to discuss the topic by stating what he intended to cover and quoting liberally from various books on the subject. The broadcast provoked considerable comment, little of it favourable. Scrimgeour left for a visit to Australia after this broadcast and, while he was away, Wilson moved to restrict Scrimgeour's right to run his programme as he saw fit. 'Upon your return from Australia, please confer with me concerning the above matter [the 'Sex and its Contagious Diseases' talk] and, in the meantime, no talk on sex, or other highly controversial subjects, must be given by those who are deputising for you during your absence.'[47]

On his return Scrimgeour asked for but was denied permission to broadcast the Auckland meeting of the Society for Closer Relations with Russia. Ignoring this denial, his response in his next 'Man in the Street' talk on 10 August 1941 was to discuss the meeting and introduce and play a recording of the speech given at it by the Rev. Percy Paris.[48] The minister regarded this action as 'deliberately evading the instructions given'.[49]

After these broadcasts the 'Man in the Street' session was again suspended. The rationale for the suspension became confused with conflicting statements that it followed from the pro-Russia broadcasts, and that it was caused by Scrimgeour's ill health. In September 1941 Scrimgeour went on sick leave for what was later reported as a mild nervous breakdown.[50] The minister instructed that any announcements 'should only intimate that the suspension of the session is due to Mr Scrimgeour's indisposition'.[51] This action was followed by the publication of another pamphlet, 'The "Uncle Scrim" Mystery. What has happened to the "Man in the Street" session?' This pamphlet, strongly pro-Scrimgeour, put the argument that he was no longer sick but was being kept off the air for his progressive outspokenness. There is no conclusive evidence that Scrimgeour was involved in the publication of the pamphlet, and he wrote that he did not know who compiled the material or published it.[52] But he did take care to inform correspondents that his suspension was not due to illness and that any resumption of the broadcasts was 'waiting Mr Fraser's decision'.[53] It was clear from this time that Scrimgeour would not submit to the usual public service discipline but was prepared to challenge the minister's authority by appealing to the general public.

The 'Man in the Street' session remained under suspension but Scrimgeour was not prohibited from broadcasting otherwise. However he remained contentious and increasingly came into conflict with his minister and with J. T. Paul, the director of publicity in the Prime Minister's Department. As the chief wartime censor, Paul was responsible for the censoring of all broadcasts.

In February and March 1942 Scrimgeour was upbraided by Wilson for broadcasting talks — first on the Red Army and then on the availability of labour to build air-raid shelters — without submitting the scripts for censorship. After the second incident Wilson noted on the file copy of his letter to Scrimgeour, 'How many times do we have to threaten him with dismissal before we do it?'[54] These incidents were followed by a further complaint from Paul. A 2ZB script 'had not only been broadcast without approval by me but had been approved by the Controller of Commercial Broadcasting. You will note that the script makes reference to the presence of Marines, ships in port and other matters, and contains also references to men returned from overseas and who have received "a period of about a week's leave". This broadcast has been made contrary to all instructions and established practice.'[55] Scrimgeour was reminded in April 1942 to submit the scripts of all his radio talks beforehand to Paul.[56] In June the minister further instructed the censor to refer all scripts received from Scrimgeour to the minister.[57]

AMALGAMATION: FIRST STEPS

The disagreements between Scrimgeour and the minister influenced the government's wartime plans for the future of the broadcasting services. Part of the war effort were attempts by the government to have both broadcasting services release radio technicians and others with radio experience so that they could be enlisted into the armed forces. Wilson presented this case to Shelley and Scrimgeour, but was unsuccessful — both men insisted that they were unable to release any skilled technicians.[58] The government then formed the Committee for the Rationalisation of Radio Broadcasting Services and Release of Radio Technicians. The committee recommended an amalgamation not just of the two engineering sections — which would mean surplus radio technicians could be released — but of the two broadcasting services. It called for a new administrative structure whereby the amalgamated services were controlled by a permanent head, assisted by an administrative panel of four: a secretary, the chief engineer and the senior officials from the commercial and non-commercial divisions. The government's desire to make more skilled radio technicians available for the armed forces was real and urgent. However, the proposed amalgamation of the two broadcasting services made it clear that the government was using the occasion as a pretext for getting rid of Scrimgeour.

According to Shelley, the report was 'a very good effort and presents no difficulties.'[59] Wilson presented the proposal to amalgamate the two services to Cabinet, which approved it on 3 December 1941. However, Scrimgeour's reaction was such that the approval was held up by the prime minister and not implemented until the second half of 1943.[60]

Scrimgeour argued that 'the essence of the success of the Commercial Service is without question its standing as a separate entity in the minds of the listeners . . . In order to maintain this attraction the Commercial Service must preserve its atmosphere distinct and separate from that of the National Service.'[61] He also attacked the proposed amalgamation as an attempt to remove himself from the control of a separate NCBS. That this was part of the proposal was verified when Wilson suggested that Shelley be the permanent head of the amalgamated services, with Scrimgeour taking a position as head of the Commercial Division at a reduced salary. In a long reply detailing his broadcasting background and relationships with the Labour Party and the NCBS, Scrimgeour maintained that 'such a proposal is an indignity — it is totally out of harmony with the spirit of the agreement entered into by the late Mr Savage, and certainly not warranted if the record of success of the commercial service under my direction is taken into consideration'.[62]

In a further interview with the minister two months later, Scrimgeour 'was emphatic in saying he would not work under Shelley . . . In a threatening manner he said that if he was not appointed he would "go to work on the wharf" also that if we wanted a fight we could have it . . . When he left the understanding was "Director or nothing".'[63] Scrimgeour applied for membership of the Auckland branch of the New Zealander Waterside Workers' Union, and his application was 'approved by an overwhelming majority'.[64] He made his reasons for the application public: 'It appeared to me I might need something to fall back on . . . I also understood that Professor

J. Shelley, Director of Broadcasting, would take control of the amalgamated services, and rather than take a secondary position in that event I would prefer to select my own avenue of service.'[65] While Scrimgeour remained in control of the NCBS, the plan for amalgamation of the services was postponed. The prime minister argued that, because of the differing natures of the two services, their proposed amalgamation was cancelled rather than postponed: 'A commercial and a non-commercial service could not be fused into one but a good deal might be achieved in a reorganisation and coordination by which better use might be made of the engineers and technicians at present engaged in the separate services.'[66] The amalgamation of the two engineering sections took place in August 1942. The NBS chief engineer, J. R. Smith, assumed overall control and his NCBS counterpart, R. E. Grainger, became a member of Smith's staff.

SCRIMGEOUR'S DISMISSAL, REINSTATEMENT AND REDISMISSAL

On 1 March 1943 Wilson and Paul met with Fraser and discussed Paul's complaint that Scrimgeour had again evaded the censorship requirements. The complaint was that he had broadcast the previous Sunday evening without submitting a script for censorship. Instead a musical continuity script had been submitted which 'bore no indication that it was to be used as a talk by Mr Scrimgeour.'[67] The minister considered that 'Scrim was tricky as usual'. Scrimgeour was suspended from his position with the suspension 'based on failure to obtain censorship'.[68]

Scrimgeour was 'curious to ascertain' whether the government distinguished between censorship for the purpose of security and censorship 'to restrain any well-known person from communication by radio . . . with his unseen public'. For that reason the director of publicity was given the 'innocuous' script of a compère's part in a musical programme and, after the script was passed, Scrimgeour took this part at the microphone. The *Press* noted that 'Scrimgeour was suspended because he took a trivial part in a trivial broadcast . . . The explanation must be assumed to lie, not in any situation that Mr Scrimgeour's broadcast created but in the situation from which it emerged; and that, plainly, was a bad one.'[69]

In a widely reported statement Scrimgeour complained that his office door was locked against him and 'he was thus deprived of the ordinary courtesy of being allowed to remove any private papers from his desk'. However Greenberg, the acting NCBS controller, held that Merton Bullivant had been instructed to meet Scrimgeour and let him know he could come and collect his effects; but that Scrimgeour had 'adopted a threatening attitude towards myself [Greenberg] and expressed his intention to commit personal violence'. Scrimgeour made it clear he would not return unless and until he was reinstated.[70]

The suspension of Scrimgeour was a matter of considerable public comment and caused enough concern within Labour Party and trade union circles for the executives of the Labour Party and the Federation of Labour to set up a joint committee in an endeavour to effect a reconciliation between Scrimgeour and the minister. The four members of the committee were James Roberts and Michael Moohan, respectively president and secretary of the party, Fintan Patrick Walsh, president of the FOL and Angus McLagan, ex-president of the FOL and at the time a member of the government with a seat in the Legislative Council. The outcome of the committee's intervention was that Scrimgeour was reinstated with an assurance that there would be no discrimination against him. For his part Scrimgeour agreed to carry out the government's broadcasting policy, submit all scripts to the censor, not broadcast without ministerial approval, work harmoniously with the minister, and cease his public attacks on the government.[71]

These conditions meant that Scrimgeour's unique broadcasting rights were formally ended, even though he was reinstated. However, his supporters argued that he had won a victory and 'backed by the FOL . . . [Scrimgeour] humbled his Minister and forced a reversal of a Cabinet decision.'[72] The reinstatement was widely regarded in the press as an instance of party and trade union 'bosses' being the real rulers in New Zealand and forcing the government to back down. The *Press* referred to the government as suffering 'political humiliation'.[73] The Opposition argued the same case. 'We have the extraordinary position of the Government delegating its powers to four trade union bosses who were the Government's accredited representatives to settle a dispute between Mr Scrimgeour and the Government.'[74]

At the same time Scrimgeour came to the notice of the military authorities and was called up for military service. The first ballot for men of Scrimgeour's age was held in November 1942. Almost forty years of age, a married man with three children, and the head of a government department, Scrimgeour could reasonably have expected to be excused from military service.

The other executive officers of the NCBS asked the minister to make a formal appeal to the Armed Forces Appeal Board on Scrimgeour's behalf. Their request only highlighted the gulf between the minister and the NCBS. The executive officers argued that the war effort had already led to a severe reduction in the NCBS executive staff. They detailed Scrimgeour's responsibilities and ventured the opinion that his abilities were unique in New Zealand, and that, 'If Mr Scrimgeour is required for military duties we have no hesitation whatsoever in stating that it will have a detrimental effect upon what is now a well-oiled administration.'[75] Wilson's reply made it clear he would not support Scrimgeour. 'There is nothing contained therein which changes the opinion I expressed to Mr Scrimgeour himself . . . to the effect that I had no intention of appealing for him as I did not think his services were essential to the success of the Commercial Broadcasting Service

. . . Very many of the statements made in your memorandum are at the most half-truths.'[76]

John A. Lee, by now Scrimgeour's only real ally in Parliament, queried 'whether certain persons will use the balloting organizations not for the purpose of recruiting people for the Army, but in order to revenge themselves against certain people . . . I felt that the Board was used as an instrument of Ministerial vengeance.'[77]

An unsuccessful appeal was lodged on Scrimgeour's behalf by the Friendly Road Church. Wilson, subpoenaed to appear at the appeal, argued that Scrimgeour's absence would not adversely affect broadcasting. He added, 'I might get more willing and loyal service from his successor.'[78] Scrimgeour was ordered into the air force, but before entering that service he made further press statements in which he argued that he was being railroaded, and that his case was one of discrimination and victimisation at its worst by the prime minister and the minister in charge of broadcasting. He also offered a new version of the settlement preceding his reinstatement, including a clause whereby any military service on Scrimgeour's part was restricted to one month's training in each year. The four members of the Labour Party and FOL committee promptly denied the accuracy of this account, and the minister described it as improper and improbable.[79] Scrimgeour's statements were a clear breach of his signed agreement and the replies to them signalled his loss of support at the higher levels of the Labour Party and the FOL.

Fraser seized the initiative. At a Labour Party caucus meeting on 18 June 1943, the Cabinet recommended that Scrimgeour be dismissed, and this was carried by 26 votes to 5. Wilson telegrammed all NCBS station directors to inform them of the dismissal, and to ensure that Scrimgeour would no longer broadcast: 'Mr Scrimgeour must not be allowed to speak over the air either from a microphone or from a record or otherwise and all such records which may be in your possession must be forwarded immediately to Professor Shelley in Wellington stop You will be held personally responsible for the prompt and strict carrying out of these instructions.'[80] Thus ended the New Zealand broadcasting career of the country's most prominent broadcaster.

It was not the end of Scrimgeour's prominence in public discussion during 1943. He stood for Parliament as an independent candidate for Wellington Central — Fraser's electorate. Soon after Scrimgeour's announcement that he would be a candidate, the prime minister referred in the House to Scrimgeour and alleged that certain communists, who had previously had the run of 2ZB and the NCBS head office, had attempted to suborn 2ZB employees and start a 'freedom' radio statio, and were now organising Scrimgeour's election campaign.[81] Fraser also produced a recording which had been made at Scrimgeour's NCBS farewell. He called it a recording of a 'drunken orgy . . . a most disgraceful thing . . . with obscene and blasphemous language . . . Anybody who could perpetuate filth on to a record was not fit to be in charge of a Government Department.'[82]

The recording was certainly unfortunate but a more temperate discussion of it was offered by the *New Zealand Observer*:

> The party was a farewell to a man who was going off on military service and his colleagues seem to have thought it would be amusing and original to make some records, which later could be given to him as a souvenir of the occasion . . . There is not one derogatory reference to the Government or the Prime Minister in the whole series. They are not even mentioned at any stage . . . Although the language in several passages is that of a smoke concert or 'buck party' at a particularly buoyant phase of the proceedings, no really objectionable language is heard from Mr Scrimgeour himself.[83]

The campaign for Wellington Central was personal and acrimonious. Scrimgeour's declared purpose was 'to make clear the extent to which Mr Savage's humanitarian programme is being sabotaged — and this deliberately by Mr Fraser'.[84] Scrimgeour, who was given no access to radio, had no electoral success. A joint statement from the FOL and the New Zealand Labour Party strongly supported Fraser and denounced Scrimgeour.[85] Scrimgeour's only satisfaction was to see Fraser's winning total drop to less than a majority of the votes cast. After the campaign Scrimgeour returned to the air force. He was pronounced medically unfit for overseas service and remained in the air force in New Zealand until November 1944, when he was discharged and left with his family to live and work in Australia.

The government investigated the NCBS and assessed the loyalty of all its major personnel. Descriptions such as 'most dangerous', 'fanatical Scrim-ite, suspected of being a member of the Communist Party', 'responsible for Scrim joining the Waterside Workers Union' and 'needs careful watching' were among the comments made.[86] However there were no wholesale changes in personnel in the NCBS and no attempts to dismiss staff who were regarded as Scrimgeour loyalists.

John A. Lee wrote of Scrimgeour, 'Scrim is a national institution. To tens of thousands he is Saint Radio and to some thousand Radio Satan.'[87] He was neither of these extremes, but presented a fascinating mixture of altruism and venality, sophistication and naivety, that combined with a broadcasting brilliance to make him a national character of compelling interest and importance. The ghost of Scrimgeour haunted successive ministers of broadcasting and kept free and open discussion out of broadcasting long after his dismissal. Scrimgeour was unquestionably the major broadcaster of his era, the only figure to rise above the general prohibition on free speech over the airwaves. Before and after the Labour victory he maintained a huge and loyal audience. The majority New Zealand audience was always determinedly lowbrow and Colin Scrimgeour was the man who, along with Michael Joseph Savage, developed the populist combination of commercial and state-controlled broadcasting that was distinctive of New Zealand radio.

Broadcasting plays its part in the war effort. Aunt Daisy pictured in 1942 opening a National Savings Investment Account. RADIO NZ

AMALGAMATION

After Scrimgeour's demise, the government moved to amalgamate the two broadcasting services. Shelley was appointed acting controller of commercial broadcasting and, on 19 June 1943, the Cabinet again approved the amalgamation. The necessary legislative action was taken in the Statutes Amendment Act 1943, the fourth clause of which repealed the separate status of the NCBS. Shelley as director of broadcasting was in charge of all New Zealand broadcasting, and Ian Mackay, with the non-descript title of co-ordinating officer, was under Shelley as the chief executive of the Commercial Division.

There were some immediate benefits from amalgamation. A single purchase committee was formed to cover both commercial and non-commercial broadcasting and, with the end of competitive bidding, the purchase cost of radio serials and other recorded programmes was lowered. Shelley also ended the practice whereby advertising agencies had chosen and imported programmes, found a sponsor and recorded the attendant commercials, and then sold the whole package, with commission, to the NCBS. From 1944, in a declaration that the broadcaster rather than the advertiser must be responsible for programme selection, only programmes

The cover of a book celebrating Daisy Basham's wartime journey to the United States. WHITCOMBE & TOMBS

selected and purchased by the broadcasting service were offered to advertisers. The managing director of Radio Features Ltd later stated that this move cost his company half its income, but that it was needed to clean up the Commercial Division, a 'somewhat odorous body'.[88]

While the amalgamation meant the end to the independent action of the Commercial Division, it brought no change to the overall political control of the broadcasting service. Although he was director of the entire broadcasting service, Shelley remained powerless on broadcasting matters when they conflicted with the political desires of the minister. Nor, in spite of the removal of Scrimgeour, were all broadcasting employees now working through departmental channels rather than private political connections. This was most evident in Daisy Basham's trips to the United States in 1944 and 1946. As Aunt Daisy she was commercial radio's best salesperson, and this status was marked by a special loading which was applied to advertising in her session.[89] Her importance to the war effort was recognised. In 1940 she was the mainstay of the successful campaign to have the country's apple crop sold and consumed at home. She took part in many subsequent appeals, and in 1944, while New Zealand was a major American base, it was decided that she should undertake a three-month goodwill visit to the United States. Difficulties arose when she proposed to record her programmes in advance so they could continue during her absence. Mackay counselled against this, arguing that it presented too many difficulties for both the service and advertisers.[90] Shelley supported Mackay, but Basham made it clear to the director that she was working directly with Walter Nash, the minister of finance, and would await a decision from him.[91] She had arranged her trip first with Mrs Fraser and then with the prime minister and minister of finance before discussing it with either Mackay or Shelley.[92] The result was a directive from Wilson to Shelley that 'everything possible is to be done to facilitate Mrs Basham's visit to the USA by keeping the "Aunt Daisy" session going during her absence.'[93] The decision led to considerable complications, but helped with her expenses. As Mackay noted, 'Radio Features [Basham's employer] received £20 per week from the Service for the Aunt Daisy session so that in effect by recording the sessions in advance this would pay for the trip and not cost the agency anything additional.'[94]

Daisy Basham's 1944 trip was seen as part of the war effort, and therefore of national importance. In 1946, after the war, similar broadcasts by her in her 'American Diary' were criticised as justifiable only in terms of their political support for the Labour Party in that year's election. Once again Shelley counselled against allowing the broadcasts, but Fraser intervened and arranged for her to give nineteen talks over the YA and ZB stations. Mackay argued that this was done 'because it was considered the material could be of some assistance in the election campaign.'[95] *Freedom*, a National Party supporting journal, cited her reports of American praise for the New Zealand rationing system and complained the broadcasts had been arranged with the Prime Minister's Department 'with a subtle bias in favour of the Government'.[96]

Amalgamation had been imposed on the NBS and NCBS, and while it brought benefits to both, it was many years before there was any unity between the commercial and non-commercial broadcasters. Particularly among the commercial broadcasters there was no doubt that the aim of the administrative reordering was to bring the Commercial Division to heel. Its title alone ranked it below the non-commercial stations, now referred to as the National Service. The NBS continued as the name of the entire government department and was now the collective title for what had been the NBS and NCBS. The senior positions were regarded as those of the National Service. Further weight was given to this in March 1944, when permission was given to establish the position of assistant director. This position went to William Yates, the National Service secretary. While there was officially only one broadcasting service, in practice the division between commercial and non-commercial broadcasting remained strong. The antagonism between the two sections precluded any attempt at cooperation; and the Commercial Division broadcasters feared that the National Service's ascendency meant an attack on the Commercial Division's resources and conditions of employment. Mackay has described the ensuing changes, as the National Service took control of and removed Commercial Division resources such as its recording equipment, took charge of future staff appointments to both the National Service and Commercial divisions, and gradually established a pattern of seniority of National Service executives over their Commercial Division counterparts.[97] Commercial and non-commercial broadcasting had distinct histories, personnel and practices. Their amalgamation into one service did not change that but added to the dissatisfaction of the Commercial Division personnel the enmity of the conquered. Rapprochement between the two broadcasting sectors was a major task requiring attention in postwar radio.

10
New Zealand Broadcasting Service

After World War II there was a renewed pride in the New Zealand identity. This was shared within broadcasting where the National Broadcasting Service was renamed so that it would be clearly known as New Zealand's. Such a change also offered a new title indicative of the combination of the commercial and non-commercial stations into one organisation rather than the takeover of the commercial stations by NBS. From 1 April 1946 the National Service and Commercial Divisions were known as the New Zealand Broadcasting Service (NZBS).

THE SHORTWAVE DIVISION OF THE NZBS

As part of the assertion of a national identity, the NZBS began a new broadcasting service — international shortwave radio transmissions. The P&T Department no longer retained exclusive control of shortwave radio and, at the end of the war, had two 7.5 kW transmitters which it no longer needed. These were bought and installed by the NZBS, and regular broadcasts began on Dominion Day, 27 September 1947. One transmitter was beamed at the eastern states of Australia, the other to the Pacific. The shortwave programmes were retransmissions of domestic programmes, usually those from 2YA, along with a special evening programme. This was mainly music but always included bulletins of New Zealand news and talks or documentary programmes on New Zealand life. Reception reports came from all around the world, but the power of the transmitters meant that adequate reception was limited to a range of 3000 miles.

THE NATIONAL ORCHESTRA

Within broadcasting, the most public expression of the national confidence that followed World War II was the establishment of an orchestra. After the war, Shelley again brought up the idea of a symphony orchestra. Nine days after the surrender of Japan he submitted a proposal to establish a national orchestra under the control of the Broadcasting Service.[1] This was accepted,

The combined 4YA/4ZB studios at the 1947 Dunedin Annual Winter Show — still not accepting the combined organisation's new title, the NZBS.

and in June 1946, Fraser announced that New Zealand would have a permanent symphony orchestra. It was an expected announcement, for Andersen Tyrer had been appointed as musical director and conductor four months previously. A sixty-five-member orchestra gave its first performance on 6 March 1947 in the Wellington Town Hall. The speed of establishment of the orchestra was due to the support given by the prime minister. Just as commercial radio came to New Zealand through the efforts of Savage, with the assistance of Scrimgeour, so was the establishment of the National Orchestra the result of Fraser's enthusiasm and Shelley's assistance.

The National Orchestra became a subject of both celebration and dissension. Its existence was greeted as evidence of a national maturity and musical accomplishment; yet it sparked off a continuing debate, centred initially on the quality of the orchestra, with much criticism of the appointment and performance of Andersen Tyrer. Parliament also came under criticism for not advertising the position of conductor. Cabinet took responsibility, but the Opposition felt that Shelley had been the main person behind the appointment.[2] The opening concert of the orchestra received mixed reviews. J. C. Beaglehole, in the *Listener*, wrote one of the more critical. As ever the orchestra was welcomed: 'The very fact that the orchestra is at last in existence and giving concerts is important. For bringing it into existence the Government deserves all our thanks.' However, Beaglehole saw 'the actual performance' as 'more notable for hearty goodwill than for refinement'.[3] In an editorial the *Listener* delivered a back-handed compliment: 'That point was reached at the first concert, which certainly proved that while there is still a long and difficult road ahead of the orchestra, it is equipped for the journey.'[4]

In the following years the debate focused less on the orchestra's competence, especially after Tyrer left at the end of 1949, and more on its cost, which rapidly exceeded Shelley's estimate of £40,000 a year. It directly impinged on the NZBS which was responsible for the orchestra. The NZBS received many hours of broadcasting from the orchestra. In the 1952–53 year, for example, the orchestra gave sixty-five evening concerts, four lunch-hour concerts and seventeen free concerts for schools. All but four were broadcast, as were a further sixteen studio performances.[5] The number of hours of broadcasting did not allay criticism that the orchestra's expense was not justified. The costs of the orchestra were a matter of public record, and engendered strong public discussion when broadcasting finances were under debate. From 1949 the NZBS began some years of deficit during which the annual costs of the orchestra exceeded the annual losses of the broadcasting service. Increasingly the orchestra was regarded as a major cause of the NZBS's financial problems.

Other criticisms followed. Sidney Smith, the member for Hobson, argued that 'People in small towns who contributed towards the cost . . . did not get an opportunity of hearing the orchestra other than over the air and in any case the best orchestras in the world could be heard over the air any day of the week.' Smith's absurd estimate that less than 1 per cent of people had an opportunity to hear the orchestra[6] could be disregarded, but it was not so easy to ignore the wider view that the orchestra, paid for by the general populace, was for the enjoyment of a highbrow minority. In the 1954 supply debate, opposing views were expressed within the ruling National Party. J. R. Hanan noted that 'the deficit on the National Orchestra had now reached £86,236 . . . such a large sum was quite out of proportion to the benefits obtained', while H. R. Lake considered the orchestra 'was a luxury New Zealand could well afford'.[7] The National Orchestra was appreciated as a cultural asset for the country, but its cost and the meeting of that by the Broadcasting Service remained contentious.

STATION CONSTRUCTION AND DELAYS

The inauguration of the National Orchestra was a development that masked stagnation in other areas of broadcasting. When the orchestra was formed broadcasting was in its tenth year since a new station had been opened. This was in spite of broadcasters' perennial desire to expand operations into other centres. Furthermore, the NZBS faced an urgent need to renew its existing stations. As Shelley acknowledged in his 1946 report, 'practically all existing plant is obsolete and requires replacement'. The difficulties extended to the NZBS accommodation. It was operating under a considerable handicap — in Wellington the head office and stations were housed in nine different buildings. None of the studios had originally been constructed with broadcasting in mind and their acoustical limitations were severe. The spread-out nature of the service made administration difficult. By the end of the war Shelley declared that the service was 'in an impossible position so far as staff and accommodation was concerned'; and he gave this as the explanation for being unable to prepare the annual accounts.[8]

Orders for new transmitters and plant were placed in 1946 with AWA in Sydney. The equipment began arriving in New Zealand from 1947, though much of it, particularly transmitter masts, was delayed because of the shortage of steel in Australia. However there was no rapid postwar expansion for broadcasting — it was 1949 before the first new station opened. The NZBS had to contend with a general shortage of resources, and with its low priority on the postwar list for the building resources of the country. It was noted in the 1947 report that the NZBS had received no allocation of building resources, and in the 1948 report that no permits at all had been granted the service for the erection of new buildings. There was no immediate relief: not until the opening of new studios for 2ZA Palmerston North in April 1956 was there what the *Listener* described as 'the first permanent new NZBS station building since 1ZB was completed in 1941'.[9] Under the circumstances the decision was taken in 1947 to find what it was hoped would be only temporary accommodation in buildings which had been erected for war purposes. This, however, was not a practical solution to the accommodation problems in Wellington. In 1947 Shelley described the completion of Broadcasting House as 'imperative'. Nothing was done. One of the last statements by Jones as minister in 1949 was that the building's construction, including the Conservatorium of Music, must wait until the building situation eased.[10]

Adding to the NZBS's difficulties was the general shortage of electric power in New Zealand. Electric power authorities wished to limit the number of broadcasting hours, and from 1944 all stations were required to reduce their hours of transmission. It was not that the broadcasting stations themselves were considerable consumers of power; but the total number of listeners represented a large power usage. If radio transmissions ceased at 10pm, listeners would turn off their radio sets and most would go to bed, which meant they also turned off their electric lights and heaters. These restrictions continued in the postwar years and were at their worst in the winter of 1947 when all stations were reduced to 6½ hours of transmission per day. Broadcasting hours that winter were 9am–11am, 1.30pm–2.30pm and 6.30pm–10pm. The effects on broadcasting were dramatic — they included a 40 per cent loss of revenue for the Commercial Division.[11] The

"I'm so glad they're going to extend the broadcasting hours—the radio means so much to Arthur"

Lodge, the *Listener* cartoonist, comments on the removal of restrictions on broadcasting hours, 27 November 1953. *LISTENER*

restrictions were carried, though not at so severe a level, into the 1950s. They were increasingly criticised: Labour MP Clyde Carr compared the broadcasting service to 'a dear old grandmother who tucked everybody into bed at ten o'clock or half past ten o'clock at night'. The daytime restrictions were relaxed gradually but the earlier night-time closedown remained. It was not until the spring of 1953 that the minister finally announced that the need to conserve power was over.[12]

In the postwar years the most capable director of broadcasting would have had difficulty advancing his department's interests, and the task was beyond Shelley. Shelley was not widely regarded as an able administrator. 'He was chafed by the details of regulation and not patient enough to master them,' wrote J. H. E. Schroeder, a later director. C. E. Beeby is recorded as stating that Shelley never learned the administrator's primary lesson, that you have to be equably prepared to argue the impregnable case over and over before you get the only possible right answer.[13] These were significant faults in the inaugural head of a government department. While there was an eightfold increase in broadcasting staff during Shelley's years as director, this was not accompanied by a commensurate development in broadcasting itself[14] The slow growth of the NBS and NZBS while he was director is in some measure due to his administrative failings. The postwar years under Shelley's directorship were years of waiting, as a lack of stations and equipment and inadequate accommodation impeded radio broadcasting's growth.

The years of waiting were accompanied by a large increase in the financial reserves of the NZBS as licence and advertising revenue accumulated but could not be spent. The mounting funds were a matter of concern to broadcasting officials, for they served as an excellent pretext for the NZBS's critics to press for a drop in the licence fee. The Marsden MP, A. J. Murdoch, argued in 1944: 'The annual licence fee of £1.5s is much too high, and I consider that it should be reduced to 10s a year, in view of the considerable credit balance in the broadcasting account.'[15] In the following years many other Opposition members joined in this argument. By 1948 the accumulated funds of the NZBS totalled just over two million pounds.

NEWS AND TALKS PROGRAMMES

It was also increasingly acknowledged in the postwar years that the method of preparing radio news was inadequate. The New Zealand content of news bulletins continued to be supplied by the Information section of the Prime Minister's Office. The Opposition had long regarded this as propaganda for the government, and now their complaints were joined by criticism from within the Labour caucus of journalistic incompetence. The dissension surfaced in Parliament after the main 9pm news bulletin on 8 September 1947 failed to mention, much less give details of, a serious rail accident that had occurred in Auckland four hours earlier. The minister explained the omission was because 'news of the accident was not available to the broadcast service'. In the following discussion it was established that radio news items were regularly broadcast more than twenty-four hours after they had appeared in newspapers. Dr A. M. Finlay, a new Labour MP, saw this as 'a very grave weakness in the broadcasting service', and there were calls for the NZBS to be allowed to establish its own news-gathering service.[16] The government failed to investigate, let alone introduce, any new system, and news bulletins continued to be prepared under prime ministerial supervision.

After World War II the NZBS talks programmes were also subject to criticism. The policy which prohibited controversial programmes was discussed soon after the war. The first intimation of change came in reply to an inquiry from Federated Farmers at the end of 1946; the minister wrote that the NZBS was investigating 'arranging broadcasts of controversial subjects additional to those already broadcast during the parliamentary debates'.[17] The next month Shelley, asked by the minister if any progress had been made, proposed broadcasting talks on controversial topics, beginning with group discussions (called 'open forums'), followed by debates and studio presentations by opposing speakers.[18] The proposal was accepted, but in a way that brought into question the government's commitment. Jones took more than four months to reply and, although he instructed Shelley to proceed, made it clear a political decision would be made for each broadcast.[19]

Jones announced in Parliament in August 1947 that there would be broadcast 'at an early date a series of discussions on selected controversial subjects by speakers representing different points of view'.[20] This postwar reconsideration of the broadcasting of controversial topics was made in

The First Citizens' Forum, 23 March 1948. The scene as one of the speakers voices his views into the 4ZB microphone.

response to criticism not just from Labour's political opponents, but more particularly from within the Labour caucus. There was a growing dissension in the caucus over the party's broadcasting policy. It surfaced in the House in 1945 when Clyde Carr asserted that 'the broadcasting service needs a great deal of cleaning up, reforming and altering, so as to make it more popular and more satisfactory'. He asked for the broadcasting of controversial subjects and was joined by another Labour speaker, Terence McCombs, who said, 'We have been too conservative in that area.'[21] A third Labour MP, Finlay, speaking publicly after change in policy was announced, went so far as to compare New Zealand radio unfavourably with Labour's old enemy, the press. 'I understand a move is to be made for . . . an open forum. That development I cordially welcome. I wish it had taken place many years ago. I wish our radio was as available for controversy as I am bound to say the press is at the present time.'[22]

The details of the new programmes were announced in September. They began on both national and commercial stations. Reporting on the national stations, Shelley noted in 1948 that 'All main stations now conduct at least one such session weekly. The aim . . . is rather to stimulate public thought and discussion than to reach conclusions. Speakers are carefully chosen to give as nearly as possible a balanced presentation of the various aspects of the question under discussion.'[23] The commercial stations' contributions to the new style of broadcasts were in weekly sessions entitled 'ZB Citizens Forum'. The sessions were held in public halls in the four main centres, with a panel of speakers supporting and attacking the subject under discussion. Topics in the first months included, 'Is our education system satisfactory?', 'Should import restrictions be continued?', 'Should married women work?', 'Do New Zealand men lack a clothes sense?' and 'Should large-scale immigration be encouraged?'.

The minister kept a close scrutiny on the discussion topics and rejected, among others, 'That equal work should be rewarded with equal pay, regardless of sex', 'Does advertising add to the cost of living?' and 'Is the watersider being unduly criticised?'[24] That the selection of each topic remained subject to ministerial approval was not made public. Even so there was criticism that the government was reluctant to allow a full and free discussion. The 'ZB Citizens Forum' programme format at first included comment from the audience. After the minister stopped this it was stated in Parliament that 'The audience gave frank expressions of opinion that were not always favourable to the Government and, apparently, the Government could not "take it"'.[25] The new programmes were a means of using broadcasting for a more open discussion, such as had been requested by broadcasters since the 1920s. However the government did not surrender its power to choose broadcasting topics and programme formats. The changes introduced were in response to criticism of the government's policy but were not a fundamental change of that policy.

In practice, some of the more open discussions on radio began in the women's sessions rather than in the official talks programmes. In 1948 on both commercial and national stations the women's sessions were coordinated into 'Women's Hour' programmes. This was the start of an expansion of these programmes through the 1950s. During that decade the various session organisers broadcast entertaining and informative material and took advantage of the relatively unscripted format of their programmes to introduce discussion topics previously disallowed as controversial.

BROADCASTS TO SCHOOLS

One programme area that was extended considerably by the Labour Government was school broadcasts. Until the 1940s these were organised separately from each of the YA stations and taken by individuals from each centre, usually from the local teachers' training college. The programmes were of varying quality and relied on the continuing goodwill of local volunteers. In 1941, with the collapse of this goodwill, the decision was taken to bring the programmes under a central control and to relay school broadcasts from one national station to the others. A full-time employee was appointed to organise the school broadcasts.[26] This was the start of the 'Broadcasts to Schools' section. In the postwar years the section was expanded and the long-term administration of the broadcasts was extended beyond the limits of individual commitment. A committee with representatives from both the NZBS and the Department of Education was in overall control of the programme. A financial contribution towards the broadcasts was made by the

Participants in a conference of National Station women's sessions organisers, 1956.

Department of Education. The NZBS continued to prepare the booklets that accompanied the broadcasts, but the Department of Education took responsibility, through its School Publications Branch, for their printing and distribution free to all schools.

There were broadcasts to secondary schools, notably for lessons in French and Music — both subjects in which New Zealand had a shortage of qualified teachers. The majority of the broadcasts were to primary schools. In the 1950s the broadcasts were for 2¾ hours weekly. All but fifteen minutes of this was aimed at the primary schools. Even at the end of the 1950s over 60 per cent of New Zealand primary schools were country schools with three or fewer teachers, and the broadcasts were directed particularly to the needs of these schools.[27]

HOW THINGS BEGAN

The Broadcasts to Schools programme was as vulnerable as any other to government intervention and censorship. One of the more celebrated examples occurred at a time when the more general question of controversial broadcasts was being considered. The programme *How Things Began* was a series of thirty-three school broadcasts produced in New Zealand from BBC scripts previously broadcast in Britain. The series, which began in March

1947 and which dealt with the early stages of life on earth, drew criticism from individuals and groups protesting at the teaching of evolution. Responding to this criticism, the minister instructed Shelley to suspend the series immediately.[28] The suspension came on the day the tenth broadcast in the series, entitled *Apes and Men* — regarded by the director of Catholic education in Wellington as 'the most dangerous of all' — was to go to air.[29] When the actors arrived for the next recording session, they were told the series was suspended.

The controversy over the *How Things Began* series was enormously increased by its suspension. The complaints (largely from religious groups) about the broadcasts were now joined by many more complaints (largely from educationalists) about their suspension. Most of the newspaper comment was also critical of the suspension. In a typical editorial the *Wanganui Chronicle* argued, 'New Zealand if it continues in this way will soon qualify for a Scopes trial such as made the United States the laughing stock of the world.'[30] The series was never resumed, even though Shelley assured inquirers it was merely suspended 'pending further investigation of its suitability'.[31] The decision of the minister of education 'that it would be unwise to have any part of the school broadcasts become the centre of a controversy or the basis of discord' meant that the suspension could be seen as a matter peculiar to school broadcasts. But within the government there was still little acceptance of any controversial broadcasting, no matter what the intended audience.

In a letter to the *Listener* which provoked much discussion, Sir Thomas Hunter, the principal of Victoria University College, referred to the suppression of the series in a criticism of broadcasting in New Zealand. He took issue with Shelley's frequent allusions to broadcasting as a democratic instrument similar to the meetings of citizens in Athens: 'Surely he [Shelley] overlooked the fact that the Greek people met its speakers face to face and could express its approval and dissent . . . Broadcasting is not democratic in nature and . . . until all shades of opinion may be heard on the air the totalitarian character of our broadcasting system will remain.'[32]

The criticism of the director of broadcasting was warranted. In spite of his Reithian attitudes, Shelley failed to follow Reith's example and seek to extend the traditional journalistic duties — to report and to discuss — into broadcasting. In the BBC Reith argued tenaciously and largely successfully for the right to report news and to debate current issues. Shelley's greatest failing was his unwillingness to stand against the government's day-to-day political control and use of radio for its own ends. Shelley made no stand: he knew it was not a contest he could win. It was one he was obliged to enter and his lack of leadership on this matter contributed to the worst feature of New Zealand radio during his years as director — the continuing inadequacies of the news and talks programmes.

SHELLEY AND THE FUTURE OF BROADCASTING

Shelley's plans for broadcasting were presented to the Cabinet late in 1943 in his 'Proposals for the Development of the National Broadcasting Service in Relation to the Cultural Life of the Community'. He proposed an increase in the number of stations and an addition to the national system of broadcasting. He called for a further fifteen to twenty small stations, each with a 250-watt power output, to be on the air for three or four hours daily in the smaller towns and populated areas. With radio previously considered from a national point of view, Shelley now asked for 'The adoption of a supplementary policy — that of using radio as a local institution to serve as an instrument for developing the cultural life, artistic endeavour and civic consciousness of towns and districts.'

The four YA stations were to be district stations. Each was to have 'small, fulltime orchestral combinations . . . to present independent programmes, to form nuclei for larger orchestral combinations, and also to provide items or help with concerts by societies at the local stations in the districts.' The development of drama was an important part of the proposal. Shelley's plan in this regard was for the full-time employment of four to six actors stationed at the Wellington head office under the supervision of a producer. Their main task would be to record plays in Wellington, though Shelley also envisaged the troupe of actors undertaking some touring.[33]

The proposal was approved by Cabinet on 6 April 1944.[34] This did not mean it would be introduced immediately or without extensive modification. Even while being approved, it was regarded as a matter to be addressed after the war.

Shelley's proposal next appeared — and for the first time publicly — in the service's 1947 report. At this stage four types of station were discussed. 2YA was the 'national' station. Next were the 'district' stations: in 1947 these were the other YA stations plus 2YC Wellington, 2YH Napier, 3ZR Greymouth and 4YZ Invercargill. These stations were to serve their districts, sometimes rebroadcasting programmes from the national station and sometimes supplying programmes to it. The third type were the 'local' stations, situated in the smaller towns and serving their immediate locality. As well as these, there were the 'alternative' stations in the main centres. The various types of station were envisaged in a hierarchy of excellence, with the best of the artists from the local stations being selected to broadcast from the district stations, and the best of these broadcasting nationally.

Shelley's development proposal represented a considerable change from the policy of centrally controlled, national broadcasting from relatively few stations. It reversed the prevailing exclusion of local performers from the airwaves. Local talent was to have the opportunity to broadcast from stations restricted in range to the local community. However the local broadcasts were to remain those selected by Reithian standards. Performers who most closely approximated the traditional high cultural standards in both content and performance would be entitled to broadcast from the more powerful stations. The proposal did not accept other standards; nor did it admit the new categories and styles of programming that had become part of broadcasting. The prime tasks of broadcasting were still to offer excellent examples of genres of entertainment that existed prior to the advent of broadcasting, and to educate an audience to appreciate such fare.

Shelley's attitude is evident in the National Service programming particularly in his lack of action with regard to sports broadcasting — one of the most popular components of YA programming. At the end of the war, as McCarthy notes, 'No station had an organised sports set-up, those interested in sport depending on players or officials at various games either phoning through results or bringing them into the station.'[35] It was an unsatisfactory situation, and one which had not changed since the repudiation by the NZBB of the position of head of sports broadcasting held by Alan Alardyce. During and after the war the individual who most attempted to change the situation was Winston McCarthy. McCarthy was first employed in broadcasting in 1937 as a programme organiser for the NBS alternative station, 2YD. He was not allowed to announce on 2YD or elsewhere, as his voice, 'more Cockney than Oxford', was unacceptable to Shelley.[36] McCarthy, however, gradually inveigled his way into sports commentating and became highly regarded for his rugby commentaries in particular. He toured Britain in 1945–46 with the 2NZEF team, the Kiwis, to record commentaries for later rebroadcast in New Zealand. His unrestrained style won over even the BBC officials, and many of his commentaries were also broadcast in Britain. In New Zealand after this tour, McCarthy, as he put it, 'began consolidating my position as Sports Supervisor'.[37] The position was essentially self-appointed, and McCarthy was never able to move beyond temporary contractual appointments with the NZBS, primarily for sports commentating. Sports commentaries remained a popular aspect of the YA stations' fare, but the whole topic of sport was of low status for Shelley in the cultural hierarchy of broadcasting programming and he was not prepared to develop it.

A notable omission from Shelley's proposals, both in 1943 and 1947, was any discussion of commercial broadcasting. In spite of their majority audiences, Shelley did not propose any additions to the five existing commercial stations. This was a major failing in broadcasting development proposals from the director who, from 1943, was head of commercial and non-commercial broadcasting. From his initial competitive, even combative, attitude towards the NCBS, Shelley became the senior executive charged with continuing and furthering commercial broadcasting.

In the postwar years many broadcasting officials modified their opposition to lowbrow culture. They were learning that there was a new sensibility — that

Winston McCarthy, sports commentator.

individuals could enjoy works of both high and low culture. The point is made by Mulgan, Supervisor of Talks:

> One couple whose interests ranged from Bach to Einstein never missed the 'Japanese House-boy' series. If there is a high-brow play . . . it is Chekov's *The Cherry Orchard*. As I was going in to a WEA performance of *The Cherry Orchard* in Auckland I overheard a conversation between a citizen and the box-office. They were sorry there was no room that evening, but they could give a seat on Monday. No, he could not come on Monday because he was going to wrestling in the Town Hall. Chekov and wrestling; that incident taught me a lot.[38]

Shelley was never personally accepting of other than traditionally defined cultural standards. However he showed more interest in commercial broadcasting towards the end of his term as director. He understood and accepted the differing nature of commercial broadcasting and worked to promote its success. He continued to pay attention to the announcing of commercials and, as was his practice with the national stations, issued regular instructions to the commercial stations on the methods of presentation.[39] Shelley was obliged by his position to support commercial broadcasting, and thus his work as director cannot be understood solely in terms of the high culture with which he is most associated. However his orientation and his loyalty lay with that culture, and under his leadership the National Service was clearly more important than the Commercial Division.

SHELLEY'S RETIREMENT

James Shelley retired in April 1949. He was knighted in June, and returned to live in England. His years as director were tempestuous ones that witnessed the separate development and then amalgamation of commercial and non-commercial broadcasting, Labour's political control of broadcasting, and the extended difficulties for broadcasting brought by the world war. In his twelve and a half years as director he oversaw the spread of radio ownership, and knew when he left that a new phase of station growth was under way. He also successfully established his two main dreams for broadcasting — the National Orchestra and an extensive drama division. These represented for Shelley the best achievements of the broadcasting service, and are the main fruits of his policy of using broadcasting as an instrument for cultural enrichment.

Shelley had his faults, but many of them stemmed from his good nature. As the *Listener* noted on his retirement, 'He has been too human, too sensitive, too kind, for any of the brutalities that usually go with efficiency, with the result that he has never been free of dead wood or of lame dogs.'[40] These human failings were allied with his inspirational qualities, which were his real gift to New Zealand broadcasting. When he left, broadcasting's National Service continued with Shelley's understanding of the importance of its work and enthused with a sense of vocation towards it.

William Yates was appointed to succeed him as director. This was an expected promotion — Yates had been assistant director since 1944. He came to broadcasting in 1934 from a position as accountant with the first Unemployment Board. Yates's other interest was drama; his 1934 appointment in broadcasting was as supervisor of plays. Most of his subsequent broadcasting experience was in administration, but he was also experienced as a producer, scriptwriter and actor in concert-party plays and reviews.

A feature of the personnel changes that followed Shelley's retirement was the continuing absence of commercial broadcasters from the senior positions within the NZBS. Yates was from the National Service. Ian Mackay, who had been the senior Commercial Division executive since Scrimgeour's ousting and who was regarded, at least by the younger staff, as the most innovative broadcaster in the NZBS, applied for but did not win the position of assistant director. He appealed, but lost when Yates stated he had advised the Public Service Commission to appoint from outside the NZBS.[41] Mackay left the NZBS the following year and went to Australia to pursue his career in commercial broadcasting. The successful applicant was J. H. E. Schroeder, associate editor of the *Press*, whose reputation was as a writer and literary critic. In 1958 Schroeder succeeded Yates as director.

X-CLASS STATIONS

In its final year in office the Labour Government introduced the local stations first proposed by Shelley six years earlier. In spite of the dominance of advocates of non-commercial broadcasting within the executive ranks of the NZBS, the Labour Government did not accept that the local stations, referred to as X-class stations, were to be the 'centres of artistic activity' Shelley had envisaged. The decision taken as to their nature is a further indication of the importance of commercial broadcasting in New Zealand.

On 18 January 1949, station 3XC Timaru began transmissions with a 2-kW transmitter. It was the first of the X-class stations. 3XC's transmitter was not placed in the same building as the studio but was erected outside the city, at Washdyke. Unattended, it was operated by remote control from the studio. This new procedure was followed at all the X-class stations. Station 3XC was joined in 1949 by 1XH Hamilton, 1XN Whangarei and 2XA Wanganui. In the same year 2XG Gisborne and 2XN Nelson were increased in power to 2 kW and also became X-class stations. In September 1953 the NZBS turned 2XP New Plymouth into a further X-class station. The final one, 2XB Masterton, was opened in 1958.

There were never more than eight X-class stations — half of the number proposed by Shelley. Their 2-kW transmitters were eight times more powerful than Shelley had proposed, and the stations did not join the local community in the manner he envisaged. The distinctive feature of the X-class stations was that they were hybrid, commercial/non-commercial stations. They operated for some of the day with advertising and for some of the day without. As of 1950, their first full year, these stations broadcast on weekdays from 7am to 10am and from 6.30pm to 10.30pm. They carried advertising in the morning sessions and for the first hour of the evening programme. Their Sunday programmes, which like all commercial stations were without advertising, were from 8am to 11am and 6.30pm to 10.30pm. The stations increasingly became centres of commercial and popular programming rather than the Reithian scheduling originally proposed. However the stations combined the resources of the National Service and the Commercial Division and acted as a lead in a slowly emerging rapprochement between the two sections of broadcasting.

THE 1949 CONSCRIPTION REFERENDUM

The government's use of radio for its own ends was most obvious in the broadcasting decisions made in its final year in office about the referendum on compulsory military service. The referendum involved the NZBS, which broadcast government members' speeches in favour of the proposal for conscription. The NZBS was also charged with arranging public address systems at the various meetings planned for the prime minister, minister of rehabilitation, minister of defence and the leader of the Opposition. This was an unusual task for the NZBS — it had to hire most of the equipment from private firms.

The prime minister made the various broadcasting decisions; and the NZBS was less considerate to groups opposed to conscription. Many requests were received from such groups for permission to broadcast their views. Application was also made from the same groups to buy advertising time to publicise their meetings. The decision in regard to advertising was that 'it will be permissible for each announcement to include the name of the speaker, auspices, time, date and place of the meeting. There must be no reference, however, to the title or subject matter. Announcements will be accepted under the usual commercial contract and at normal rates.'[42] While these were the conditions that applied at election times to political meetings, the refusal of permission to advertise the topic of meetings hindered the publicising of the case against the government's proposal.

Fraser's initial intention was not to allow groups opposed to conscription to broadcast at all. He eventually decided to allow some airtime for his opponents, but he imposed stringent conditions. In particular, he nominated those groups that could broadcast, and the broadcasts went ahead only after being vetted by him. Four organisations were given permission to broadcast, each for a talk not exceeding twenty minutes in length. They were the New Zealand Peace and Anti-Conscription Federation, the 2nd NZEF Association, the New Zealand Christian Pacifist Society, and the Peace Committee of the Society of Friends (Quakers) in New Zealand.[43] One of the groups chosen, the 2nd NZEF Association, was an unlikely opponent to conscription and had little in common with the others. As their dominion secretary explained to Yates, they were 'not opposed to the principle of conscription' but they felt that the government's proposal was not 'the best method of military preparedness available to us'.[44]

The speech of the Anti-Conscription Federation was the one most altered by Fraser. As an illustration, the following, already a rewriting of an earlier rejected draft, was itself rejected:

> Women make no mistake.
> Conscription will not save your sons from becoming casualties. Conscription does mean the needless sacrifice of young lives, for it will not give New Zealand either a well-trained or a well-equipped Army. For centuries the nations of Europe have employed conscription without preventing one war. Conscription does not mean real defence preparedness — in World War Two over two million armed French conscripts passed straight into prisoner-of-war cages without firing a shot.

Eventually the following was approved:

> Women make no mistake.
> Conscription will not give New Zealand either a well-trained or a well-equipped Army. For centuries the nations of Europe have employed conscription without preventing one war. Conscription does not mean real defence preparedness.[45]

FROM LABOUR TO NATIONAL

The Labour Government's broadcasting decisions at the time of the conscription referendum were among the final actions of its fourteen-year term of office. They were a reversal from its announcement that it was opening radio to a more free discussion, but were in accord with its long-term policy to forge the medium as an arm of government. After the death of Savage, the Labour Government lost its broadcasting vigour. Apart from the dismissal of Scrimgeour and the amalgamation of the NBS and NCBS into the NZBS, Labour did not give much consideration to the broadcasting portfolio. While this could in part be explained by the wartime stringencies (which for instance prevented the construction of new stations), there was an overall reluctance to allow any change. With the exception of Peter Fraser's support for the National Orchestra, Labour in its declining years tried to avoid publicity, and showed little inclination to expand the role of broadcasting.

Even Savage's great publicity innovation — the broadcasting of Parliament — turned against Labour. Initially opposed to the change, National eventually applauded the measure, with Sidney Holland stating in 1944 that it 'was one of the things the present Government had done which had turned out to be right'.[46] Labour at first benefited from the parliamentary broadcasts. It was introducing an innovative legislative programme, and it had the better radio speakers. National speakers soon adapted to radio broadcasting, however. Broadcast debates favour the forceful side, and in the late 1940s the National Opposition, increasingly on the attack, scored against a defensive government.

The broadcasting of election addresses during Labour's years remained evenly divided between the two major parties. However, for Labour's two final elections as government there was one important change which was strongly criticised by National. In the 1938 and 1943 elections the Labour and National leaders both broadcast their final addresses the night before the election. In 1946 and 1949 Fraser took the final night for himself, forcing Holland to give his concluding address a day earlier. National saw this as a lack of fair play from a declining government attempting to bolster its electoral chances.[47]

Labour's fourteen-year hold on power ended at the general election of 1949. The new National Government, which appointed Frederick Doidge as minister in charge of broadcasting, did not introduce any sudden changes in New Zealand's broadcasting arrangements. It soon became obvious that the government had a tit-for-tat view of access to the airwaves and would not reverse its predecessor's policy and allow its political opponents equal access. The following parliamentary exchange came eight months after Labour's defeat:

Walter Nash: 'We have been denied completely access to the radio since the House rose in November last.'
Frederick Doidge: 'We were denied it for fourteen years.'[48]

In 1952 Holland, the prime minister, stated after complaints from Nash about his radio time, 'I shall always be happy to afford the Leader of the Opposition radio facilities no less generous than those granted to me by his Government when I was Leader of the Opposition.'[49]

The more basic question of broadcasting policy was whether National would continue to run broadcasting as a government department. The new government's political philosophy favoured private enterprise, and there was speculation that some radio stations might be sold, or that private firms would be permitted to start new stations. In his first year in office the minister announced that this was not so. Private stations would not be started and allowed to compete with the NZBS.[50] In spite of its many years of opposition to Labour's political control of broadcasting, the National Government, once the two political parties' positions were reversed, saw no reason to change the existing system of control.

National also continued Labour's postwar broadcasting policy of concentrating development in commercial broadcasting. Over the years the X-class stations gradually increased their hours and their proportion of advertising time. By late 1958 they were broadcasting from 6am to 10.30pm, and all the additional hours, apart from those on Sunday, included advertising. In the mid 1960s they became fully commercial stations.

In November 1955, 1XH Hamilton had its hours extended to those of the ZB stations and joined them as a fully commercial station. Other fully commercial stations opened in the 1950s were 4ZA Invercargill in 1956, 2ZC Napier in 1957, and the 100-kilowatt 1ZC Rotorua in 1959. These commercial stations were joined by 1YD and 2YD which began broadcasting advertisements in 1957 after the minister's announcement that more advertising was offered than could be taken solely on the ZB stations and that the continuing NZBS deficit demanded increased revenue.[51]

Thus the increase during the 1950s was almost entirely of commercial stations. The one exception was 2YX Wellington, opened in 1950. This small 1-kW station used the original 2ZB transmitter and operated only when 2YA was broadcasting Parliament. The postwar development programme also included far fewer stations than originally announced. From 1949 to 1961 only eleven new stations were started; five of these in a short-lived flurry of activity in 1949.

The purchase of new equipment, which had started during Labour's final years in office, was continued under National, and benefited both commercial and non-commercial stations. The first of the new transmitters, of 10-kW power, was installed in Greymouth in November 1948. With its increased power, station 3YZ became one of the district-grade of stations. It was joined in April the following year by 1YZ Rotorua. By mid 1951 2YA was operating with a new 60-kW transmitter, and 1YA, 3YA and 4YA and the four ZB stations had new 10-kW transmitters. A new class of station — the YC stations, one in each of the four main centres — had been designated by Labour in 1948. These were not new stations but new call-signs, given to B stations purchased in the 1930s and operated since then within the government service. National continued upgrading them, and by mid 1951, 1YC, 3YC and 4YC operated with 10-kW transmitters. 2YC operated with the original 2YA 60-kW transmitter.

National's readiness to control and use the broadcasting service in a manner comparable to that of its predecessor was made obvious in 1951, when it introduced emergency regulations to counter the effects of the 151-day Waterfront dispute. This was one of New Zealand's most bitter industrial confrontations, and the regulations included measures to deny any publicity to the watersiders and their supporters. One Opposition member noted: 'By the Emergency Regulations the press was practically blacked out except in relation to Government statements. The radio was used only by those who were in support of the Government.'[52]

However, the 1950s witnessed an emerging willingness on the part of the government to change the system of control of broadcasting. After National's electoral victory the New Zealand component of the radio news bulletins continued to be prepared under ministerial supervision. The unease with this practice, which had been apparent from Labour's backbenchers during that government's final years in office, was now seen among members of the new government. The first change, made towards the end of National's first year in office, was to transfer the responsibility for the preparation of the New Zealand component of the news bulletins from the Information section of the Prime Minister's Office to the equivalent section within the Tourist and Publicity Department. It was a signal of intent rather than a change of substance, for the bulletins remained, of course, under political control.

In the next years the decision was made to seek a non-governmental source of New Zealand news. There was only one existing alternative — the New Zealand Press Association (NZPA), a cooperative owned by the proprietors of the various New Zealand newspapers. It had been renamed in 1942 from the previous United Press Association. The NZBS and its predecessors had long had an arrangement whereby they purchased the rights to use NZPA cables of overseas news, but these rights were limited in the amount of material that could be used and, particularly, in the time when the material could be broadcast. Application was made to the NZPA to purchase fuller rights to its news service, both national and international. The application came to nothing: it was turned down by the newspaper organisation, which saw no reason to assist a potential competitor. However, for the first time the NZBS acknowledged the inadequacy of its news. It referred to the attempts to 'remedy known weaknesses in its present news service' and admitted it was 'without an adequate source of non-official New Zealand news'.[53] The Labour MP, P. N. Holloway, called for the NZBS to develop its own news service and compete with the NZPA 'monopoly'. The minister in charge of broadcasting, by this time R. M. Algie, continued to hope the NZPA would eventually accept the NZBS application, He acknowledged the news service 'was a weakness but it represented the best that could be done with the resources available'. Speaking against setting up a distinct NZBS news service, Algie argued that the NZBS would need to master its deficit 'before it could suggest to the Government the setting up of an activity that might easily cost another £50,000 or £60,000 a year'.[54] The eventual solution was to take the large step of beginning a new national news organisation but that remained a task for the next decade.

The National Government also accepted a need for change in policy on controversial broadcasts. In 1954 Algie announced for New Zealand what had long been the BBC policy. He referred to the recently released Beveridge Report on the Broadcasting Service of Great Britain, and took its principle regarding the broadcasting of controversial material as the policy for New Zealand: 'The Service must be impartial . . . but . . . was not obliged to show that the people who spoke over the air were equally impartial . . . The Service had to be fair in seeing that partisans of different shades of thought had the opportunity to go before the microphone.' He explicitly repudiated the practice of ministerial control: 'There were two ways of handling the matter. One was for the Broadcasting Service, being operated as a public concern to be run by a minister who, with his officers, would say what he thought should go on the air. That would be a shocking system . . . A better way was for the Service to operate against a background of principle . . . the principle laid down in the Beveridge report.' Against the background of that principle two rules were laid down: 'The first was that the Service was to guarantee a reasonable equality of opportunity, and next, against that background of reasonable equality of opportunity, speakers were to be free to express their own opinions, provided that they paid attention to a proper understanding of fact, that they observed the rules of good taste, that they showed evidence of common fair play, and that they observed the rules of defamation.'[55]

The provisos were at times criticised as interpreted and applied in a manner which denied the application of the background principle. Nor was it always accepted that the Service, which remained under ministerial control, was considering the broadcasting of controversial material without recourse to a specific decision from the minister. In 1956, for example, Algie acknowledged that he had stopped a broadcast of a panel discussion on the Cook Strait cable plan, 'which would, in my view, only stir up a controversy in a field in which there was no need for it'.[56] As long as the NZBS remained under direct ministerial control, it would be difficult for a minister to resist asserting his authority, and impossible for the broadcasters to deny the presence of that authority. But Algie's 1954 announcement was the first

1ZB's Phil Shone conducting his breakfast session in 1949, though here in quest of seagulls rather than wasps.

significant step taken towards an independent broadcasting system since it had been placed under ministerial control eighteen years previously.

HOAX BROADCASTS

From the end of World War II there was much debate as to the nature and effect of radio programmes, and the responsibilities of broadcasters. Radio was in nearly all homes; the medium was an important source of information, and represented a major leisure-time activity for the population. In the circumstances the NZBS was rightly regarded as a significant social institution and was under constant public scrutiny. Part of the response to this was a reluctance to allow a light-heartedness that had been inherited from the B stations and was often seen in the ZB programmes. The most dramatic instance of this sobriety was the response to a 1949 April Fool's Day broadcast by the 1ZB breakfast-show announcer, Phil Shone. It was the most successful and most contentious hoax made on New Zealand radio.

At the time the Department of Agriculture was running an anti-wasp campaign. Shone, taking advantage of the topicality of the subject, announced that a swarm 'no less than a mile wide and several hundred yards deep' was nearing Auckland, and he warned citizens to close all doors and windows. Listeners were asked 'to assist the authorities by preparing strips of paper smeared with either jam or honey; these strips should be left outside, preferably in some open space.' Instructions were offered as to the best attire: 'Trousers should be tied at the ankle and the socks pulled over the trouser ends; gloves should be worn inside shirt sleeves and tied at the cuffs; a wide brimmed hat fitted with a fine net should be worn and the net should be worn inside the shirt and sewn to it.' Further reports were given of the wasps' progress and advice was offered on how to deal with any smaller swarms that settled: 'Blowlamps are not advised . . .' Telephone lines to the Central Police Station were jammed with enquiries about the swarm. At 9.30am, when it became apparent that many mothers were anxious about sending their children to school, Shone apologised for any anxiety he may have caused.[57]

The official reaction was antagonistic: the minister expressed strong disapproval of those responsible.[58] Shelley also condemned the broadcast, 'not only for its lack of propriety but because of its infringement of the principles of good broadcasting. Apart from its immediate effect on a section of the listeners, such a broadcast could produce doubts as to the authenticity of information broadcast from the station.'[59]

It soon became clear that the majority had thoroughly enjoyed the joke. The *New Zealand Herald* rebuked the minister for his reaction:

> Mr Jones is supposed to be a Ministerial purveyor of entertainment. If he had to disapprove could he not have shown trust in his subordinates and merely said: 'Good joke — but be more careful next time.'[60]

No further action was taken but it was made clear there were to be no 'references which are likely to create a hazard in any respect whatsoever for the Service or the listeners'.[61] The following year, after 2XA Wanganui broadcast an April Fool's Day hoax message regarding the opening of a swinging span on the town bridge, Yates, the new director, issued an instruction to all stations making it clear that responsibility took priority over lightheartedness:

> The deliberate broadcasting of information to mislead the public on April 1st or any other day is strictly prohibited . . . The Service may be charged with lacking a sense of humour. That is beside the point. The main thing the Service must have is a sense of responsibility . . . The public places great reliance on pronouncements from broadcasting stations. We must do nothing deliberately to destroy this public confidence in the Service.[62]

Not all regarded broadcasting pronouncements with the same seriousness. The poet A. R. D. Fairburn, himself a frequent broadcaster and one-time broadcasting employee, noticed that the mere fact of being broadcast gave pronouncements a status often quite beyond that merited: 'When one does broadcast, one descends the ivory substitute staircase in a seer's robe, and one's lightest words of nonsense are regarded as droppings from the mouth of the Ancient of Days, crumbs of imperishable wisdom. Droppings they are, I admit. But not of that kind.'[63] Fairburn felt that humour was the best weapon against this new pomposity, and he welcomed broadcasts such as the April Fool's Day hoaxes. His view was not shared by broadcasting's controllers, who saw it as a disciplinary matter to ensure a continuing public confidence in the organisation. The broadcasts were a manifestation of an inherent good humour and sense of entertainment within New Zealand broadcasting that continued in spite of the often dampening effects of control.

RADIO SERIALS AND MORAL CONCERN

Concern over the hoax broadcasts came largely from within the NZBS. Wider concern was voiced over the nature and effect of radio serials which, after music, took up the largest portion of broadcasting time. Broadcasting executives disliked the American serials, and from 1946 no new American features were purchased for ten years. British serials also presented difficulties. Private British productions were scarce, and the BBC productions were not available for use with advertising. Thus from the end of the war right through the 1950s, most radio serials were purchased from Australia.[64] The serials remained firm favourites for evening listening until the advent of television. Daytime serials also appeared and continuing sagas such as *Doctor Paul* and *Portia Faces Life* were avidly followed by the (predominantly female) daytime audience.

Some of the concern about radio serials stemmed from a realisation that many listeners were confusing the real and the fictional — serial characters were being regarded as actual human beings, and the serials not as fiction but as real-life models. Selwyn Toogood, an actor in *One Man's Family*, the first New Zealand-produced serial, reported an early example: when a character in the serial had a baby in 1937, the station playing the serial was showered with layettes and bootees sent in by listeners for the fictional baby. Nor was this the only example of such confusion for the actors in *One Man's Family*.[65] Hall suggested this proved the uncanny power of the radio serial to enthral households unused to live theatre or dramatic literature.[66] But such confusion continues into the third and fourth generation.

Broadcasting authorities were themselves ready to exploit this confusion — as had been obvious since the visit of the cast of the serial *Fred and Maggie Everybody*, one of the first Australian serials to be broadcast in New Zealand. The serial was written by Edward Howell and Therese Desmond Howell who, with their daughter, Madeline, were also the actors. The visit of the Howells to New Zealand over the Christmas–New Year period of 1939–40, two years after their programme had begun relaying their fictional lives, indicated that they were established, not as writers and actors, but as their characters. This confusion of character for actor was assisted by the extensive publicity they received. Generally the trio were noted as being Mr and Mrs Howell and daughter — but were then referred to as Fred, Maggie and Daisy Sproggins

The Howell Family, alias Fred and Maggie Everybody and Daisy Sproggins pictured in Dunedin during their New Zealand tour, 1939–40.

throughout the coverage.[67] Their visit, which culminated in a farewell function at a packed Majestic Theatre in Wellington at which the band of the RNZAF played, was a display both of the high public regard for radio personalities and of the acceptance of the existence of their characters.

From the end of the war much criticism was aimed at the radio serials, which were widely regarded as having a malevolent influence, particularly on the young. Alec O'Donoghue's criticism of them is strong, but not untypical:

> What frightful things the majority of the recorded plays are! portraying as they do all the fundamental details of every desperate crime imaginable. Through the broadcast of these unwholesome serials the young, impressionable listener is professionally coached for his graduation as a professional criminal. He obtains a complete course of instruction in safe-cracking, house-breaking, highway robbery, blackmail and how to apply it; as well as in all the subtle forms and practices resorted to by the professional crook in the pursuit of his nefarious calling. He is coached in the devastating art of seduction and is taught all the finer points applicable to the alienation of a woman's love for her husband. And sometimes as many as three or four of these ghastly serials are broadcast in one evening from a single station; and they may run from ten to forty episodes.[68]

Concern over the effects of the radio serials centred not just on their nature, but on the fact that, by being broadcast, they were heard by larger audiences than had been reached by any previous method of entertainment. With the home as the listening point, broadcasters had a special responsibility to ensure their wares did not give offence. Parliamentarians also gave voice.

"It all started over whether or not radio serials would tend to make me wild and disorderly"

Lodge's contribution to the discussion of the ill effects of radio serials. *New Zealand Listener*, 4 July 1952. LISTENER

They argued that the supposed ill-effects of serials should be taken into account when programmes were being scheduled. Terence McCombs suggested that 'serials portraying the careers of gangsters and those featuring violent family quarrels' should not be on early when children were listening.[69] Hilda Ross complained about the 'rubbishy serials' that were being broadcast when children should be doing their homework.[70]

Not all shared this concern. In an editorial, the *Listener* noted: 'It is one thing to send children to bed to make sure they get sleep and rest, and another thing altogether to say that if they stay up and listen to the things their parents are listening to they will be started on the road to ruin.'[71] The minister, supporting the serials in 1945, saw the complaints as part of an already long litany against the elements of lowbrow culture.

> I recall the complaint made in my youth that the boys of that day would be led astray by 'penny-dreadfuls' and 'Deadwood Dick' novels, but I do not think that we are any worse than our parents have been. We have only to consider the exploits of young New Zealanders in the war to realise that neither the films nor broadcasting have done anything to undermine their manliness and womanliness.[72]

The topic of radio programmes in general and serials in particular featured annually in parliamentary debate on broadcasting. It came to prominence again in 1954–55 at the time of the investigation of the Mazengarb Committee, a group appointed by the government to investigate moral delinquency in children and adolescents. The NZBS, 'as perhaps the principal agency of entertainment in the country', was involved in the inquiry. The Committee decided that neither the serials nor the NZBS were particularly at fault; but it did suggest that features considered unsuitable for children were not always broadcast late enough to hinder children from hearing them. It also recommended that the concept that 'crime must never pay' needed to be featured more prominently in crime serials and that, while the NZBS's selection standards for serials was adequate, a married woman should be appointed to the selection panel. The NZBS reported that these recommendations had been 'pursued'.[73]

DIVISION OF STATIONS INTO PROGRAMME TYPES

The major programming change of the 1950s was the division of stations into programme types, and with this the extension of the coordination of programmes to include the commercial stations. Schroeder, as assistant director, was the principal architect of the new ordering. He invited a submission from Sydney Musgrove, professor of English at Auckland University College. The submission, which reflected Schroeder's thinking as much as Musgrove's, was greatly influential. Musgrove contended that the problem with New Zealand radio was the large number of stations transmitting different programmes. He proposed that stations be linked either nationally or regionally, with a certain amount of time set aside for local programming; and that there be three types of stations distinguished by their types of programmes: 'The obvious arrangement is that of the three basic programmes — a "light", a "middlebrow", and a "third" as adopted by the BBC.'[74] Schroeder agreed with this plan, and added: 'I see no better basis than ZB light, YA mixed, and YC sustained while X and YZ will be broadly regulated by their predominant functions.'[75]

By April 1950 the plan was accepted. It was set in place in Auckland in June 1950, in what was described as a 'trial'. It was more an introduction than a trial, and by the end of the month Wellington, Christchurch and Dunedin were ordered to follow suit. Yates instructed that the three national-service types of stations — the YAs, YCs and YDs — 'should constantly and carefully be drawn to provide effective alternatives', and that the ZB programme 'be associated with the coordinated planning as fully as possible'.[76]

Two new elements were introduced into the long-standing coordination of programmes: the commercial stations were included, and the entire programme of a station was to be of one type. The four types of station were to concentrate on distinct programmes. The YA stations were to offer a 'variety of appeal; standard — sound, upwards of average, popular taste in all programme fields, music, talks, drama, sports, religion, educational, etc., all news announcements, etc.; Parliament'; the YC stations 'sustained quality programmes of all kinds; weather reports, news, sports, etc., excluded so far as is possible'; the YD stations 'the non-commercial light programme', and the ZB stations a 'light' programme.[77]

In this New Zealand version of the BBC 'light, middlebrow and third'

Molly McNab, of the 3ZB Women's Hour, chooses table birds, the prize for the winner of a chicken recipe contest.

division, the YC stations became the custodians of the Reithian tradition as introduced by Shelley. This was a dramatic elevation for these stations, and it confirmed that Reithian programming was of interest to a minority only. Rather than a handicap to the new network, this became its freedom from the unacceptable assertion of the Reithian tradition: that tuition in high culture was *the* purpose of broadcasting.

In light entertainment, a further decision was made giving the ZB programmes primacy. Other stations were to note what was programmed on the commercial stations and design their schedules accordingly. Yates explained:

> As the commercial (ZB) stations are repowered you should gradually revise your day time programmes using contrasting material, especially in the breakfast session. All swing and blatant dance recordings and sentimental crooners should be excluded from this session . . . Other sessions throughout the day should be contrasted according to the local ZB programmes. The lighter programmes should not be eliminated entirely, but discreetly timed to provide an alternative to serials and other spoken sessions from the ZB stations.[78]

Over the following years it became clear that non-commercial stations were not to compete with their commercial counterparts. The policy was first made explicit with regard to the breakfast sessions, which attracted large audiences and high advertising income. Yates directed, 'Do not attempt to compete with the commercial stations in programme types they favour.'[79] During the early 1950s there developed a general understanding that the commercial stations had first call on light-programming material, and that non-commercial light programming was to be made up of what was left.

The YA stations were greatly affected by these decisions, even though much of their programming remained as light entertainment. As Yates noted with regard to their breakfast sessions: 'The general object should be to present cheerful music in all forms and all moods of cheerfulness from gay to rousing.'[80] However, having first call on light material, naturally the ZB stations chose the most popular of this material. In practice the YA stations were forced to choose their light programming from non-contemporary, less popular material. Their audiences, already smaller than those of the commercial stations, declined even further. The YA stations experienced a similar difficulty at the other end of their programming range, where highbrow programming became the province of the YC stations. The demarcation was in an understanding of classical music. As Schroeder instructed:

> Classical music — this is a field shared by YA and YC. The differentiation, though, is that while YC will range it completely taking familiar and unfamiliar, traditional and experimental, popular and controversial, YA will generally depend on the familiar, the traditional and the popular . . . YA will take an easier more catholic view of what is classical.[81]

Thus during the early 1950s the YA stations, which since the start of the RBC had been promoted as the central feature of New Zealand broadcasting, began to lose both their popular and their Reithian programming to other stations. They occupied a nebulous middle range which was defined only by not being at either extreme. By 1952 the director and assistant director agreed that 'the programme status and purpose of all YAs are a real problem'. The problem was 'about it [YA programming] having few listeners, about the waste of money on fees for studio programmes nobody listens to and about the inevitable frustrating effect on programme staff'.[82]

Facing page: A section of the audience in Christchurch during the first presentation of the 3ZB Colgate Palmolive Amateur Talent Show. *Above:* Grace Green, long-time announcer first on 3ZM and then on 3ZB, celebrates her birthday (*left*) and demonstrates one of the occupational hazards of broadcasting (*right*).

The YA stations were not left without programming. But what they had was a residual category of material that was unsuitable for either of the clearly defined ZB or YC groups. Some of this material was popular and attracted relatively large audiences — sports commentaries, for instance. These were not played on the ZB stations because they required a long period of uninterrupted time during which advertisements could not be run. The YA stations also ran the BBC serials, which were sold on condition that they would not be broadcast by commercial stations. Parliamentary broadcasts also remained with 2YA; but the great legislative changes of the 1930s did not continue beyond that decade, and the parliamentary broadcasts had not retained their original audiences. Specialist programmes such as the religious broadcasts remained with the YA stations, but these also appealed to only minority audiences. Yates, and particularly Schroeder, considered having the YA stations offer predominantly talks programmes. This was not adopted — it was never a realistic method of raising the stations' popularity while broadcasting was without its own news service and the talks programme remained under political control.

NETWORKING

The technology that was most useful to the NZBS's policy of dividing stations into three main types was networking, whereby one programme was broadcast simultaneously from two or more stations. The system had first been suggested by the RBC in 1929, when it wanted to connect its YA and proposed relay stations by landlines — standard telephone lines. Stations could rebroadcast one another's programmes; that is, they could tune into the broadcast signal and rebroadcast it from their own transmitter. But the quality of such rebroadcasts was not as high as signals carried by landlines. From 1929, New Zealand broadcasters were keen to use networking, but the P&T Department was unable to provide them with permanent landlines. Their complaint occasionally surfaced in public.[83] Even when landlines became available, they were acceptable only for voice transmission, and did not afford a quality transmission of music.

In 1953 the P&T Department brought into service wide-band lines which were permanently leased to the NZBS. The wide-band lines had a 10-kilohertz band (the telephone lines had only 4 kilohertz) and they were capable of transmitting music and giving it a quality reproduction. The royal tour in 1953 provided the impetus for the introduction and use of the lines. The four main centres plus Invercargill and Greymouth were connected. Thousands flocked for a glimpse of the Queen as she travelled through their area, but the tour was pre-eminently a radio occasion, with the medium uniting the people in their enjoyment of the first New Zealand visit by a reigning monarch. Eventually all stations were in a wide-band network. By November 1953 1YA and 2YA were linked for a joint breakfast session, and later more stations were included. In theory full networking was now available for the NZBS, but in practice the wide-band lines did not give good service. They were copper-wire lines which suffered interference from the considerable amount of other wiring, particularly that used for telephones, which the P&T Department system also carried. The problems continued through the 1950s, and in 1960 the P&T Department acknowledged there was no hope for improvement until the different technology of microwave links was in operation.[84] Thus whereas networking was introduced in the 1950s, the experiments were not fully successful and full networking of a number of stations to a single programme output did not occur until the following decade.

TELEVISION

New Zealand broadcasting was radio. Looming over broadcasting in the 1950s was the spectre of television, a long-delayed medium. In 1928 the *Radio Record* first discussed television, under the heading 'Television in Five Years'.[85] It was almost thirty-five years before television stations were broadcasting regularly in New Zealand. The introduction of the medium was delayed in other countries besides New Zealand in those years by the technical difficulties and the expense associated with it, and by external events including world war. The first British television station was broadcasting by 1937, though the BBC closed it down when the war began. Television stations were set up in the United States in 1946, in Canada in 1952 and in Australia in 1956. New Zealand, a country where radio gained unusually rapid acceptance, was obviously seeking to delay the introduction of television.

In July 1949 the Labour Government set up a departmental committee to advise on the establishment of television.[86] Labour showed no haste, and seven years later the National Government's minister in charge of broadcasting acknowledged that he also was proceeding 'slowly . . . by deliberate design'.[87] Cost was one reason for the delay. Labour's minister had indicated this.[88] In a country with a policy of conserving its overseas funds this was a strong reason, but in the 1950s New Zealand was a comparatively rich country which deliberately chose not to include television among its priorities. The dominant reason was political. Since the death of Savage, successive politicians of both political parties had maintained his control over broadcasting and hindered any further broadcasting innovations. A similar unwillingness to countenance change in broadcasting was seen in the dismissal of suggestions that frequency modulation (FM) broadcasting be considered.[89] The political reluctance which was the brake on the development of an independent system of broadcasting news and discussion also retarded the adoption of new broadcasting technology.

In the 1950s broadcasting became a curious mixture of development and stagnation. There was a postwar re-equipping of almost all existing stations and the founding of the X-class stations. By 1951 this coverage plan was described as 'virtually complete', and only towards the end of the decade was there any renewed attention to starting more stations.[90] During the decade the hours of transmission were gradually increased until radio was an all-day presence. But there was still resistance to any change which might threaten the practice of using radio as a presenter and supporter of government policy. Broadcasting was considered fraught with difficulties rather than possibilities. New methods and different styles of broadcasting were available but were disregarded. It was not until the following decade that New Zealand broadcasting was remade in a manner more fundamentally distinct than anything since its advent.

Broadcasting and New Zealand

Bill Richards, author of a wonderful recollection of Waikato pioneering life, gives an account of the arrival in 1925 of the first radio in his district. It was a grand and highly enjoyable social occasion attended by all the people of the district even though radio reception was such that not a voice or note of music was heard. It was the start of years of transformation as radio radically altered peoples' interests. As Richards explained, 'We knew little, and cared less, about the "outside" Our world revolved around the surrounding districts and its inhabitants. Within a 20-mile radius were your neighbours. News was passed by word of mouth. We made our own fun.'[1] Richards and his neighbours did not consider themselves isolated, whatever the judgement of later generations. But perceptions have changed much since those pioneering days.

Radio has been instrumental in shifting our attention away from an overwhelming focus on the local community. News and talks programmes, music and serials have become part of our lives, and we now take them for granted. Broadcasting has continued a process of change started by print and continued by the telegraph, whereby our understanding of 'community' has outgrown that area encompassed by word-of-mouth communication. Broadcasting is not the only twentieth century innovation which has enlarged our interests and changed our sense of community. But it is a major one, and in New Zealand the medium has been deliberately fostered as a means of national cohesion. From the time of the first Labour Government, this took the form of a (largely successful) attempt to unite the country in an understanding and acceptance of its government's policies and actions.

With the banning of controversial material and in the absence of a news and current affairs section, broadcasting was much confined in what it could investigate and document in New Zealand. Much of the postwar focus of New Zealand programming was on recording and broadcasting the stories of individuals. This policy was seen in the work of the mobile disc recording unit which toured in the late 1940s. Its work was hampered by the fact that it used disc recorders, just as this technology was about to be replaced by tape recorders, and by its lack of knowledge of the local communities.

A joint for the weekend. Maurice King, left, and Simon Devore entertain a large crowd inside and outside the studio, during a 3ZB Friday night novelty quiz.

These faults were countered in the 1950s by new technology and by local stations with personnel who had local understanding.

Such celebration of the individual does not obscure the orientation of broadcasting to the nation rather than the local community. This came both from the New Zealand use of broadcasting to promote national cohesion and from international broadcasting practices, whereby broadcasting was aimed at the widest possible audience. In social thought in the 1950s, this widest audience was increasingly regarded as a 'mass' audience encompassing the majority of the population. The broadcasting audience was never treated entirely as one undifferentiated mass. In New Zealand the audience was divided by the different types of station and further divided in terms of an understanding of sectional interests. The sports programmes, for instance, were seen as essentially a male preserve and paid little attention to women's sport, while much daytime programming was oriented towards a female, home-bound audience. But the growth of broadcasting and the fact that its audience was spread over the entire population led broadcasters to aim their programmes at as wide an audience as possible. This policy ignored the differences between people, and the ones who lost out most from it were the Maori — the country's largest minority group. Maori were never totally ignored, however, and especially on ceremonial occasions Maori items were broadcast as an expression of a particularly New Zealand programming. There were some programmes made for Maori. The Maori news began in 1942, a weekly programme in Maori was broadcast in 1950 over the North Island YA and YZ stations, in 1957 a weekly programme was broadcast from Napier for Maori listeners in the area. Such examples are few, though. Radio was not oriented to the Maori in general, let alone to the various iwi. Broadcasting may have been a major influence in the establishment of a New Zealand nationality, but it defined that nationality by declining to focus on the differences among New Zealanders.

George Boyle interviews a tobacco worker in 1950 and demonstrates the ease of operation of the newly available tape recording equipment.
RADIO NZ

Broadcasting also aligned New Zealanders with an international culture promoted by broadcasting. Many argued that this made New Zealanders mere recipients of programmes produced elsewhere. Certainly much of the programming, such as the great majority of the music and serials, was purchased from other countries, but there was considerable New Zealand programming as well. Some of this was local versions of international types of programmes. Thus 'hit-parades' came to New Zealand radio in the 1940s and flourished through the 1950s. Quiz-shows were an American innovation with local varieties being produced in many countries. Notable among the New Zealand versions were 'The Quiz-Kids', compered by Jack Maybury, which opened in 1948 and Selwyn Toogood's 'It's in the Bag', which started in 1954. These programmes were not original to New Zealand but were never slavish imitations. Instead they were adaptations which achieved great popularity for their promotion of ordinary New Zealanders as much as for their sense of fun. Among the new programmes were some which were starting coincidentally in different countries rather than originating in one and being exported to others. The women's sessions in the 1950s were a prime example. The New Zealand women's sessions were skilful local programming developments, were widely listened to and gave women's issues and women broadcasters greater prominence than usual.

Some of the versions of international programming were not so much New Zealand adoptions of others' innovations as broadcasting types to the development of which New Zealand contributed as much as any other country. Children's sessions, religious broadcasting and sports commentaries

Wiremu Parker, doyen of the Maori announcers.

are instances. While never among the major countries exporting its broadcasting programmes New Zealand has not been without influence on international broadcasting practices.

In spite of this there was a growing concern that New Zealanders needed a stronger presence on their nation's radio stations. This surfaced first with regard to music. The type of music broadcast varied with the nature of the stations, but was the mainstay of all programming. The majority of the music broadcast was from records released by the major international gramophone companies. In the 1950s the technical quality of these recordings improved with the introduction in 1952 of LPs — long playing 33⅓ rpm discs, and in 1955 of 45 rpm single discs. Music was the major product of an international culture that spread to many countries through broadcasting. The majority of this was popular music, but the broadcasting of music was never one-dimensional. The different types of stations offered a wide selection. Musicians and the public in general in New Zealand gained much of their musical appreciation and understanding from radio.[2]

There had always been demands that New Zealand broadcasting should take a greater account of New Zealand singers and musicians. In 1948 this view was coupled for the first time with a call for the NZBS to be required to broadcast a quota of New Zealand talent. The Labour MP, Dr A. M. Finlay, noting that the federal government of Australia then required 2½ per cent of broadcast time to be devoted to Australian compositions, asked the minister to consider a similar New Zealand requirement. The Minister declined. He stated that the NZBS was always willing to broadcast work

Jack Maybury with a group of Dunedin's *Quiz Kids* in 1948 at the Centennial Industries Fair.

of acceptable standard and type by New Zealand authors and composers, and there was no difficulty in finding broadcasting time for the amount of such work available.[3] The minister's reply continued the understanding that broadcasting should be the vehicle to bring excellence to the attention of listeners and that such excellence was not readily available locally. In spite of the minister's reply, the concern remained in following years that insufficient opportunity to broadcast was given to New Zealand composers, authors and performers.

Broadcasting was an innovation that was accepted with alacrity and enthusiasm by New Zealanders. By 1950 most households had owned a radio set for more than a decade. During the 1950s transmission hours were further increased. By the end of the decade 2YA and the commercial stations, including the X-class stations, were beginning transmissions by 6am and carrying on right through the day until late evening; 2YA to 11.20pm and the ZB stations to midnight. The other national stations were on the air from around 9am until the 2YA close at 11.20pm. Occasionally one or more of the YC stations would start transmissions at 11am but usually they broadcast from 5 or 6pm to 11pm. 1YD and 2YD broadcast from three to five hours an evening. The short transmission hours of broadcasting's early years had been greatly increased and, for many, radio had become a constant part of the daily routine.

New Zealand broadcasting has been much subject to standards imposed from above on its practitioners and recipients. These have been such as to rob New Zealand of the chance to use radio broadcasting to its full advantage. In particular the decision, upheld by successive governments, not to allow New Zealand broadcasting the right to report and discuss freely, emasculated the medium for the entire period covered by this volume. Similarly, the view that broadcasting should be a medium for the cultural uplifting of a people meant that much of broadcasting's orientation has been towards programming that was unacceptable to the majority and ultimately wasteful of broadcasting's resources.

The commercial stations remained overwhelmingly the listeners' favourites up to and during the 1950s. A graphic illustration of this preference was provided one night in 1949 by Grace Green, a 3ZB announcer. Broadcasting from high up a Christchurch radio mast she asked her listeners to turn off their house lights — and the city turned to near darkness. In the absence of detailed research, exact statistics about listeners and listening habits are not available. The National Service accepted that at least two thirds of listeners tuned to the commercial stations, while a pro-commercial broadcasting estimate found that 85 per cent of listeners favoured the commercial programmes.[4] The resounding and long-lasting preferences of the New Zealand audience are not so much an appreciation of commercial broadcasting *per se*, as an affirmation that the majority of listeners, at least in the era of politically controlled broadcasting, preferred contemporary, light entertainment.

During the 1950s radio sets moved beyond the family living-room. By the start of the 1950s portable sets were below 10 lbs in weight, and in that decade they became truly portable. Radio kept its dominant focus within the family home but also began to incorporate other audiences. In particular a growing youth culture received increasing recognition, and indeed definition, via broadcasting. Musically this was expressed especially on the new 45 rpm records but to these was added a flamboyance epitomised on radio by the DJs, or disc-jockeys — the compères for the new popular music sessions. Many listeners derived an understanding that they were not just members of their families and the local community — they were also part of a culture not circumscribed by geography but sustained by the various communication media of which radio broadcasting was preeminent.

During the years covered by this history broadcasting became an enduring institution in New Zealand society. It altered existing public tastes, added new ones, changed the nature of political communication between the electorate and their representatives, and enlarged people's understanding and acceptance of the community in which they lived.

In spite of this there was a growing understanding in the 1950s that more should be available from broadcasting. This understanding was at odds with the political definition of the proper organisation for broadcasting, but

Selwyn Toogood shows some of the prizes available on *It's in the Bag*. The programme started on New Year's Day 1954 and became the country's best-known quiz show. RADIO NZ SOUND ARCHIVES

was shared by the public and by broadcasters. While the political controllers of broadcasting were wary of broadcasting innovations, broadcasters had lost their earlier apprehension about the genres of radio entertainment. Appreciative of their medium, they were desirous of independence and the chance to extend the nature of New Zealand broadcasting and its place within New Zealand society. It was a desire that would be unleashed with the great changes to the control and nature of New Zealand broadcasting that came in the 1960s.

Appendix One

RADIO LICENCE OWNERSHIP STATISTICS

As at 31 March	Total licences	Licences per hundred of population
1925	4702	0.34
1930	8597	3.59
1935	152,808	9.79
1940	345,682	21.07
1945	379,189	22.57
1950	449,347	23.86
1955	508,943	23.88
1960	577,403	24.36

Appendix Two

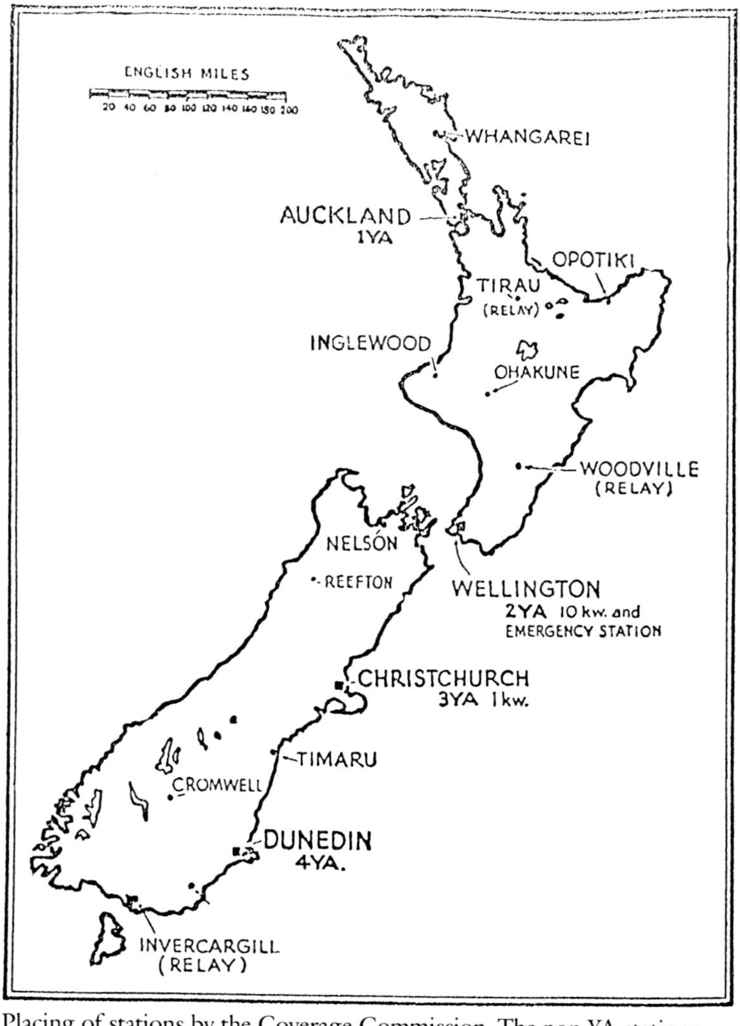

Placing of stations by the Coverage Commission. The non-YA stations were to be relay or assisted stations. *Radio Record*, 29 July 1932.

Appendix Three

STATIONS LICENSED 1922-1961

All stations were private stations for the first years but became divided into what were known as the A stations, the YA stations run by the RBC and then the NZBB, and the B stations. From 1933 no more private stations were licensed. From 1936 broadcasting became a government service and all stations started were part of either the NBS or NCBS. The two were amalgamated in 1943 under the NBS title. This was changed to the NZBS in 1946 and remained so until 1961.

1922

International Electric Co. Wellington. Originator Charles Forrest. Began broadcasting February. Co-proprietor Hope Gibbons. Call-sign 2YB allocated August 1923. Incorporated as Wellington Broadcasters Ltd, August 1923. Hope Gibbons sole proprietor, late 1923. Merged into 2YK November 1923.

Wellington Broadcasters Ltd. Originator Arthur McClay. Associates Bob Apperley, Hugh Simpson, Clive Drummond. Began broadcasting July. Call-sign 2YK allocated 1923. Station sold to Dominion Radio Co Ltd, November 1923. Managing director Gordon Harcourt. Ceased transmissions August 1925. Sold to RBC and recommenced transmissions December 1925. Call-sign changed to 2YA June 1927.

Radio Society of Christchurch. Began broadcasting intermittently late 1922. Call-sign 3AC allocated August 1923. April 1924 reregistered in name of J. I. Smail and call-sign changed to 3AQ. Sold to RBC March 1926 and call-sign changed to 3YA.

Radio Service Ltd, Auckland. Began broadcasting late 1922. Call-sign 1YA allocated 2 May 1923. Sold to RBC November 1925.

British Electrical Engineering Co. Dunedin. Proprietor Fred O'Neill. Started broadcasting 29 October 1922. Call-sign 4YA allocated 13 August. Sold to RBC November 1925 and call-sign changed to VLDN. Reverted to 4YA 1926.

British Radio Supply Co. Dunedin. Proprietor Norman Arundel. Started broadcasting late 1922. Call-sign 4YO allocated 13 August 1923. Ceased transmissions late 1925.

Otago Radio Association, Dunedin. Started broadcasting 15 November. Call-sign 4AB allocated August 1923. Changed to 4ZB, 23 September 1926. Changed to 4ZD 24 July 1937. Changed to 4XD 1 September 1948.

1923

2YA Nelson. Owned by Wilkins and Field Ltd. Started broadcasting July 1923. Ceased transmissions 1923.

1YB Auckland. Licensed 20 September. Proprietor Charles Pearson. 15 September 1925, licence reissued to La Gloria Gramophone Co., 9 August 1926. Call-sign changed to 1ZB, December 1933, licence taken over by Fellowship of the Friendly Road Inc. Opened transmissions for the NCBS 29 October 1936.
2YM Gisborne. Owned by Gisborne Radio Company. Licensed 27 October. Proprietors Percy Stevens, Ivan O'Meara, Bob Patty. Call-sign changed to 2ZM, 9 August 1926. Licence reissued to Atwater Kent Radio Service, 4 July 1928, same operators. Call-sign changed to 2XM 1 September 1948. Ceased transmissions May 1963.

1925
2YF Palmerston North. Owned by Palmerston North Radio Club. Licensed 8 December. Call-sign changed to 2ZF 9 August 1926. February 1928, licence reissued to Manawatu Radio Club. Purchased by government 11 August 1937 and ceased transmissions.

1926
1YD Auckland. Licensed to L. R. E. (Roy) Keith, 11 January. Call-sign changed to 1YQ, 1 June 1926, 1ZQ 9 August 1926. 19 September 1932, license reissued to Keith and Atwaters Piano Co. Licence revoked and station closed down 31 March 1933.

1927
3ZC Christchurch. Licensed to Messrs Finemore Strachan Ltd, 1 February. Licensee name amended to Home Recreations Ltd, December 1927. Station sold to F. R. Plominski and name changed to 3ZC Broadcasting Service Ltd, 15 October 1929. Station sold to C. F. Woodhead, December 1931. Relicensed to New Zealand Farmers Co-op Association, May 1932. Purchased by government and ceased transmissions 31 December 1933.
4ZM Dunedin. Licensed to J. D. McEwen, 8 October. Licence transferred to Messrs McCracken and Walls, 3 October 1932. Purchased by government and ceased transmissions March 1939.
4ZL Dunedin. Licensed to Radio Service Ltd, 12 October. Purchased by government and ceased transmissions 28 June 1937.

1928
2ZK Wanganui. Licensed to D. A. Morrison and Co, 20 June. Ceased transmissions 11 November 1932.
4ZO Dunedin. Licensed to Barnetts Radio Supplies Ltd, 24 October. Purchased by government and ceased transmissions July 1937.

1929
2ZQ Masterton. Licensed to N. R. Cunningham Ltd, 1 April. Ceased transmissions November 1930.
2YB New Plymouth. Licensed to North Taranaki Radio Society, 27 April. Contract station for NBS from 30 November 1938. Joined NBS 1944. Call-sign changed to 2XP 1 September 1948.
2ZH Napier. Licensed to C. B. Hansen Ltd, 8 August. Purchased by government

17 September 1937. Call-sign changed to 2YH and opened as an NBS station 17 November 1938. Call-sign changed to 2YZ, 1 September 1948.
2ZD Masterton. Licensed to W. D. Ansell, 9 August. Purchased by government and ceased transmissions 29 July 1937.
2ZE Eketahuna. Licensed to G. R. S. Allen, 18 August. Ceased transmissions December 1932.
2ZG Wanganui. Licensed to G. C. Carrad, 28 August. Ceased transmissions December 1929.
4ZI Invercargill. Licensed to G. C. Batchelor, 19 November. Ceased transmissions 6 April 1932.

1930
4ZP Invercargill. Licensed to Parsons Radio Supplies, 29 January. Purchased by government September 1937. Call-sign changed to 4YZ and opened as an NBS station 2 August 1938.
2ZP Wairoa. Licensed to G. A. Perry, 6 April. Purchased by government and ceased transmissions June 1937.
2ZX Wellington. Western Electric Co. (NZ) Ltd's shortwave experimental broadcasting station. Transmissions commenced April and ceased 25 June 1932.
2ZR Wanganui. Licensed to Radio Specialities Ltd, 10 May. Transmissions ceased April 1931.
1ZH Hamilton. Licensed to G. S. Anchor, 19 May. Purchased by NZBB and ceased transmissions 10 August 1935.
2ZI Hastings. Licensed to A. Simmons and Co. , 10 July. Sold to Sutcliffe's Music Store, 5 December 1930. Ceased transmissions 30 September 1932.
1ZM Auckland. Licensed to W. W. Rodgers Ltd, 27 August. Purchased by government September 1937 and opened as an NBS station 20 December 1937. Call-sign changed to 1YD, 1 September 1948.
2ZL Hastings. Licensed to Lockyer's Piano Stores, 12 September. Licence reissued to J. W. Holden, 19 February 1933. Purchased by government and ceased transmissions 24 November 1937.
1ZJ Auckland. Licensed to Messrs Johns Ltd, 13 November. Purchased by government and ceased transmissions December 1938.
2ZU Dannevirke. Licensed to Gramophone and Radio Co Ltd and A. R. C. Claridge, November. Ceased transmissions February 1931.
2ZJ Gisborne. Licensed to C. T. C. Hands, 17 December. Contract station for NBS from 6 September 1937. Joined NBS 1944. Call-sign changed to 2XG, 1 September 1948.
2ZO Palmerston North. Licensed to J. V. Kyle, 19 December. Purchased by government 30 September 1937 and ceased transmissions.
1ZR Auckland. Licensed to Lewis Eady Ltd, 20 December. Purchased by government and ceased transmissions 4 December 1933.

1931
1ZS Auckland. Licensed to McCabe's Radio, 19 January. License transferred to Radio Ltd, 17 August 1933. Ceased transmissions 2 June 1934.
5ZA Apia. Licensed to F. R. Plominski, 24 January. Relicensed to Apia Radio Corporation, 26 June 1931. Ceased transmissions 31 March 1933.

3ZR Greymouth. Licensed to Greymouth Radio Society, 29 January. Relicensed to West Coast Radio Society later in 1931. Purchased by government August 1937. Opened as an NBS station, 1 December 1938. Call-sign changed to 3YZ, 1 September 1948.
2ZW Wellington. Licensed to L. E. Strachan of Hamilton Nimmo and Sons Ltd, 21 May. Relicensed to 2ZW Broadcasting Service Ltd, June 1931. Purchased by government and ceased transmissions 4 December 1933.
4ZW Dunedin. Licensed to Radio Sales and Exchanges Ltd, 13 June. Relicensed to 4ZW Broadcast Service, 8 February 1933. Relicensed to A. E. E. Clarke, 3 May 1934. Ceased transmissions 2 May 1935.
4ZF Dunedin. Licensed to Laidlaw and Gray Ltd, 26 June. Ceased transmissions 31 July 1934.
4ZR Inchclutha. Licensed to B. R. Renton, 26 June. Relicensed to B. R. Renton and C. G. Clark, 28 January 1932. Purchased by government and ceased transmissions April 1937.
2ZB Napier. Licensed to T. H. Boston, 15 August. Ceased transmissions 30 September 1932.
3ZE Greymouth. Licensed to Schaef's Garage Broadcasting Service, 16 September. Ceased transmissions 11 January 1935.

1932
2ZR Nelson. Licensed to N. D. Braithwaite, 11 July. Relicensed to 2ZR Radio Club, November 1932. Purchased by government September 1937. Call-sign changed to 2YN and opened as an NBS station 30 November 1937. Call-sign changed to 2XN, 1 September 1948.
4ZC Lowburn Ferry, Cromwell. Licensed to J. C. Bilton, September. Purchased by government and ceased transmissions early 1938.

1933
2YC Wellington. Opened by NZBB, 24 February.
3ZM Christchurch. Licensed to W. J. Green and J. Younger, 27 March. Purchased by government April 1937. Call-sign changed to 3ZB and opened as an NCBS station 28 September 1937.
1YX Auckland. Opened by NZBB, 28 July. Call-sign changed to 1YC, 1 September 1948.

1934
3YL Christchurch. Opened by NZBB, 1 January. Call-sign changed to 3YC, 1 September 1948.
4YO Dunedin. Opened by NZBB, 26 March. Call-sign changed to 4YC, 1 September 1948.

1937
2YD Wellington. Opened by NBS, 25 April.
2ZB Wellington. Opened by NCBS, 28 April.
4ZB Dunedin. Opened by NCBS, 12 October.

1938
2ZA Palmerston North. Opened by NCBS, 10 October.

1939
5ZB Mobile station opened by NCBS 4 April. Transmitted from Centennial Exhibition, 8 November 1939. Ceased transmissions April 1940.

1949
3XC Timaru. Opened by NZBS, 18 January.
1YZ Rotorua. Opened by NZBS, 27 April.
1XH Hamilton. Opened by NZBS, 2 July.
1XN Whangarei. Opened by NZBS, 27 August.
2XA Wanganui. Opened by NZBS, 22 October.

1950
2YX Wellington. Opened by NZBS, 29 June.

1956
4ZA Invercargill. Opened by NZBS, 1 October.

1957
2ZC Napier. Opened by NZBS, 1 October.

1958
2XB Masterton. Opened by NZBS, 17 March.

1959
1ZC Rotorua. Opened by NZBS, 1 July.

1961
1ZD Tauranga. Opened by NZBS, 25 February.

NOTES

1 RADIO AND STATE CONTROL

1 Appleton E. V. 'The Young Rutherford', in *Rutherford* (1962), p.17.
2 Eve states Rutherford 'could detect wireless waves passing through a distance of sixty feet . . . and through opaque obstacles', though he later shows some scepticism. 1939, p.9. Wilson concurs but also expresses a point of doubt. 1983, p.58.
3 Rutherford to Mary Newton, 25.1.1896, in Eve, 1939, p.24.
4 Wilson, 1983, p.87.
5 In Chapter 6 of *The History of the Cavendish Laboratory*, quoted in Eve, 1939, p.20.
6 Rutherford to Mary Newton, 15.1.1896, Eve, 1939, p.22.
7 Wilson, 1983, pp.95-96.
8 The quotation is from Eddington, Eve, 1939, p.vii.
9 Wilson, 1983, p.94.
10 Appleton, The Young Rutherford, Rutherford, 1962, p.19.
11 Douglas, 1987, p.16.
12 Douglas, 1987, pp.18-19.
13 Donaldson, 1962, p.12.
14 Donaldson, 1962, p.12.
15 Alex Harris Ms.
16 Pocock and Garratt, 1972.
17 J. E. Strachan to A. R. Harris, 3 July 1970, Alex Harris Ms.
18 *Otago Daily Times*, 28 April 1930.
19 C. M. Turner to Seddon, 16 February 1902, Seddon papers, NA.
20 1985, p.10.
21 NZPD, Vol. 125, p.65.
22 *NZP* v.125, pp.289-90.
23 NZPD, Vol. 124, p.171.
24 Siegfried, 1982, p.52.
25 NZPD, Vol. 125, p.66.
26 NZPD, Vol. 125, p.289.
27 Ormsby Hamilton is the author of *The Commercial Development of Radio Telegraphy, Telephony and Broadcasting.* I have been unable to locate the work, and am dependent here on extracts from the text in the L. H. Steele diary in the Alex Harris manuscripts.
28 AJHR, F1, 1904.
29 NZPD, v.135, p.823.
30 AJHR, F1, 1904.
31 AJHR, F1, 1907.
32 AJHR, F8, 1908.
33 AJHR, F1, 1908.
34 AJHR, F1, 1909.
35 AJHR, F8, 1910.
36 AJHR, F1, 1910.
37 Deakin to Ward, 29 October 1909; Ward to Deakin, 29 November 1909, AJHR, F8, 1910.
38 Primrose to Ward, 25 November 1910, Ward to Primrose, 7 February 1911, AJHR, F8, 1911.
39 J. E. Strachan to A. R. Harris, 24 September 1970, Alex Harris Ms.
40 11 September 1908.
41 AJHR, F8, 1910.
42 *Break-In*, December, 1970.
43 RNZSA, T949.
44 McDonald, 1958, pp.185-6.
45 Clarkson, 1970.
46 J. E. Strachan to A. R. Harris, 3 July 1970, Alex Harris Ms.
47 AJHR, F1, 1911.
48 RNZSA T2780.
49 Clarkson in Alex Harris Ms.
50 AJHR, F1, 1912.
51 NZPD, Vol. 160, pp.111-2.
52 AJHR, F1, 1913.
53 *Evening Post*, 15 October 1912.
54 AJHR, F1, 1913.
55 NZPD, Vol. 164, p.558.
56 Hes, 1973, p.21.
57 Clarkson, 1973, p.14.

58 Clarkson in Alex Harris Ms.
59 RNZSA D899.
60 RNZSA T2780.
61 NZPD, Vol. 163, p.852.
62 Hall, 1980, p.4.
63 NZPD, Vol. 164, pp.128, 569–70.
64 NZPD, Vol. 164, p.558.
65 NZPD, Vol. 167, p.128.
66 NZPD, Vol. 171, p.553.
67 Hall, 1980, pp.9–10.
68 See Curnow, 1963, Chapter Two, for details of the origins of AWA.
69 RNZSA D899.
70 Hall, 1980, p.4.
71 RNZSA D1032.

2 The Birth of Broadcasting

1 Quoted in Barnouw, Vol. 1, 1966, p.25.
2 Quoted in Douglas, 1987, p.172.
3 Barnouw, Vol. 1, 1966, pp.69–70, 78–9.
4 Briggs, 1961; Burns, 1977, Chapter 1.
5 Hall, 1980, p.4.
6 Douglas, 1987, p.299.
7 See the announcement by Ward, Postmaster-General, NZPD, Vol. 183, p.999.
8 NZPD, Vol. 184, p.608.
9 NZPD, Vol. 185, p.709.
10 NZPD, Vol. 187, pp.117–18.
11 NZPD, Vol. 192, p.890.
12 Post Office Chronological Record Covering Regulatory Control of Amateur Radio In New Zealand, *Break-In*, January–February 1973, p.3.
13 RNZSA T949.
14 RNZSA T951.
15 RNZSA T949.
16 Ibid.
17 Brenda Bell, quoted in *Break-In*, January–February, 1973, p.42.
18 Alex Harris Ms., p.147.
19 23 August 1921.
20 RNZSA T949.
21 RNZSA T1907.
22 Hall, 1980, pp.9–10.
23 RNZSA T1905.
24 RNZSA.
25 RNZSA T949.
26 Bell to Jack, quoted in Hall, 1980, p.9.
27 *Otago Witness*, 14 April 1922.
28 6 April 1922.
29 Quoted in Downes and Harcourt, 1976, p.10.
30 Hall, 1980, p.11.
31 RNZSA T570.
32 RNZSA T1907.
33 RNZSA T950.
34 Drummond Ms.
35 RNZSA T949.
36 Spectrum 25, RNZSA D1032.
37 RNZSA T2781.
38 RNZSA T950.
39 RNZSA T949.
40 Hall, 1980, p.11. RNZSA T950, T949.
41 RNZSA T950.
42 Chronological Record Covering Regulatory Control of Amateur Radio in New Zealand, Post Office Headquarters, *Break-In*, January-February, 1973.
43 Alex Harris Ms. p.147.
44 quoted in Hall, 1980, p.5.
45 NZPD, Vol. 195, p.873.
46 NZPD, Vol. 196, p.97.
47 In particular, see in this regard Strachan's full listing, on behalf of the Canterbury Progress League, of matters needing attention. J. E. Strachan to J. G. Coates, 9 August 1922, in Alex Harris Ms, p.133.
48 NZPD, Vol. 197, pp.194–5.
49 Alex Harris Ms.
50 Hall, 1980, p.6.
51 RNZSA T949.
52 RNZSA T950.
53 Drummond Ms.
54 RNZSA T2782.
55 Drummond Ms.
56 5 January 1925.
57 RNZSA T2981.
58 RNZSA T951.
59 *New Zealand Wireless & Broadcasting News*, 20 December 1923, quoted in Hall, 1980, p.16.
60 *Evening Post*, 31 December 1925, 21 January, 26 February 1926.
61 RNZSA T950.
62 *Evening Post*, 10 November 1923.
63 AAFK890/12b NA.
64 RR, 21 December 1928.
65 NZPD, Vol. 204, pp.1075–8.
66 NZPD, Vol. 208, p.853.
67 NZPD, Vol. 208, p.850.
68 *New Zealand Gazette*, 22 June 1925.
69 Collins, 1967, pp.25–6. Mackay, 1953, p.23.
70 *Evening Post*, 21 November 1923.

3 The Radio Broadcasting Company

1 RNZSA T570.
2 quoted in Hall, 1980, p.17.
3 RNZSA T570.

4 RNZSA T570.
5 NZPD, Vol. 208, p.583.
6 However in 1934, after the end of the RBC, though more for matters to do with the marketing of primary produce exports than broadcasting, Goodfellow formed the Democrat Party, a political grouping to the right of the then governing Forbes Ministry.
7 RR, 7 October 1927.
8 A dealer's fee was £10 per year in the four main centres, £5 in towns having a population above 5000 and £2 elsewhere.
9 NZPD, Vol. 208, pp.850–1.
10 RNZSA D2068.
11 Harris to Acting-Secretary, P&T Department, 5 April 1929, HC.
12 RR, 22 July 1927.
13 *Evening Post*, 29 October 1925.
14 Drummond Ms.
15 *Evening Post*, 3 and 23 September 1925.
16 RR, 22 July 1927.
17 NZPD, Vol. 208, pp.852–3.
18 NZPD, Vol. 209, p.977.
19 AJHR, F1, 1928.
20 Gallagher, 1987.
21 RR, 11 October 1929.
22 3 August 1928.
23 NZPD, Vols. 213, pp.682–94, 214, pp.156–63.
24 RNZSA D1032.
25 RNZSA D2145.
26 RNZSA D1032.
27 Gordon, 1981, p.40.
28 RNZSA D2145.
29 RNZSA D2068.
30 Hall, 1980, p.23.
31 Harris to McNamara, 13 June 1927, HC.
32 RNZSA T1044.
33 Gordon, 1981, p.40.
34 RNZSA T1044.
35 22 July 1927.
36 29 July 1927.
37 McNamara to Harris, 10 June 1927, HC.
38 Harris to McNamara, 13 June 1927, HC.
39 RNZSA T570.
40 NZPD, Vol. 212, p.720–1.
41 Quoted in Hall, 1980, p.20.
42 RR, 16 August 1929.
43 Hall, 1980, p.22.
44 RNZSA D2145.
45 RNZSA T1015.
46 *Christchurch Star*, Sports Edition, 9 July 1927.
47 3 August 1928.
48 3 August 1928.
49 30 September 1927.
50 See, for example, Hall, 1980, p.25.
51 Coming of Age; The Friendly Road; 21 Today; 1929–1950, ATL.
52 11 March 1926.
53 RR, 18 November 1927.
54 RR, 18 November 1927.
55 McNamara to Harris, 13 November 1931, Harris to McNamara, 14 November 1931. HC.
56 NZPD, Vol. 204, pp.236–7.
57 Marsden to Gibbs, 12 July 1926, E 8/4/33, NA.
58 For example, see Gibbs to Marsden, 5 August 1926, E 8/4/33, NA.
59 Strong to Minister, 20 July 1927, E 8/4/33, NA.
60 RR, 23 December 1927.
61 Strong to Minister, 20 February 1929, E 8/4/33, NA.
62 Strong to Minister, 26 July 1929, E 29/71/1, NA.
63 Strong to Minister, 4 August 1930, E 29/71/1, NA.
64 For example see the material in E 29/71/1, NA.
65 17 April 1931.
66 22 July 1927.
67 Harris to Station Directors, 8 December 1927, HC.
68 28 October 1927.
69 RR, 6 April 1928.
70 RR, 3 August 1928; see this issue also for a survey of the various musicians then employed by the RBC.
71 RR,16 September 1927.
72 Harris to Station Directors, 8 December 1927, HC.
73 Bellingham to Harris, 6 December 1928, HC.
74 *Evening Post*, 2 July 1931.
75 Harris to McNamara, 7 March 1927, HC.
76 McNamara to Harris, 10 March 1927, HC.
77 Harris to McNamara, 17 March 1927, HC.
78 NZPD, Vol. 212, p.977.
79 NZPD, Vol. 213, p.690.
80 NZPD, Vol. 218, p.167.
81 NZPD, Vol. 213, p.693.
82 NZPD, Vol. 213, p.690.

4 Disputes and Growth

1 See Appendix One for radio licence statistics.
2 The fullest study of the history of New Zealand radio manufacturing is that by Stokes, 1986.
3 See, for example, RR, 12 August 1927.
4 RNZSA T2300.
5 RR, 22 November 1929.
6 RNZSA T1015.
7 RR, 19 December 1930.
8 28 March 1925.
9 *Evening Post*, 28 April and 2 May 1925.
10 *Evening Post*, 24 April and 9 July 1925.
11 Hall, 1980, p.58.
12 RR, 22 July 1927.
13 NZPD, Vol. 214, pp.159–60. RR, 7 October 1927.
14 McLean to Harris, 15 November 1927, HC.
15 Harris to McLean, 16 November 1927, HC.
16 30 March 1928.
17 RR, 30 March 1928.
18 RR, 30 March 1928.
19 For a full listing of stations see Appendix Three.
20 RNZSA T53.
21 Collins, 1967, pp.13–16.
22 NZ Yearbook, 1929.
23 15 February 1929.
24 RNZSA T298.
25 RR, 19 April 1929.
26 RR, 15 November 1929, and see Hayden to Goodfellow, 8 August 1930, Goodfellow to Hayden, 11 August 1930, and Hayden to Carruth, undated but circa 8 August 1930, HC.
27 Robinson, 1964, Chapter 18.
28 RR, 10 April and 31 July 1931.
29 Mackay papers.
30 L. E. Strachan, press statement, January 1929, HC.
31 RR, 22 February 1929.
32 17 December 1925.
33 4 January 1929.
34 Harris to McNamara, 30 January 1929, HC.
35 Harris to Ball, 21 January 1929, HC.
36 Harris to Goodfellow, 26 April 1929, HC.
37 See E. H. Culver, Hastings Radio Society, to Postmaster-General, 9 August 1929, HC.
38 Harris to Tyle, 9 July 1929, HC.
39 Kent to Harris, 9 December 1929, HC.
40 Harris to McNamara, 12 February 1930, HC.
41 Harris to Goodfellow, 26 September 1927, HC.
42 RNZSA T570.
43 NZPD, Vol. 219, pp.786–7.
44 RR, 30 January 1931.
45 *Evening Post*, 30 October 1930.
46 RR, 20 and 27 February and 6 March 1931.
47 RR, 30 January 1931.
48 *Evening Post*, 31 October 1930.
49 Goodfellow to Harris, 12 May 1930, HC.
50 Goodfellow to Harris, 15 May 1930. Harris to Kent, 12 September 1930, HC.
51 Mackenzie to Harris, 14 July 1930, HC.
52 *Evening Post*, 20 November 1930.
53 *Evening Post*, 7 November 1930.
54 11 November 1927.
55 RR, 17 and 24 February 1928.
56 RR, 22 March 1929.
57 RNZSA T1015.
58 Hall, 1980, pp.36–37.
59 20 July 1928.
60 NZPD, Vol. 218, pp.165–6.
61 NZPD, Vol. 218, pp.161, 168.
62 RR, 26 October 1928.
63 See the Drummond and Mackay manuscripts for an appreciative description and O'Donoghue, 1946, pp.37–46 for a more antagonistic appraisal.
64 RNZSA T1015.
65 Harris to Heighway 17 April 1930, HC, and see the RR editorial, 24 April 1930.
66 RR, 13 September 1929.
67 RR, 13 September 1929.
68 RR, 28 February 1930.
69 11 April 1930.
70 4 April 1930.
71 RR, 17 April 1930.
72 RR, 17 April 1930.
73 Harris to Secretary, Wellington Rugby Union, 1 June 1931, HC.
74 Hall, 1980, p.47.
75 See, for example, the report of that year's International Press Conference, held in Melbourne, *Evening Post*, 2 October 1925.
76 See, for example, the *Radio Record* editorial on the strength of press news as opposed to radio news, acknowledging 'The ear will never replace the eye', 7 December 1928.
77 Ball to Harris, Heighway to Harris, 18 January 1929, HC.

78 Harris to Heighway, Harris to Ball, 21 January 1929, HC.
79 1 February 1929.
80 Clarke to Harris, 22 December 1930, HC.
81 Harris to Clarke, 30 December 1930, HC.
82 Harris to Goodfellow, 14 January 1930. Goodfellow to Harris, 3 February 1930, HC.

5 The Extension of Broadcasting and The Fall of the RBC

1 28 March 1929.
2 RR, 8 November 1929.
3 Hall, 1980, p.25.
4 Hegan, 1980, p.53.
5 Gordon, 1981, p.28.
6 Collins, 1967, p.25.
7 Hegan, 1980, p.53.
8 17 February 1928 and see the issues for the four previous weeks.
9 RR, 16 August 1929.
10 RR, 13 June and 4, 18 and 25 July 1930.
11 Collins, 1967, pp.30–3.
12 Hall, 1980, p.25.
13 RR, 28 December 1928.
14 RR, 7 December 1928.
15 1 February 1929.
16 RNZSA T583.
17 RR, 17 April 1930.
18 2 January 1931.
19 2 January 1931.
20 Harris to Goodfellow, 12 May 1930, HC.
21 Goodfellow to Harris, 18 November 1930, HC.
22 Harris to Goodfellow, 24 December 1930, HC.
23 Collins, 1967, p.40.
24 NZPD, Vol. 230, p.667.
25 Frame, 1982, p.106.
26 22 July 1927.
27 Quoted in Hall, 1980, p.28.
28 See Harris's statements in RR, 16 August and 6 September 1929.
29 RR, 8 and 15 February 1929.
30 RNZSA T841.
31 RNZSA D2069.
32 Drummond MS.
33 RR, 21 March 1930.
34 RR, 28 March 1930.
35 1976, p.56.
36 RR, 13 February 1931.
37 13 February 1931.
38 Quoted in RR, 13 March 1931.
39 RNZSA T2331 and T2335.
40 RNZSA T434.
41 RNZSA T583.
42 A Visit to the Home of 2ZW, ATL.
43 RNZSA T1044.
44 NZPD, Vol. 222, pp.1069, 1071.
45 NZPD, Vol. 223, p.675.
46 NZPD, Vol. 226, p.274.
47 NZPD, Vol. 226, p.399.
48 NZPD, Vol. 226, pp.286–7.
49 NZPD, Vol. 226, p.403.
50 Quoted in RR, 24 October 1930.
51 Quoted in RR, 17 October 1930.
52 Goodfellow to Harris, 10 November 1930, HC.
53 For example, in Harris to Chappell, 20 December 1930, HC.
54 *Evening Post*, 16 October 1930.
55 19 September 1930.
56 26 June 1931.
57 16 October 1930.
58 Parliamentary Committee Submission, 29 June 1930, HC.
59 Quoted in RR, 28 August 1931.
60 *Evening Post*, 21 August 1931.
61 Harris to Goodfellow, 22 August 1931, HC.
62 Heighway to Harris, 13 July 1931, HC.
63 RR, 6 July 1931.
64 *Evening Post*, 20 August 1931.
65 RR, 17 July 1931.
66 Quoted in RR, 17 July 1931.
67 RR, 4 September 1931.
68 RR, 25 September 1931.
69 Hamilton to Harris, 21 October 1931, HC.
70 Goodfellow to Harris, 4 December 1931, HC.
71 Goodfellow to Harris, 8 December 1931, HC.
72 Goodfellow to Harris, 10 December 1931, HC.
73 *Evening Post* and most dailies, 17 August 1931.
74 RNZSA T570.
75 18 August 1931.
76 24 August 1931.
77 NZPD, Vol. 230, p.804.
78 RNZSA T570.

6 New Zealand Broadcasting Board

1 NZPD, Vol. 230, p.657.
2 The phrases are from Burns, 1977, p.21.
3 Burns, 1977, p.22.

4 NZPD, Vol. 230, p.657.
5 NZPD, Vol. 230, pp.656–7.
6 NZPD, Vol. 230, p.663.
7 NZPD, Vol. 230, p.803.
8 NZPD, Vol. 230, p.661.
9 NZPD, Vol. 230, p.658.
10 NZPD, Vol. 230, p.660.
11 NZPD, Vol. 230, p.663.
12 NZPD, Vol. 230, p.659.
13 Mackay, 1953, p.40. NZPD, Vol. 249, p.140.
14 13 January 1933.
15 Mackay, 1953, p.40. NZPD, Vol. 249, p.140.
16 NZPD, Vol. 230, p.802.
17 NZBB minutes, 23–24 January 1932. 23 January 1935.
18 For example, see Hands to Hutchinson, 15 and 16 May 1933, AAFK 890/5a, NA.
19 Hands to Hutchinson, 3 November 1932, AAFK 890/4c, NA.
20 Mackay, 1953, p.47.
21 AADL 890/2d, NA.
22 Hands to Station Director 4YA, 27 June 1932, AADL 890/2d, NA.
23 Hands to Macfarlane, 19 September 1932, AAFK 890/6d, NA.
24 Hutchinson to Hands, 16 June 1932, AAFK 890/4c, NA.
25 NZBB minutes, 17 November 1932.
26 AAFK 890/7j, NA. NZBB minutes, 14 December 1932.
27 The report is printed in full in RR, 29 July 1932.
28 Interview with Tom Clarkson, 12 March 1991.
29 NZPD, Vol. 237, p.744.
30 Clarkson, 1981.
31 NZBB minutes, 28–29 April 1932.
32 RR, 24 April 1932.
33 NZBB minutes, 16 March 1933.
34 RR, 18 March 1932.
35 Annual Report, NZBB, AJHR, 1933.
36 Analysis of NZBB expenditure, NZ Year Book 1939, p.786.
37 The Story of 2ZR, M3.
38 RR, 23 February 1934.
39 Annual Report, NZBB, AJHR, 1935.
40 Annual Report, NBS, AJHR, 1937.
41 RR, 24 February 1933.
42 NZBB minutes, 3–4 February 1932.
43 A full review of the plebiscite is in RR, 9 September 1932.
44 NZBB minutes, 27 March 1934.
45 RR, 23 December 1932.
46 Hands to Macfarlane, 10 January 1934, AAFK 890/6f, NA.
47 RNZSA D2069.
48 Collins, 1967, p.25.
49 RNZSA T841; and see the *Radio Record* of 2 August 1929 account of the first Wellington Radio Exhibition and the celebrity status given to Drummond, with hundreds of people waiting to the 10.30pm close down to hear his 'G-o-o-d Night'.
50 Hands to Macfarlane, 19 January 1932, AAFK 890/6d, NA.
51 Hands to all Station Directors, 1 February 1932, AAFK 890/15c, NA.
52 Hands to all Station Directors, 30 May 1932, AAFK 890/15c, NA.
53 Hands to all Station Directors, 12 March 1932, AAFK 890/15c, NA.
54 Ball to Hands, 12 April 1932, AAFK 890/15c, NA.
55 RR, 21 October 1932.
56 Hands to Station Directors, 26 August 1932, AADL 890/2d, NA.
57 22 February 1935.
58 See the correspondence between Macfarlane and Hands in AAFK 890/6d and AAFK 890/6e, NA.
59 Hands to Macfarlane, 29 February 1932, AAFK 890/6d, NA.
60 NZBB minutes, 23–24 June 1932.
61 NZBB minutes, 16 August 1935.
62 RNZSA T1015.
63 NZBB minutes, 28–29 April 1932.
64 NZBB minutes, 29 May 1934.
65 AAFK 890/10c, NA.
66 NZBB minutes, 28 August 1932.
67 RR, 1 September 1933.
68 Macfarlane to Hands, 28 November 1932, AAFK 890/6d, NA.
69 RR, 1 April 1932.
70 30 June 1933.
71 For example, see J. E. Polson, Principal, Teachers' Training College, Christchurch, to Director of Education, 16 October 1933 and Director to Polson, 18 October 1933, E29/71/1, NA.
72 NZBB annual report, AJHR 1936.
73 AJHR, E1, 1933–36.
74 Hands to Macfarlane, 22 March 1934, AAFK 890/6f, NA.
75 NZBB Annual Report, AJHR, 1936.
76 Leys to Hands, 6 June 1932, AAFK 890/4c, NA.
77 See the 1932 correspondence between Hands and Sir Cecil Leys, Managing Director of the *Auckland Star*, AADL 890/2d, NA.
78 NZBB minutes, 26 October 1934.

79 Hands to Hutchinson, 21 February 1935, AAFK 890/5c, Hands to Macfarlane, 21 February 1935, AAFK 890/7a, NA.
80 Hutchinson to Hands, 4 November 1932, AAFK 890/4c, NA.
81 Annual Report, AJHR, 1933.
82 'Newspapers must be made to realise that if they attempt to stifle publicity concerning broadcasting, we have in our hands a wonderful organisation to use in the interests of broadcasting.' Hutchinson to Hands, 17 May 1932, AAFK 890/4c, NA.
83 NZBB minutes, 23 December 1931.
84 Hands to Macfarlane, 9 February 1932, AAFK 890/6d, NA.
85 C. S. McNulty to Hamilton, 12 December 1933, AAFK 890/10e, NA.
86 22 February 1935.
87 NZBB minutes, 15 February 1933.
88 NZBB minutes, 9 August 1933.
89 See, for example, RR, 2 and 30 August 1935.
90 Thanks to the collected papers of Ian K Mackay, the secretary of the 2ZR Radio Club during the years 1933 to 1935 and a man who had a long and formidable career in broadcasting, 2ZR is one of the private stations about which information remains available.
91 The Story of Station 2ZR, M3.
92 *Nelson Evening Mail*, 27 April 1933.
93 Mackay to Secretary, Station 1ZB, 22 August 1935, M3. *Nelson Evening Mail*, 15 April 1933.
94 'The Story of Station 2ZR', M3.
95 'Microphones and Men', M1.
96 'Microphones and Men', M1.
97 NZBB minutes, 24 August 1932.
98 NZBB minutes, 17 November 1932.
99 *Nelson Evening Mail*, 15 April 1933.
100 AAFK 890/4a, NA.
101 NZPD, Vol. 234, pp.646, 651.
102 RR, 3 August 1934.
103 RR, 12 October 1934.
104 RR, 23 June 1933.
105 RR, 12 January 1934.
106 RR, 15 June 1934.
107 Coming of Age; The Friendly Road; 21 Today; 1929-1950, ATL, and see RR, 19 December 1930.
108 'The Scrimgeour Story', Spectrum Documentary, Part 1.
109 Scrimgeour, Lee, Simpson, 1976, p.7.
110 *John A Lee's Weekly*, 10 December 1941.
111 Simpson, 1974, pp.154-6.
112 Ibid, p.154.
113 Don to Nimmo, 24 June 1932, MSY1926, S papers, ATL.
114 25 June 1932, MSY1926, S papers, ATL.
115 NZPD, Vol. 234, pp 642-3.
116 NZPD, Vol. 234, p.643.
117 Vickery to Hamilton, 3 November 1932, AADL 890/2d, NA.
118 Hall, 1980, p.55.

7 Victory for the B Stations and Labour

1 3 March 1933
2 RNZSA T1247.
3 Hall, 1980, p.55.
4 Hands to Hutchinson, 2 November 1933, AAFK 890/4c, NA.
5 *New Zealand Herald*, 13 June 1933.
6 RR, 16 June 1933.
7 *New Zealand Herald*, 12 June 1933.
8 Hands to Hutchinson, 12 June 1933, AAFK 890/5a, NA.
9 Edwards, 1971, p.58.
10 16 June 1933.
11 Hutchinson to Hands, 9 June 1933, AAFK 890/5a, NA.
12 Hutchinson to Hands, 11 July 1933, AAFK 890/5a, NA.
13 M2, ATL.
14 NZPD, Vol. 237, pp.146-7.
15 Hands to Macfarlane, 23 August 1933, AAFK 890/6e, NA.
16 NZPD, Vol. 237, p.147; see also the earlier report in RR, 15 September 1933.
17 Macfarlane to Hands, 3 March 1933, AAFK 890/6e, NA.
18 Hall, 1980, p.56.
19 *Listener*, 17 July 1964.
20 The Scrimgeour Story, Part 1, Spectrum.
21 Edwards, 1971, p.60.
22 Vickery to Hamilton, 19 December 1933, AAFK 890/9e, NA.
23 Coming of Age; The Friendly Road, ATL.
24 NZPD, Vol. 237, pp.944-5.
25 Edwards, 1971, p.60. The names of the leading citizens of Auckland who supported the Friendly Road were later listed in NZPD, Vol. 249, p.264.
26 NZPD, Vol. 237, p.944.
27 RR, 23 March 1934.
28 Hands to Macfarlane, 9 March 1934, AAFK 890/6f, NA.
29 RR, 25 May 1934.

1935, AAFK 890/6e, NA.
30 RR, 19 April 1935.
31 Fry, 1957, p.83.
32 RNZSA T1247.
33 In Mackay papers, Box 1, ATL. And see the debate in NZPD, Vol. 238, pp.878–80.
34 NZBB minutes, 21 June, 16 August 1935.
35 Hands to board members, 24 July 1935, AAFK 890/4a, NA.
36 Macfarlane to Hands, 17 December 1934, AAFK 890/6f, NA. RR, 6 July 1934.
37 Hutchinson to Hands, 4 July 1934, AAFK 890/5b, NA.
38 AADL 890/2e, NA.
39 24 November 1932. AAFK 890/7j, NA.
40 WEA Dominion Broadcasting Committee report for winter of 1933, p.4, AAFK 890/13b, NA.
41 Shuker, 1984, p.96.
42 NZBB minutes, 16 March 1933.
43 Hands to N M Richmond, 18 September 1933, AAFK 890/13b, NA, NZBB minutes, 14 September 1933.
44 NZBB minutes, 27 March 1934.
45 NZBB minutes, 21 February 1936.
46 Hands to Macfarlane, 1 May 1933, AAFK 890/6e, NA.
47 NZBB minutes, 24 January 1934.
48 Hamilton to Hands, 31 January 1934, AAFK 890/10e, NA.
49 Hall, 1980, p.68.
50 NZPD, Vol. 238, p.57.
51 Shaw, 1934, p.25.
52 Hands to Macfarlane, 5 April 1934, AAFK 890/6f, NA.
53 The full script is in RR, 20 April 1934.
54 Edwards, 1971, p.65.
55 Shaw, 1934, pp.8–9.
56 AAFK 890/7j, NA.
57 Hamilton to Vickery, 21 May 1934, AAFK 890/9e, NA.
58 NZPD, Vol. 238, pp.837–9, 868.
59 NZPD, Vol. 238, pp.903–4.
60 *Auckland Star*, 4 July 1934 and see RR, 13 July 1934.
61 NZPD, Vol. 241, p.234.
62 NZBB minutes, 27 March 1935.
63 Vickery to Hamilton, 6 March 1935, AAFK 890/9e, NA.
64 RR, 7 April 1933, Shuker, 1984, p.95.
65 12 April 1935.
66 12 April 1935.
67 NZPD, Vol. 243, p.98.
68 NZBB minutes, 11–12 April 1935.
69 For example, see Hands to H. R. Cathie, Secretary of 1ZM Association, 5 August 1935, AAFK 890/6e, NA.
70 NZBB minutes, 11–12 April 1934.
71 NZBB minutes, 11–12 April 1934.
72 14 June 1935, AADL 890/2d, NA.
73 *New Zealand Gazette*, 1935, Vol. 1, p.781, 28 March 1935.
74 Garland to Hands, 20 June 1935, AAFK 890/12c, NA.
75 Hands to Garland, 27 June 1935, AAFK 890/12c, NA.
76 AAFK 890/11a, NA.
77 McNamara to Hands, 16 July 1935, AAFK 890/11a, NA.
78 Hands to McNamara, 25 July 1935, AAFK 890/11a, NA.
79 O'Donoghue to Hands, 15 September 1932, AAFK 890/11b, NA.
80 Simpson, 1974, p.154, Scrimgeour, Lee, Simpson, 1976, pp.40–1.
81 RR, 24 August and 9 November 1934.
82 RR, 7 December 1934.
83 10 August 1934.
84 Edwards, 1971, p.70.
85 Hands to Hutchinson, 3 May 1934, AAFK 890/5b, NA.
86 RR, 22 February 1935.
87 M2.
88 RR, 19 April 1935.
89 For a public discussion of the issues see the article by Canon Percival James, RR, 1 March 1935.
90 Hands to Station Directors, 24 May 1935, AAFK 890/12b, NA.
91 4YA Station Director to Head Office, 16 July 1935, AAFK 890/12b, NA.
92 21 and 28 May 1935.
93 Hutchinson to Hands, 29 July 1935, AAFK 890/6a, NA.
94 O'Donoghue to Hands, 11 July 1935, AAFK 890/12b, NA.
95 23 May 1934.
96 Hutchinson to Hands, 26 May 1935, AAFK 890/5b, NA.
97 11 July 1935, AAFK 890/12b, NA.
98 Hutchinson to Hands, 25 September 1935, AAFK 890/6a, NA.
99 *Auckland Star*, 3 October 1935.
100 NZPD, Vol. 243, pp.98–103; Hands argued that this was the final act of an employee who had previously been found unsatisfactory, Hands to Phillips, 3 October 1935, AAFK 890/6b, NA.
101 Mackay, Secretary NZ Federation of B Station Proprietors, to John Bilton, Proprietor 4ZC Cromwell, 3 October 1935, M2.
102 *Auckland Star*, 4 October 1935.
103 Scrimgeour to Mackay, 23 September

1935, M2.
104 O'Donoghue to Hands, 18 November 1935, AAFK 890/12b, NA.
105 Much has been written of this incident. Along with this material I am also indebted to Clarkson's own account, Clarkson, 1981, and to an interview with him, 12 March 1991.
106 Clarkson, 1981, p.8.
107 Scrimgeour, Lee, Simpson, 1976, pp.47–50. Clarkson, 1981, p.12.
108 Scrimgeour, Lee, Simpson, 1976, p.46.
109 25 November 1935.
110 Clarkson, 1981, p.10.
111 *Auckland Star*, 27 November 1935.
112 Scrimgeour, Lee, Simpson, 1976, p.51.
113 *Auckland Star*, 25 November 1935.
114 NZPD, Vol. 245, p.760.
115 NZPD, Vol. 245, pp.807–8.
116 O'Donoghue to Hands, 25 November 1935, AAFK 890/12b, NA.
117 Clarkson, 1981, p.6–7.
118 20 August 1938, p.371.
119 Jones to Vickery, 26 June 1936, AAFK 890/9e, NA.
120 Day, 1990, p.238.
121 T2332 RNZSA. Ingram, 1984, p.359.
122 RR, 23 February 1934, and see the issue of 9 February 1934.
123 NZPD, Vol. 244, pp.345–6.

8 Government Controlled Broadcasting

1 NZPD, Vol. 230, p.661.
2 Quoted in Hall, 1980, p.83.
3 Vickery to Jones, 12 February 1936, AAFK 890/9e, NA.
4 Cabinet schedule, 18 February 1936, N2028.
5 NZPD, Vol. 244, p.56.
6 Frame, 1982, p.131.
7 Gustafson, 1986, p.141.
8 RR, 18 September 1936.
9 Scrimgeour to Wilson, 7 December 1941, N2029.
10 NZPD, Vol. 263, pp.640.
11 NZPD, Vol. 244, pp.338, 349.
12 Scrimgeour to Savage, 6 April 1936, N2281.
13 Cabinet schedule 15 June 1936, N2028.
14 Scrimgeour to Savage, 6 June 1936, N2281.
15 Bell to Scrimgeour, 20 March 1936, N2281.
16 Scrimgeour to Savage, 6 April 1936, N2281.
17 RR, 31 July 1936.
18 NZPD, Vol. 245, pp.750–1; see also Scrimgeour to Wilson, 7 December 1941, N2029.
19 NZPD, Vol. 245, p.780.
20 M1, 1952, p.97.
21 NZPD, Vol. 245, p.818.
22 *Standard*, 1 July 1936.
23 O'Donoghue, 1946, pp.15–16. Savage's first choice was not Henry's nephew Rudall Hayward, the film-maker, as Mackay told Gustafson. 1986, pp.194, 331.
24 A full discussion and appraisal of Shelley is offered in Ian Carter's recent biography which, like this work, was commissioned by the Broadcasting History Trust.
25 Quoted in Hall, 1980, p.87.
26 RR, 4 September 1936.
27 Quoted in Hall, 1980, p.88.
28 *New Zealand Herald*, 20 October 1936.
29 *New Zealand Herald*, 26 October 1936.
30 NZPD, Vol. 249, pp.87–8.
31 Labour Caucus Minutes, 25 November 1936, NA.
32 RR, 12 March 1936.
33 Scrimgeour to Wilson, 7 December 1941, N2029.
34 RR, 28 August 1936.
35 *Auckland Star*, 24 August 1936.
36 T2730 RNZSA.
37 Algie to Marshall, 11 July 1956, AAFL 563/4b, NA.
38 Hall, 1980, p.91.
39 Calculated from NCBS returns in Annual Reports, AJHR, 1938–1939.
40 T2730 RNZSA.
41 NZPD, Vol. 247, p.1071.
42 NZPD, Vol. 248, pp.429, 546–7.
43 Scrimgeour to Wilson, 7 December 1941, N2029.
44 In particular see the long campaign by *Truth*, and the questioning by the Opposition, during the debate on the Broadcasting Amendment Act 1937, of the activities of Stannage Radio Ltd and Radio Vision Ltd.
45 Cabinet schedule, 28 April 1937, N2038.
46 Lee to Savage, 10 August 1938, S19.
47 9 November 1936, S19.
48 Minutes of Executive meeting, North Canterbury LRC, 12 November 1942, N2029.
49 RR, 5 June 1936.
50 9 June 1936.
51 Mackay to Minister in Charge of

Broadcasting, 25 May 1937, AADL 564/105c, NA. M1 1952, pp.93-7.

52 See the report of the B Station Protestation Committee delegation meeting with Peter Fraser, Acting PM, on 12 June 1937, AADL 564/105c, NA.

53 M1, 1952, p.94.

54 *Auckland Star*, 2 July 1936.

55 Cabinet schedule 15 June 1936, N2028.

56 NZPD, Vol. 249, pp.267-8.

57 NZPD, Vol. 249, p.1120.

58 Annual Report, AJHR, 1938.

59 NZPD, Vol. 249, p.272.

60 See, for instance, the statements by J. V. Kyle, the 2ZO proprietor, in *New Zealand Herald*, 27 September 1937. Also NZPD, Vol. 248, p.447, Vol. 249, p.164.

61 AADL 919/1a, NA.

62 Acting-Minister in Charge of Broadcasting to Otago Radio Association, 12 April 1937. President, ORA, to Minister in Charge of Broadcasting, 28 April 1937, AADL 564/106h, NA.

63 Mitchell to Minister, 16 October 1937. Shelley to President, ORA, 13 December 1937, AADL 564/106h, NA.

64 President, ORA, to Shelley, 20 December 1937, AADL 564/106h, NA.

65 Mitchell to Minister, 24 February 1942, AADL 564/106h, NA.

66 Shelley to Minister, 21 June 1944, AADL 564/106h, NA.

67 AADL 564/632a, NA.

68 AADL 546/620d, NA.

69 Shelley to DG, P&T Department, 2 July 1948, AADL 564/620d, NA.

70 Scrimgeour to Savage, 26 July 1937, N2281.

71 Du Maurier to Savage, 9 November 1936, N2281.

72 Shiel to Scrimgeour, 4 November 1936, N2281.

73 T2730 RNZSA.

74 Annual Report, AJHR, 1937.

75 Annual Reports, AJHR.

76 Annual report 1938, AJHR.

77 Annual report, 1938, AJHR.

78 T252 RNZSA.

79 1967, p.119.

80 RR, 4 September 1936.

81 1967, pp.63-4.

82 Collins, 1967, pp.119-20.

83 8 April 1942. Drummond joined the NZBS after WWII and finally retired in 1955.

84 C. E. Beeby, who knew Shelley earlier in Canterbury and throughout his term as director, described him as detesting economics and having no interest in news or current affairs. Interview 10 April 1991.

85 AADL 564/287g, NA.

86 Sullivan, 1987, pp.30-1.

87 See the *New Zealand Herald* and the *Auckland Star* of 5 April 1937 for an initial reaction to Scrimgeour's 'outburst'. The debate continued throughout the New Zealand press for much of April. Fraser's statement is in the Auckland newspapers of 24 April.

88 Edwards, 1971, pp.103-6.

89 NZPD, Vol. 249, p.149-50.

90 See, for example, NZPD, Vol. 249, p.194 and Vol. 252, p.442.

91 NZPD, Vol. 249, p.100.

92 NZPD, Vol. 249, p.153.

93 NZPD, Vol. 248, p.455.

94 NZPD, Vol. 250, p.269.

95 Minister to A. P. O'Shea, 16 December 1946, AADL 564/299a, NA.

96 NZPD, Vol. 250, p.135.

97 NZPD, Vol. 249. p.349-51.

98 Minister to Mason, MP, 8 July 1938, AADL 564/84a, NA.

99 AADL 564/343b, NA.

100 Lee to Savage, 18 May 1938, Shelley to Savage, AADL 564/343b, NA.

101 AADL 563/4a, NA.

102 NZPD, Vol. 252, pp.99-124.

103 RR, 5 February 1937.

104 Hall, 1980, p.95.

105 M1, Commercial Broadcasting and the Announcer.

106 quoted in Hall, 1980, p.114.

107 Gordon, 1981, pp.43, 64.

108 NZPD, Vol. 248, p.933.

109 Annual Report, 1938, AJHR.

110 *NZ Observer*, 3 March 1943.

111 S50.

112 NZPD, Vol. 255, pp.478, 490.

113 Gordon saw the two as 'always at loggerheads', 1981, p.70. C. E. Beeby considered the two 'detested one another', interview 10 April 1991.

114 M1, 1952, p.112.

115 RR, 7 May 1937.

116 AADL 564/295a, NA.

117 T294 RNZSA.

9 Broadcasting in the War Years

1 RR, 15 July and 19 August 1938.
2 RR, 28 October 1938.
3 RR, 23 December 1938.
4 See also Shelley's comments in RR, 20 August 1937.
5 Jensen, 1966, p.16.
6 NZBB minutes, 26 October 1934.
7 T583 RNZSA.
8 1958, p.193.
9 Annual Report, 1939, AJHR.
10 Toogood, 1979, p.36. Hall, 1980, p.115.
11 Shiel to Scrimgeour, 2 February 1939, AADL 563/19a, NA.
12 Mackay's report, AADL 563/19a, NA.
13 Laurenson's report, AADL 563/19a, NA.
14 NZPD, Vol. 258, p.154.
15 NZPD, Vol. 257, p.823.
16 NZPD, Vol. 257, p.825.
17 *Otago Daily Times*, 5 March 1966.
18 NZPD, Vol. 252, p.310.
19 Shelley to Duff, 3 March 1937, AAFT 794/6b, NA.
20 Duff to Shelley, 30 March 1939, AAFT 794/6b, NA.
21 1958, p.191.
22 See M2, 'The Professional Years', pp.47–9, for an NCBS view of the relationship with Duff and Holcroft; and Holcroft, 1969, pp.18–21 for an editor's view.
23 NZPD, Vol. 256, p.29.
24 NZPD, Vol. 257, p.592; see also M2, 'Broadcasting and the Politicians', pp.2–3, for a different reasoning for the completion of this building.
25 27 September 1940.
26 NZPD, Vol. 257, pp.16–17; see also Collins, 1967, pp.86–8 for a discussion of this incident.
27 NZPD, Vol. 257, p.17.
28 Collins, 1967, pp.96–8.
29 The work of the Broadcasting Unit is well covered in Palmer's technical and general reports, AADL 564/109a, NA.
30 Palmer's report on technical services, 29 June 1944, AADL 564/109a, NA.
31 1958, p.191.
32 Shelley to Beeby, 13 January 1943, AADL 564/109b, NA.
33 Palmer to Director, 14 June 1943, AADL 564/109b, NA.
34 Report of Royal Commission on the Loss of certain Vessels by Enemy Action and alleged Leakage of Information, 21 March 1941, AADL 563/21a, NA.
35 25 July 1941, AADL 563/21a, NA.
36 Director, Meteorological Office to Controller, NCBS, 7 July 1941, AADL 563/21a, NA.
37 7 April 1941, AADL 563/21a, NA.
38 Fry, 1957, p.107.
39 AADL 563/21a, NA.
40 Hall, 1980, p.142.
41 Scrimgeour to Minister, 20 May 1942, AADL 563/21a, NA.
42 Hall, 1980, p.139.
43 5 April 1940.
44 3 March 1943.
45 The actual broadcast is not recorded but this extract is reported in various places including M1, Appointment with Yesterday, p.119, *New Zealand Observer*, 3 March 1943, and Hall, 1980, pp.107–8.
46 M1, Appointment with Yesterday.
47 Wilson to Scrimgeour, 15 May 1941, AADL 563/11a, NA and N2029.
48 The talk was published in a pamphlet entitled 'Rev. Percy Paris praises Russia in the Famous Man in the Street Talk', Russian Relations Pamphlet No 3.
49 Minutes of Executive meeting of North Canterbury L R C, 12 November 1942, N2029.
50 *New Zealand Observer*, 17 December 1941.
51 Wilson to Acting Controller, NCBS, 12 September 1941, AADL 563/11a, NA.
52 In the frontispiece of Scrimgeour's copy, S131.
53 For example, Scrimgeour to Nicholls, 4 November 1941, AADL 563/11a, NA.
54 Wilson to Scrimgeour, 17 February 1942, 10 March 1942, N2029.
55 Paul to Wilson, 1 March 1942, N2029.
56 Wilson to Scrimgeour, 15 April 1942, N2029.
57 Statement by Wilson to Prime Minister, 1 March 1943, N2029.
58 Wilson to Fraser, 3 March 1941, N2029.
59 10 October 1941, N2029.
60 N2029.
61 quoted in Hall, 1980, p.103.
62 Scrimgeour to Wilson, 7 December 1941, N2029.
63 Wilson: notes made immediately after interview, 26 February 1942, N2029.
64 Secretary, Auckland Branch, New Zealand Waterside Workers' Union, to

Scrimgeour 13 March 1942, S papers, MSY1929.
65 *Evening Post*, 12 March 1942.
66 NZPD, Vol. 261, p.471.
67 Paul to Wilson, 2 March 1943, N2029.
68 Statement by David Wilson to Prime Minister in the presence of J T Paul, 9pm 1 March 1943, N2029.
69 8 March 1943.
70 Greenberg to Wilson, 4 March 1943, N2029.
71 Labour Party Caucus minutes, 4 March 1943, NA. Also NZPD, Vol. 263, p.507.
72 M2, Broadcasting and the Politicians, p.33.
73 8 March 1943.
74 NZPD, Vol. 262, p.953.
75 12 February 1943, AAFL 563/4b, NA.
76 15 February 1943, AAFL 563/4b, NA.
77 NZPD, Vol. 262, p.133.
78 Quoted in Hall, 1980, p.103.
79 *Evening Post*, 15 June 1943.
80 18 June 1943, N2029.
81 See M2, Broadcasting and the Politicians, for a more innocent version.
82 NZPD, Vol. 263, pp.639–40.
83 28 July 1943.
84 From a Scrimgeour election pamphlet, N2029.
85 N2029.
86 N2029.
87 'The "Uncle Scrim" Mystery'.
88 Sinclair Carruthers, 20 May 1946, M3.
89 Hall, 1980, p.119.
90 Mackay to Shelley, 11 February 1944, AADL 563/3a, NA.
91 Basham to Shelley, 10 February 1944, AADL 563/3a, NA.
92 M3.
93 24 February 1944, AADL 563/3a, NA.
94 M3.
95 M3.
96 13 November 1946.
97 1953, pp.90–1, 102–3.

10 New Zealand Broadcasting Service

1 The history of the orchestra is well documented. See, in particular, Jensen, 1966. Tonks, 1986.
2 NZPD, Vol. 273, pp.501, 516–25.
3 21 March 1947.
4 14 March 1947.
5 Annual Report, AJHR, 1953.
6 NZPD, Vol. 298, p.1692.
7 NZPD, Vol. 304, pp.1551–62.
8 NZPD, Vol. 269, p.807.
9 22 October 1954.
10 NZPD, Vol. 287, p.2268.
11 Annual report, AJHR, 1947.
12 NZPD, Vol. 300, pp.1645–47.
13 Director's newsletter to Station Managers, May 1961, M3.
14 By 1949 the NZBS had 809 employees. *New Zealand Gazette*, 1949, vol. III, pp.1903–1912.
15 NZPD, Vol. 264, p.333.
16 NZPD, Vol. 278, pp.366–72.
17 Jones to General Secretary, 16 December 1946, AADL 564/299a, NA.
18 Minister to Director, Director to Minister, 13 January 1947, AADL 564/299a, NA.
19 20 May 1947, AADL 564/299a, NA.
20 NZPD, Vol. 277, p.468.
21 NZPD, Vol. 272, pp.42–3.
22 NZPD, Vol. 277, p.757.
23 Annual Report, AJHR, 1948.
24 Hall, 1980, p.165.
25 NZPD, Vol. 280, p.586 and see pp.576, 582.
26 See Director to Minister of Education, 15 September 1941 and minutes of meeting of 16 September 1941, AADL 564/375B, NA.
27 AADL 564/383B, NA.
28 Jones to Shelley, 27 May 1947, AADL 564/394b, NA. For a full discussion of this incident see McGeorge, 1990.
29 Quoted in McGeorge, 1990, p.8.
30 15 October 1947.
31 Shelley to H. Blackburn, Head Teacher, Mangapehi School, 25 June 1947, AADL 564/394b, NA.
32 1 July 1949.
33 Shelley to Minister, 15 November 1943, N2029.
34 N2029.
35 McCarthy, 1973, p.91.
36 Ibid, p.58.
37 Ibid, p.91.
38 1958, pp.192–3.
39 Examples include, 'We must ensure at all times that the listener is very conscious of the commercial being broadcast in as effective a way as possible. To help this effectiveness the programme announcement should not be spoken as though it was a continuation of a commercial announcement. It is a question of inflection, of pause or the personality of the announcer involved.

I would be glad if you would adhere to this at all times, and instruct the announcers accordingly.' — Shelley to J. W. Griffiths, 1ZB Station Manager, 10 February 1948, AADL 563/1a, NA; 'You will recall that originally a rigid instruction was in effect that after a commercial the following record was to be brought in unannounced, on the theory that listeners would remember last words spoken and if the record was announced would be inclined to forget the commercial and concentrate on the title of the record. It is noted on some stations that announcers are continuing to broadcast a commercial and then introduce the recording. Please ensure that the original instruction is enforced.' — Shelley to all commercial station managers, 15 April 1948, AADL 563/1a, NA.

40 22 April 1949.
41 M1.
42 Yates to J. Young, Head Office Commercial Division, 13 July 1949, AADL 564/343e, NA.
43 Fraser to Secretary of The Peace Union, 14 July 1949, AADL 564/343e, NA.
44 H. W. Bullen to Yates, 5 July 1949, AADL 564/343e, NA.
45 AADL 564/343e, NA.
46 NZPD, vol. 266, p.505.
47 NZPD, Vol. 288, p.3024–5.
48 NZPD, Vol. 294, p.99.
49 NZPD, Vol. 297, p.749.
50 NZPD, Vol. 293, p.4012.
51 NZPD, Vol. 312, pp.1283, 1477.
52 NZPD, Vol. 294, p.65.
53 Annual Report, AJHR, 1955.
54 NZPD, Vol. 307, pp.1949, 1953.
55 NZPD, Vol. 303, p.405.
56 NZPD, Vol. 308, p.632.
57 J. W. Griffiths, Station Manager 1ZB to Shelley, 1 April 1949, AADL 564/539a, NA.
58 *New Zealand Herald*, 5 April 1949.
59 Shelley to Griffiths, 7 April 1949, AADL 564/5393a, NA.
60 6 April 1949.
61 Shelley to Griffiths, 14 April 1949, AADL 564/539a, NA.
62 AAFT 794/6b, NA.
63 A. R. D. Fairburn to Ian Mackay, 9 April 1954, M3.
64 Hall, 1980, p.172.
65 1979, p.36.
66 1980, p.121.
67 Witness the *Listener* from 29 December 1939 to 19 January 1940.
68 1946, pp.22–3.
69 NZPD, Vol. 266, p.251; see pp.269–271 for further parliamentary discussion on radio serials.
70 NZPD, Vol. 270, p.139.
71 6 October 1944.
72 NZPD, Vol. 272, p.47.
73 Annual Report, AJHR, 1955.
74 Musgrove to Schroeder, 7 December 1949, AADL 564/559b, NA.
75 Assistant Director to Director, 28 November 1950, AADL 564/559b, NA.
76 Director to Station Managers, 13 April 1950, AADL 564/34a, NA.
77 Yates to Station Managers, 13 April 1950, AADL 564/34a, NA. The same instruction was repeated in numerous memoranda during 1950.
78 Yates to Station Manager, 2YA, 10 November 1950,AADL 564/34b, NA.
79 Yates to Station Manager, 2YA, 27 February 1952, AADL 564/34b, NA.
80 Yates to YA Station Managers, 27 February 1952, AADL 564/560a, NA.
81 Schroeder to Station Manager 2YA, 8 November 1950, AADL 564/34a, NA.
82 Schroeder to Station Manager 1YA, 26 June 1952, AADL 564/560a, NA.
83 NZPD, Vol. 272, p.46.
84 May 1960, AAFL 564/4b, NA. Annual Reports, AJHR, 1954–5. Downes, undated. N R. Palmer interview, 13 September 1991.
85 9 March 1928.
86 Annual Report,1950.
87 NZPD, Vol. 306, p.1213.
88 NZPD, Vol. 285, p.781.
89 NZPD, Vol. 273, pp.105–6.
90 Annual Report, AJHR, 1951.

BROADCASTING AND NEW ZEALAND

1 1986, p.81.
2 For instance, I am indebted to Rick Bryant and Midge Marsden for bringing to my attention the influence that Arthur 'Turntable' Pearce and his jazz sessions had on them and other musicians of their generation.
3 NZPD, Vol. 283, p.2452.
4 Peter Downes interview, 17 October 1991. NZPD, Vol. 293, p.4014.

Bibliography

BOOKS

Barnouw, Erik, *A History of Broadcasting in the United States*, New York, Oxford University Press.
(1966) v.1, *A Tower in Babel.*
(1968) v.2, *The Golden Web.*
(1970) v.3, *The Image Empire.*

Barnett, Stephen & Jim Sullivan (compilers) (1988) *In Their Own Words: From the Sound Archives of Radio New Zealand*, Wellington, Government Printing Office.

Briggs, Asa, *The History of Broadcasting in the United Kingdom*, London, Oxford University Press.
(1961) v.1, *The Birth of Broadcasting.*
(1965) v.2, *The Golden Age of Wireless.*
(1970) v.3, *The War of Words.*
(1979) v.4, *Sound and Vision.*
(1985) *The BBC: The First Fifty Years*, London, Oxford University Press.

Burdon, R. M. (1965) *The New Dominion*, Wellington & London, Reed/Allen & Unwin.

Burns, Tom (1977) *The BBC: Public Institution and Private World*, London, Macmillan.

Carter, Ian (1993) *Gadfly: The Life and Times of James Shelley*, Auckland, Auckland University Press.

Collins, Ken G. (1967) *Broadcasting Grave and Gay*, Christchurch, Caxton Press.

Day, Patrick (1990) *The Making of the New Zealand Press*, Wellington, Victoria University Press.

Donaldson, Frances (1962) *The Marconi Scandal*, London, Hart-Davis.

Douglas, Susan J., (1987) *Inventing American Broadcasting: 1899-1922*, Baltimore, John Hopkins University Press.

Downes, Peter & Peter Harcourt (1976) *Voices in the Air. Radio Broadcasting in New Zealand: A Documentary*, Wellington, Methuen/Radio New Zealand.

Edwards, Les (1971) *Scrim: Radio Rebel in Retrospect*, Auckland, Hodder & Stoughton.

Eve, A. S. (1939) *Rutherford*, Cambridge University Press.

Frame, Janet (1982) *To the Is-land*, New York, Braziller.

Fry, A. S. (1957) *The Aunt Daisy Story*, Wellington, Reed.

Gordon, John (1981) *All the World's a Stage*, Wellington, Mallinson Rendell.

Gregory, R. J. (1985) *Politics and Broadcasting: Before and Beyond the NZBC*, Palmerston North, Dunmore Press.

Gustafson, Barry (1986) *From the Cradle to the Grave: A Biography of Michael Joseph Savage*, Auckland, Reed Methuen.

Hall, John Herbert (1980) *The History of Broadcasting in New Zealand: 1920–1954*, Wellington, Broadcasting Corporation of New Zealand.

Hamer, David (1988) *The New Zealand Liberals: The Years of Power*, Auckland University Press.

Hegan, Eddie (1980) *No Choice: My Years in Vaudeville and Variety Around New Zealand*, Auckland, Hodder & Stoughton.

Hight, James & Alice M. F. Candy (1927) *A Short History of the Canterbury College*, Christchurch, Whitcombe & Tombs.

Holcroft, M. H. (1969) *Reluctant Editor: The* Listener *Years*, 1949–67, Wellington, Reed.

Inglis K. S. (1983) *This is the ABC: The Australian Broadcasting Commission, 1932–1983*, Melbourne University Press.

Jensen, Owen (1966) *The NZBC Symphony Orchestra*, Wellington, Reed.

Lee, John A. (1973) *Political Notebooks*, Wellington, Alister Taylor.

Lee, John A. (1975) *Simple On A Soapbox*, Whitcombe & Tombs. (first published 1963).

Lee, John A. (1981) *Diaries 1936–1940*, Christchurch, Whitcoulls.

Mackay, Ian K. (1953) *Broadcasting in New Zealand*, Wellington, Reed.

McDonald, K. C. (1958) *A History of Waitaki Boys' High School 1883–1958*, Whitcombe & Tombs.

Mulgan, Alan (1958) *The Making of A New Zealander*, Wellington, Reed.

O'Donoghue, Alec F. (1946) *The Rise and Fall of Radio Broadcasting in New Zealand: The Tragedy of Government Control and Management*, Auckland, A. F. O'Donoghue.

Oliphant, Mark (1972) *Rutherford: Recollections of the Cambridge Days*, Amsterdam, Elsevier.

Pocock, R. F. & G. R. M. Garratt (1972) *The Origins of Maritime Radio: The Story of the Introduction of Wireless Telegraphy to the Royal Navy between 1896 and 1900*, London, Her Majesty's Stationery Office.

Richards, Bill (1986) *Off the Sheep's Back*, Auckland, Lindon Publishing.

Robinson, Howard (1964) *A History of the Post Office in New Zealand*, Wellington, Government Printer.

Rutherford, Ernest (1962) *The Collected Papers of Lord Rutherford of Nelson*, v.1, London, Allen & Unwin.

Sanders, James (1979) *Dateline-NZPA: The New Zealand Press Association, 1880–1980*, Auckland, Wilson & Horton Ltd.

Scrimgeour, C. G. (1937) *Chats: Talks On the Friendly Road*, Auckland, Unity Press.

Scrimgeour C. G., John A. Lee, Tony Simpson (1976) *The Scrim-Lee Papers*, Wellington, Reed.

Shaw, George Bernard (1934) *What I Said in New Zealand: The Newspaper Utterances of Mr. George Bernard Shaw, March 15 to April 14, 1934*, Wellington, The Commercial Printing & Publishing Company of NZ Ltd.

Shuker, Roy (1984) *Educating the Workers? A History of the Workers' Education Association in New Zealand*, Palmerston North, Dunmore Press.

Siegfried, Andre (1982) *Democracy in New Zealand*, Wellington, Victoria University Press (first published 1904, Paris).

Simpson Tony (1974) *The Sugarbag Years*, Wellington, Alister Taylor.
Stokes, John W. (1986) *The Golden Age of Radio in the Home*, Invercargill, Craigs.
Sullivan, Jim (1987) *A History of Broadcasting News: 1921–1962*, Timaru, Radio New Zealand Sound Archives.
Tonks, Joy (1986) *The New Zealand Symphony Orchestra: The First Forty Years*, Auckland, Reed Methuen.
Toogood, Selwyn (1979) *Out of the Bag*, Auckland, Methuen.
Walker, R. R. (1973) *The Magic Spark: 50 Years of Radio in Australia*, Melbourne, Hawthorn.
Wilson David (1983) *Rutherford: Simple Genius*, Cambridge, Massachusetts, MIT Press.
Wilson Ormond (1946) *What's Wrong With Broadcasting?* Hamilton, Pauls.

MANUSCRIPTS

Broadcasting Archives, NA.
Clarkson, T. R. (1981) Reminiscences of Radio, Telecom Museum & Archives.
Drummond, D. A. V. C., Collected Papers, Peter Downes Collection.
Hall, John Herbert, Collected Papers, RNZSA.
Harris Collection, RNZSA.
Harris, Alex Rowan (1975) 'History of Wireless in New Zealand Before the Broadcast Age 1902–1923', ATL.
Mackay, Ian Keith, Collected Papers, ATL.
Nash, Walter, Collected Papers, NA.
Sargeson, Frank, Collected Papers, ATL.
Scrimgeour, C. G., Collected Papers, ATL.
Seddon, R. J., Collected Papers, NA.
Tyrer, T. G., Collected Papers, ATL.

PAMPHLETS

'A Visit to the Home of 2ZW. The Voice of the Capital City', Commercial Printing & Publishing Co. of NZ Ltd, 1931, ATL.
'Coming of Age. The Friendly Road, 21 Today: 1929–1950', ATL.
'Reverend Percy Paris Praises Russia in the Famous Man in the Street Talk', Sunday August 10, 1941, Russian Relations Pamphlet No 3, ATL.
'The Scandal of New Zealand Broadcasting', 1ZB Radio Club, Unity Press, Auckland, 1935, ATL.
'More Scandal', 1ZB Radio Club, Unity Press, Auckland, 1935, ATL.
'The Uncle Scrim Mystery. What Has Happened to the Man in the Street Session?', C&G Print, 1942, ATL.

PAPERS

Clarkson Tom (1973) 'Radio Telegraphy in New Zealand in the "Spark" Era', *Break-In*, January-February, pp. 12–18.
Curnow, Ross (1963) 'The Origins of Australian Broadcasting,1900–1923', in *Initiative and Organisation*, (eds) Ian Bedford & Ross Curnow, Cheshire.

Downes, Peter (undated, unpublished) 'The Long Road to Networking'.

Gallagher, Brian (1987, unpublished), 'A technical discussion of 2YA to commemorate the 60th anniversary of the station'.

Hes, R. L. (1973) 'Sixty Years of Achievement; The Story of Amalgamated Wireless (Australasia) Ltd', *Break-In*, January–February, pp. 20–26.

McGeorge, Colin (1990, unpublished), 'Why *How Things Began* Ended'. New Zealand Association for Research in Education Conference Paper.

Index

NAMES

Algie, R. M., 302, 336
Allardyce, Alan, 79-81, 96, 113-14, 167-8
Allen, G. R. S., 325
Allen, James, 31, 59
Anchor, J. S., 153, 325
Anderson, George, 29
Ansell, N. D., 325
Apperley, Bob, 44, 323
Arundel, Norman, 43, 55, 323
Atkinson, Karl, 122, 231
Atmore, Harry, 175

Ball, John, 77, 91, 103, 120, 129, 131, 166, 331-3
Ball, Robert, 8
Bansgrove, R., 48
Barnard, William Edward, 261
Barnett, Frank, 224
Barnett, M. A. F., 156-9
Basham, Maud Ruby (Aunt Daisy), 188, 239-41, 267, 269, 279-81, 339
Batchelor, G. C., 325
Baume, Eric, 201
Beaglehole, J. C., 284
Beeby, Bernard, 230, 250-2, 337-8
Beeby, C. E., 212, 287, 337
Bell, Brenda, 27, 32, 329
Bell, Culford, 77, 81
Bell, E. J., 81
Bell, Francis, 30
Bell, Frank, 27, 40
Bell, Harry, 216-18, 336
Bellingham, W. J., 88-90, 330
Bickerton, A. W., 8
Bilton, J. C., 326
Bingham, John M., 21, 52, 67, 71-72, 77-78, 103, 157-8
Bingham, Mrs John M., 72
Birkinshaw, Frank, 211
Blair, Archibald, 143
Booth, H., 153
Boston, T. H., 326
Boucher, E. R., 185
Braithwaite, N. D., 174, 176, 326
Brandon, Cyril, 19
Branly, Edouard, 7, 12
Brew, F. C., 170
Broadfoot, Walter, 234
Bryant, Rick, 340
Bullivant, Merton, 216, 222, 229, 275
Bunce, W. K., 48
Burrell, Bob, 47
Bush King, C. J., 83
Byrd, Richard, 130-1

Campbell, Hugh, 28
Carr, Clyde, 77, 136-7, 165, 287, 289
Carrad, G. C., 325
Carrick, 228
Catton, S., 188
Chapman, Annette, 166
Chappell, A. B., 153, 332
Clare, Maurice, 250, 255
Clark, G. C., 326
Clark, Edward, 26
Clarke, A. E. E., 326
Clarke, Basil, 260
Clarkson, T. E., 22, 156, 204-6, 208, 328-9, 333, 336
Clifford, George, 114
Coates, Gordon, 36, 48, 57-59, 62, 64-65, 71, 123, 125, 183, 187, 197, 206, 214
Cocker, W. H., 197
Coe, H. V., 47
Collins, Ken, 101, 165, 229-30, 261, 329, 331-3, 337-8
Condor, W. T., 201
Coney, Barry, 112
Cook, A., 154
Cooper, John Sissons St George, 11-12
Coughlin, Fr., 187
Coutts, Morton, 55

Crossley Clitheroe, J., 154
Crowther, Frank, 135
Curry, A. L., 74, 78-79, 130, 145, 262, 265

Davies, Jack, 52
Davis, E. T., 153
Deakin, Alfred, 18, 328
De Forest, Lee, 34
Dickinson, H. D., 195
Doidge, Frederick, 300
Donald, J. B., 125, 137-42, 149
Douglas, Major, 193, 195
Downes, Peter, 131, 329, 340
Dowsett, H. M., 17
Drake, Mrs, 74
Drummond, Clive, 32, 41, 44-45, 53-54, 68, 74, 77, 81, 130, 165, 231-2, 271, 323, 329, 331-3, 337
Duff, Oliver, 257-8, 338

Eady, Forbes, 202
Eady, Ray, 136, 184
Earle, Charles, 172-3, 256
Edison, Thomas, 34, 63
Edwards, Jim, 180
Endean, William, 235
Eve, A. S., 8, 328

Fairburn, A. R. D., 304-5, 340
Falconer, W. C., 32
Fessenden, Reginald Aubrey, 34
Finlay, A. M., 289, 319
Fisk, Ernest, 59
Fleming, John Ambrose, 34
Forbes, George, 105, 151, 186-7, 234, 237
Forrest, Charles, 43-44, 46-47, 53, 56, 238, 323
Frame, Janet, 127, 215, 332, 336
Fraser, Peter, 218, 220, 233, 255-6, 261, 272-3, 275, 277-8, 284, 298-9, 337-8, 340
Freeman, L. G., 48

Garland, Thomas, 82-83, 178, 182-3, 185-9, 199, 335
Gibbons Hope, 43, 53, 59, 238, 323
Gibbs, Arthur, 43, 156-60
Gilchrist, Geoff, 205
Giles, Violet, 44
Godley, General,22
Goodfellow, William, 62-69, 73, 77-78, 90, 98, 100, 104, 107, 109, 111, 120, 126-7, 138, 143-7, 191, 330, 332
Gordon, John, 74-75, 122, 242
Grainger, R. E., 275
Green, Grace, 311, 320
Green, W. J., 326
Greenberg, L. J., 267-8, 275, 339
Grennell, Airini, 242
Griffiths, J. W., 340

Haggitt, Ray, 53
Hakopa, Hamiora, 123
Hall, Anna Rose, 82
Hall, John Herbert, 305, 329-32, 334-40
Halsey, Captain, 22
Hamilton, Adam, 142, 149, 168, 180-1, 195, 197, 207, 214, 332, 334-5
Hanan, J. R., 285
Hands, C. T. C., 325
Hands, E. C., 152, 154-5, 164, 166-7, 171, 176, 180, 183-4, 190, 193, 199-202, 208, 211, 242, 245, 333-6
Harcourt, Gordon, 44, 53, 59
Harcourt, Peter, 131, 323, 329
Hargest, James, 234
Harris, Alex, 50, 328-9
Harris Ambrose Reeves, 62-69, 71, 73, 76-77, 79, 84, 87-91, 98-100, 104, 107-8, 110-11, 113-14, 119-20, 122, 125-7, 130, 138, 141, 143-7, 164, 166-7, 211, 328, 330-32
Harrison, W. L., 134, 158
Hathaway, Alfred, 20
Hayden, S. J., 78
Hayward, Henry, 220
Heenan, J. W., 256
Hegan, Eddie, 95, 122-3, 332
Heighway, Arthur J., 78, 172-3, 256, 331-2
Henderson, A. G., 119
Henderson, Jim, 247
Henney, Tom, 130
Hertz, Heinrich 1-2, 12
Hickey, Pat, 179
Hicks, Stanton, 19
Hindman, W., 188
Hislop, T. C. A., 139
Hodgekinson, E., 162
Holcroft, M. H., 257-8, 338
Holden, J. W., 325
Holland, H. E., 92, 192
Holland, Sidney George, 80, 299-300
Holloway, P. N., 302
Holyoake, Keith Jacka, 175
Hood, George, 128
Howard, E. J., 70, 92, 110
Howell, Edward, 305-6
Howell, Madeline, 305-6
Howell, Therese Desmond, 305-6
Huggins, Bill, 37, 52, 67, 72, 75, 77, 103
Huggins, W. P., 12
Hughes, Stan, 39
Humphries, Percy, 205
Hunter, Thomas, 292
Hutchinson, G. R., 150-1, 172, 184, 197, 333-5

Hutter, Gordon, 80, 115, 170

Islington, Lord, 24

Jack, Robert, 37-43, 55-56, 61, 127, 156, 329
James, Percival, 335
Jenner, Ernest, 170
Johnston, N. B., 262-3
Jones, A. W., 153
Jones, Frederick, 207, 244, 271, 286, 288, 304, 336, 339

Keith, Roy, 48, 101, 183, 324
Kelvin, Lord, 9
Kent, George, 108
Kingi, Mete, 123
Kingsford-Smith, Charles, 129-30, 216
Krishnamurti, Jidda, 192-3, 195
Kyle, J. V., 325, 337

Laing, R. M., 8
Lake, H. R., 285
Lampburgh, Charles, 80
Laurenson, M. D., 254, 262-4, 338
Lawrence, Ralph, 38
Lee, John A., 92, 179, 203, 205-6, 217, 223, 236, 271-2, 277-8, 334, 336-7
Leech, J. T., 154
Lilburn, Douglas, 256
Livingstone, H. G., 197
Lodge, Oliver, 7-8
Lodge Nevile, 287, 307
Lynn, Vera, 245

Macfarlane, L. R. C., 150-1, 167, 170, 185, 191, 332, 334
Mackay, Ian K., 4, 151, 175, 201, 203, 216, 218, 241, 245, 254-5, 279-81, 296, 331, 333-5, 338-40
Marconi, Guglielmo, 9-10
Markman, A. T., 98
Marsden, Midge, 340
Mason, P. H., 204-5
Maxwell, James Clerk, 6
Maybury, Jack, 316
McCarthy, C., 188
McCarthy, Winston, 294-5, 339
McClay, Arthur, 28, 44-47, 53, 56, 323
McCombe, Jack, 80
McCombs, Terence, 289, 307
McEwen, J. D., 324
McIntyre, W. H., 145
McKeen, Robert, 150, 195
McKellar, Arnaud, 66, 145
McKenzie, Whang, 80
McLagan, Angus, 276
McLean, J. M., 98
McNab, Molly, 309-10
McNamara, John, 198, 217, 230, 250
McNamara, George, 76, 107-8, 137, 141, 143, 184, 200, 206-8, 330, 335
Meinung, O. E., 20, 42
Meredith, Bill, 80
Minhinnick, Gordon, 215
Mitchell, D. G., 20-21, 39, 42, 225-6, 337
Moncrieff, J. R., 128
Montague, J. F., 75, 124
Moohan, Michael, 276
Mulgan, Alan, 198, 251, 257, 263, 270, 295
Munns, G. C.,126
Murdoch, A. J., 288
Murray, Haydn, 250
Musgrove, Sydney, 308, 340

Nash, J. A. 137
Nash, Walter, 177, 214, 218, 220, 225, 236, 280, 300
Natzke, Oscar, 255
Neale, Leslie, 201, 225, 238
Nimmo, R. H., 141-2, 184, 334
Nimmo, Robert Rutherford, 39-40
Nosworthy, William, 70

O'Donoghue, A. F., 154-5, 202, 306, 331, 335-6
O'Meara, Ivan, 31, 51, 55, 324
O'Neill, Fred, 43, 55, 62, 67-68, 83, 108, 323
Orbell, J. A., 4
Orbell, R. J., 38, 41
Ormsby Hamilton, J. W., 15, 328
Owen, J. H., 153

Page, Bernard, 135
Palliser, Ernest, 197
Palmer, N. R., 262-4, 338, 340
Paora, Uramo, 242
Paris, Percy, 272
Parr, James, 65, 149, 152
Passmore, James L., 11-12, 42, 197
Patterson, A. S. 117, 240
Patty, R. J., 31, 48, 51, 55
Paul, J. T., 273, 275, 338
Payne, E. J. L., 102-3
Pearce, Arthur, 340
Pearson, Charles, 52, 185, 324
Perry, G. A., 325
Phillips, Kenneth, 197, 335
Pitama, Te Ari, 242
Pitt, Albert, 13
Plinge, Walter, 231
Plominski, F. R., 324-5
Power, Justin, 237
Prentice, J. M., 77

Reith, John, 35, 90, 130, 149, 234, 292
Renton, B. R., 326
Rhodes, Robert, 23, 26, 29-30
Richards, Bill, 314
Roberts, James, 276
Robinson, A. E., 235-6
Robinson, Brian, 28
Ross, Hilda, 307
Rushworth, H. M., 214
Russell, Cecil, 21
Russell, C. R., 153
Rutherford, Ernest, 7-11, 328

Salt, Gordon McBride, 97
Sarnoff, David, 34
Savage, Michael Joseph, 92-93, 150, 180-1, 203, 207, 213-19, 221-5, 233-4, 236, 238, 241-2, 249, 256, 271-2, 278, 284, 299, 313, 336
Schaef, Len, 133-4
Schroeder, J. H. E., 287, 296, 308, 310-11, 340
Scott, Robert, 12
Scrimgeour, Colin Graham, 166, 177-80, 182-90, 193-4, 200-7, 212, 216, 218, 221-4, 228-9, 233, 237, 241-5, 271-7, 279-80, 284, 296, 299, 334-9
Seddon, Richard, 12, 14, 328
Sewell, W. A., 195
Shaw, George Bernard, 192-5, 335
Shaw, Ian, 188
Shelley, James, 171, 211, 219-22, 229-34, 236, 245-6, 249-52, 255-7, 262, 264, 266, 274-5, 279-81, 283-9, 292-7, 304, 337-9
Shepherd, Gwen, 83
Shiel, Beaumont, 216, 222, 241, 337-8
Shipherd, Douglas, 47, 52
Shone, Phil, 303-4
Shrimpton, E. A., 17, 43-44, 50, 67
Simpson, A. E. H., 20, 43-44, 323
Sinclair, W. J., 48
Slade, Lionel, 42, 67
Smail, J. I., 62, 67, 323
Smith, A. Lee, 14
Smith, J. R., 158, 160, 208, 275
Smith, R. W., 114
Smith, Sidney Walter, 285
Smith, Sydney George, 235
Spackman, Bernie, 36, 55
Spackman, L. S., 36-37, 40
Spiers, Mick, 133-4, 211
Stannage, John, 216
Stark, Rawson, 19
Statham, Charles Ernest, 45
Sterling, H. H., 117
Stevens, P. R., 31, 55, 225-6
Stout, Robert, 42
Stowell, Henry H., 124
Strachan, J. E., 12, 19-20, 29, 35, 37, 48, 328-9
Strachan, L. E., 106, 135, 158, 238, 326
Strong, T. B., 85, 330
Sullivan, D. G., 48, 203, 220
Sutherland, Jack, 38-40

Tahiwi, Kingi, 242
Talbot, Rod, 188, 203, 242
Thacker, Henry, 157, 205
Thomson, J. C., 35
Thomson, J. J., 8-9
Tinney, Guy, 28
Tirikatene, Eruera, 242
Tollerton, George, 268
Toogood, Selwyn, 305, 316, 319, 338
Trim, Colin, 75, 77
Tuckwell, E. S., 200
Tunney, Gene, 130
Twain, Mark, 75
Tyrer, Anderson, 255, 284-5

Ulm, Charles, 190

Vause, Sid, 46
Venables, Drayton, 67, 75, 77
Vickery, H. D., 150-1, 157, 168, 186, 196-7, 208, 213, 334-6
Vincent, Stan, 55

Wall, Arnold, 211, 246-7
Walsh, Fintan Patrick, 276
Walters, Theo, 244
Ward, A., 114,
Ward, C., 48
Ward, Joseph, 13-16, 18, 73, 105, 128, 328-9
Waters, W. A., 153
Webb, J. W., 77
Weeks, Ernest, 83
Wells, R., 224
Whitaker-Wilson, C., 164
Whitley, John, 149
Wilkie, Alan, 128
Wilson, David, 328
Wilson, David, 272-7, 336, 338-9
Woodhead, C. F., 324
Wragge, Harold, 21
Wrathall, Dudley, 75, 78, 188, 268

Yates, William, 231, 281, 296, 304, 309-11, 340
Younger, J., 326

RADIO STATIONS

1XH Hamilton, 297, 300, 327
1XN Whangarei, 297, 327
1YA Auckland, 1, 51-52, 62, 67, 69-70, 72, 74, 77-78, 80-82, 88, 98, 102-3, 108-9, 124, 128, 137, 153, 155-6, 158, 160-1, 164, 169-70, 178, 186, 190, 193, 200, 208, 235, 301, 312, 323
1YB Auckland, 52, 100, 102, 185, 324
1YC Auckland, 301, 326
1YD Auckland, 101, 300, 319, 324-5
1YQ Auckland, 101-2, 324
1YX Auckland, 162, 185, 326
1YZ Rotorua, 301, 327
1ZB Auckland, 136, 157, 185, 187-8, 190-1, 193, 199-207, 212, 216, 218, 222, 224-5, 227-9, 239, 241, 242, 249, 251, 259, 262, 270, 286, 303, 324, 334, 340
1ZC Rotorua, 300, 327
1ZD Tauranga, 327
1ZH Hamilton, 104, 153, 159, 190, 325
1ZJ Auckland, 225, 235, 325
1ZM Auckland, 190-1, 195, 218, 235, 246, 270, 325, 335
1ZQ Auckland, 102, 183, 324
1ZR Auckland, 136, 177-9, 182-5, 325
1ZS Auckland, 185, 325
2AZ Taihape, 55
2GI Palmerston North, 101
2XA Wanganui, 304, 327
2XB Masterton, 297, 327
2XG Gisborne, 297, 325
2XM Gisborne, 226, 324
2XN Nelson, 297, 326
2XP New Plymouth, 297, 324
2YA Nelson, 54, 323
2YA Wellington, 70-74, 77-81, 83-85, 87, 91, 102-4, 122-4, 127-30, 134, 144, 146, 155, 158, 160, 162, 164, 166, 170, 193, 197, 213-4, 234-6, 249, 258-60, 262, 283, 293, 300-1, 311, 319, 323, 340
2YB Wellington, 53, 56, 59, 246, 323
2YB New Plymouth, 102, 103, 158, 324
2YC Wellington, 162, 293, 301, 326
2YD Wellington, 245-6, 294, 300, 319, 326
2YF Palmerston North, 101-2, 324
2YH Napier, 246, 293, 325
2YK Wellington, 28, 53-54, 59, 62, 68, 323
2YM Gisborne, 55, 100, 102, 324
2YN Nelson, 246, 326
2YX Wellington, 300, 327
2YZ Napier, 325
2ZA Palmerston North, 227, 246, 286, 327
2ZB Wellington, 238-9, 242, 245-6, 267, 269-70, 273, 277, 300, 326
2ZC Napier, 300, 326
2ZD Masterton, 104, 159, 325
2ZE Eketahuna, 102, 325
2ZF Palmerston North, 104, 111, 159, 324
2ZG Wanganui, 102, 325
2ZH Napier, 159, 195, 324
2ZI Hastings, 112, 325
2ZJ Gisborne, 159, 246, 325
2ZK Wanganui, 102, 324
2ZL Hastings, 199, 325
2ZM Gisborne, 102, 225-6, 324
2ZO Palmerston North, 199, 325, 337
2ZP Wairoa, 199, 325
2ZQ Masterton, 102, 111, 324
2ZR Nelson, 4, 159, 174-6, 201, 216, 246, 326, 333-4
2ZR Wanganui, 325
2ZU Dannevirke, 111, 325
2ZW Wellington, 84, 134-5, 137, 139, 158, 177, 179-80, 184-5, 238, 326, 332
2ZX Wellington, 325
3AC Christchurch, 55, 67, 81, 323
3AQ Christchurch, 62, 67, 323
3XC Timaru, 297, 327
3YA Christchurch, 1, 70, 72, 77-78, 80-81, 88, 102, 108-9, 118, 122-3, 128-30, 136-7, 144, 155, 158, 160, 162, 166, 168, 170, 193, 203, 220, 251, 301, 323
3YC Christchurch, 301, 326
3YL Christchurch, 162, 326
3YZ Greymouth, 300, 326
3ZB Christchurch, 242-3, 268, 309-11, 315, 320, 326
3ZC Christchurch, 101, 106, 111, 135, 184, 324
3ZE Greymouth, 134, 326
3ZM Christchurch, 174, 177, 189, 194, 211-2, 216, 238, 311, 326
3ZR Greymouth, 134-5, 159, 211, 246, 293, 326
4AB Dunedin, 55, 100, 102, 323
4XD Dunedin, 20, 42, 48, 226, 323
4YA Dunedin, 1, 55, 62, 67-68, 72, 74, 77-78, 103, 108-9, 118, 128, 130, 137, 154-5, 158, 160, 171, 200, 242, 284, 301, 323, 335
4YC Dunedin, 301, 326
4YO Dunedin, 55, 162, 185, 323, 326
4YZ Invercargill, 246, 293, 325
4ZA Invercargill, 300, 327
4ZB Dunedin, 102, 111, 238, 242, 269, 284, 289, 323, 326
4ZC Lowburn, 102, 159, 174, 326, 335
4ZD Dunedin, 225-6, 239, 323
4ZF Dunedin, 326
4ZI Invercargill, 103-4, 325
4ZL Dunedin, 102, 324

4ZM Dunedin, 102, 177, 201-2, 224-5, 238, 324
4ZO Dunedin, 102, 224, 324
4ZP Invercargill, 103, 159, 200, 246, 325
4ZR Inchclutha, 225, 326
4ZW Dunedin, 190, 326
5ZA Apia, 134, 325
5ZB Mobile Station, 253-5, 327

GENERAL

advertising, 46-47, 51, 56, 106, 108, 111, 137-9, 148, 171, 175, 180-1, 183, 196, 199, 204, 217, 219, 227-9, 237, 239-40, 243, 257, 279, 290, 300, 311
Advisory Council, 148, 153, 155-6, 159, 176, 191, 196
Amalgamated Wireless Co. of Australasia Ltd, AWA, 31, 53, 57, 59, 66-67, 104, 141, 160-2, 169, 184, 286
amateur radio, 11-12, 20-22, 27-30, 35-36, 50-51
Anzac Wireless Squadron, 32
Auckland Radio Association, 97
Auckland Radio Listeners' League, 98, 216
Auckland University College, 97, 195, 197, 308
audience, broadcasting, 3, 94, 97, 208
Australasian Performing Rights Association, APRA, 109-11, 125, 173-4, 209, 226
Australia, 13-26 *passim*, 73, 104, 109-10, 124, 126, 130-1, 137, 148, 165, 189, 200-3, 218, 229, 238, 241, 244, 252-3, 283, 286, 296, 312
Australian government, 15-16, 18

B stations, 100-6, 108, 110-12, 133-9, 141-2, 149-50, 153, 158-9, 164, 174-91, 198-204, 211, 216-19, 223-7, 238, 242, 245, 323
Beveridge report, 302
British Broadcasting Company, 35, 58, 90
British Broadcasting Corporation, 58, 86, 90, 98, 130, 140-2, 149, 163-4, 170, 181, 192-3, 196-7, 223, 230, 233, 259-60, 262, 264, 291-2, 294, 302, 305, 308, 311-2
Broadcasting Act 1931, 148
Broadcasting Act 1936, 208, 214, 219, 224
Broadcasting Amendment Act 1934-5, 195-6, 199
Broadcasting Amendment Act 1937, 221
Broadcasting Corporation of New Zealand, 4

cable companies, 17-19
camp concerts, 270
Canterbury Philosophical Institute, 41
Canterbury University College, 7, 11, 30, 38, 41, 135, 171, 220, 246
Cavendish laboratory, 8-9
censorship, 93, 200, 223, 261-2, 267, 273, 275, 291
Chatham Islands, 17, 25, 37
children's programmes, 55, 78, 81-83, 121, 155, 178, 266, 316
Christchurch International Exhibition 1906, 16, 21
Church of the Helping Hand, 201, 225, 238
Conscription referendum, 297-8
controversial material, broadcasting of, 51, 91-92, 127, 146, 163, 166, 172, 177, 191-9, 212, 223, 235-6, 270, 272, 288-92, 302, 314
Copyright Act 1913, 109-10
Copyright (Temporary) Amendment Act 1928, 110
crystal sets, 57, 95, 134, 136

Democrat Party, 330
Dominion Alliance of B Class Stations, 184, 201
Dominion Radio College, 31
drama production, 251-3, 293, 296

elections, & broadcasting, 45, 68, 236-7, 299
electric power & broadcasting, 79, 95, 157, 286

Federated Farmers, 288
First New Zealand Expeditionary Force, 32
First New Zealand Wireless Troop, 32
frequency modulation, 313
Friendly Road, Fellowship of the, 182-90, 200, 228, 238, 277

gramophone record companies, 111, 125, 159, 173-4, 226

hoax broadcasts, 303-5
horse racing, broadcasting of, 79-80, 167-8, 210, 213

International Radio-Telegraphic Convention 1912, 25

Labour Party, 3-4, 73, 105, 148, 150, 182, 191-2, 202, 204, 206, 208, 212-13, 218, 234, 237, 274, 276-8, 288-9, 299,
Listeners' Leagues, 97-98, 142, 153, 176

listeners' licence, 49, 57-58, 60, 144, 146, 148, 288
listeners' questionnaire 1932, 163-4
listeners' representation, 97-100
local talent, broadcasting of, 61, 122, 125, 317-18

magnetic detector, 7, 8, 10
Manawatu Radio Society, 153
Maori broadcasting, 123-4, 242-3, 315
Marconi Wireless Telegraph Company, 10, 13, 15-16, 18, 22, 31, 35
marine communication, 10, 15-17, 19, 23, 36
Mazengarb Committee, 307
Mobile Broadcasting Unit, 262-6
Musical and Dramatic Committees, 99, 122, 146, 153-6, 164, 197
musicians, employment of, 87-88, 112-13, 122, 244, 249, 250

NBS String Orchestra, 249-50, 255-6
Napier earthquake, 131-3
National Broadcasting Service, 160, 167, 219, 221, 223, 229-36, 238, 241, 244-6, 249-52, 255-6, 258-65, 269-71, 275, 279, 281, 283, 287
National Centennial Symphony Orchestra, 255
National Commercial Broadcasting Service, 188, 221-3, 225, 227-9, 233-4, 237, 239-46, 249, 251-3, 258-9, 267, 269, 270-81, 294, 299
National Council of Women, 196
National Orchestra, 249-50, 283-85, 296, 299
National Party, 285, 299-301
networking, 311-12
New Zealand and South Seas Exhibition, 67-68, 108, 112
New Zealand Broadcasting Board, 4, 147-77, 181, 183-5, 189, 192-219, 226, 233, 245, 251, 256
New Zealand Broadcasting Service, 1, 2, 167, 283, 285-8, 291, 296-7, 300-3, 307, 311-12, 317-18
New Zealand Cooperative Dairy Company, 63-64, 66, 78
New Zealand Farmers' Union, 235
New Zealand Federation of B Station Proprietors, 201, 204, 218
New Zealand Listener, 230, 256-8
New Zealand Press Association, 301
News & talks programmes, 4, 90-92, 119, 121, 146, 171-2, 233-4, 259-60, 262-4, 283, 288-90, 301, 308, 311
Newspaper Proprietors' Association, 90, 104, 119, 196, 228
North Taranaki Radio Society, 102-3

Otago Radio Association, 42, 55, 100, 102
outside broadcasts, 75-76, 80, 127-31

Pacific Islands, 17-19, 71, 100, 123
Palmerston North Radio Club, 101
parliamentary broadcasts, 4, 213-15, 234, 299-300, 308, 311
Post & Telegraph Act 1908, 28
Post & Telegraph Amendment Act 1913, 28, 30
Post & Telegraph Amendment Act 1920, 36
Post & Telegraph Amendment Act 1924, 57, 66
Post & Telegraph Amendment Act 1933, 157
Post & Telegraph Department, 4, 15-17, 21-22, 28, 32, 36, 43, 47, 51-60 *passim*, 65, 68, 73, 76, 80, 84-85, 90-91, 97-107 *passim*, 125, 127, 132-59 *passim*, 176, 196, 199-200, 205-8, 283, 312, 181-4, 188-9
programming, conflict of 'popular' and 'quality', 88-90
  coordination of, 86-87, 166-9, 260, 309-10
  musical, 2, 86-90, 238, 309-11
  NCBS, 238-42
  NZBB, 153-4, 164-9, 209, 211
  RBC, 73-77, 121-7, 145-7
  station division into types, 308-11
  wartime, 266-71

Radio Apia, 26, 32
Radio Auckland VLA, 24-25, 37
Radio Awanui, 26, 32, 37
Radio Awarua, 26
Radio Broadcasting Company, 1, 4, 62, Chapters 3-5 *passim*, 148, 153, 156, 159, 164, 166-7, 172-3, 177, 191-2, 233, 256, 260, 311
Radio Chathams VLC, 25, 37
radio clubs, 175-7, 184
Radio Corporation of America, 35
Radio Coverage Commission, 156-9
radio receivers, 94-96
radio reception, 40-41, 70, 72-73, 156
Radio Record, 78-79, 172-3, 256-7
Radio Regulations 1923, 49-51, 55-56
Radio Regulations 1925, 60-61, 91, 125, 146, 162-3
Radio Regulations 1932, 162-3, 191-2, 196, 199
radio serials, 3, 126-7, 145, 165, 238-9, 245, 252-3, 279, 305-8
Radio Society of Christchurch, 42, 55, 59, 62, 67, 153
radio telephone appeals, 269
Radio Wellington VLW, 23-24, 26, 28, 37, 43, 122
recording, introduction of, 202, 250-1

Reform Party, 65, 92, 142, 148, 151, 197
Reithian standards, see Reith, John
religious broadcasting, 51, 60-61, 78, 83-84, 99, 113, 121, 153, 177-9, 182, 187, 201, 308, 316
Royal Navy, 12, 15, 20-21
royal tour 1953, 312
rugby, 53-54, 79-80, 115-8, 127, 168-9, 210, 235, 294

school broadcasts, 84-86, 170-2, 210, 213, 220, 285, 290-2
shortwave radio, 104, 130-1, 283
Southland Radio Society, 153
sports broadcasting, 79-81, 113-18, 121, 127, 130, 145, 167-70, 188-9, 210, 294, 311, 315-16
Statutes Amendment Act 1943, 279
sub-Antartic islands, 17, 25

Taranaki Radio Society, 153
Telefunken Company, 22-23, 26, 31
television, 162, 312-13

United Party, 105, 138, 148
United Press Association, 171, 233, 260, 301
University of Otago, 39-40

Victoria University College, 292,

Waitaki Boys' High School, 21-22, 38
Wanganui Maori Concert Party, 123-4
Waterfront dispute 1951, 301
weather forecasts, 23, 122, 267
Wellington Radio Station, 23
Wellington Radio Society, 97
West Coast Radio Society, 133
wireless telegraphy, 1, 10-15, 17, 23, 31, 34, 37, 49
Wireless Telegraphy Act 1903, 13-15, 19, 28, 33
Workers' Educational Association, 135, 171, 192, 195-6, 220
Women & broadcasting, 238-41, 290-1, 315-16
World War One, 30-33
World War Two, 254, 258-75, 283

X-class stations, 296-7, 300, 313